CODES AND MODULAR FORMS

A Dictionary

CODES AND MODULAR FORMS

A Dictionary

Minjia Shi
Anhui University, China

YoungJu Choie
Pohang University of Science and Technology (POSTECH), South Korea

Anuradha Sharma
Indraprastha Institute of Information Technology Delhi, India

Patrick Solé
I2M, (CNRS, Centrale Marseille, University of Aix Marseille), France

World Scientific

NEW JERSEY • LONDON • SINGAPORE • BEIJING • SHANGHAI • HONG KONG • TAIPEI • CHENNAI • TOKYO

Published by

World Scientific Publishing Co. Pte. Ltd.
5 Toh Tuck Link, Singapore 596224
USA office: 27 Warren Street, Suite 401-402, Hackensack, NJ 07601
UK office: 57 Shelton Street, Covent Garden, London WC2H 9HE

British Library Cataloguing-in-Publication Data
A catalogue record for this book is available from the British Library.

CODES AND MODULAR FORMS
A Dictionary

ISBN 978-981-121-291-8

For any available supplementary material, please visit
https://www.worldscientific.com/worldscibooks/10.1142/11628#t=suppl

Foreword

The dictionary between codes and modular forms has a long and distinguished history going back to the 1970's (work of Broué and Enguehard). It was built on an even older dictionary between lattices and modular forms via theta series that dates from the times of Jacobi and his quartic identity of Lie group fame. Then, in the 1960's at the time of the discovery of the Leech lattice via the so-called Construction A, a third dictionary, between codes and lattices, this time, connected the two precedent dictionaries.

In the 1990's a paradigm shift occurred in Coding Theory: the migration of alphabets from finite fields to more general rings. It is my pleasure to see my former student (Minjia Shi) and my good friend (Patrick Solé), two famous experts on codes over rings as co-authors of this book. The other two authors (Choie and Sharma) are number theorists with an expertise in modular forms. The strong argument for using these rings lie in the successful treatment of various classes of modular forms ranging from elliptic, to Jacobi form, to Siegel.

The book you are holding in hands, dear reader, is a good introduction to this rich and varied domain, at the crossroad of number theory and combinatorics, written by a quatuor of experts. It should serve the novice like the aficionado equally well.

Shixin Zhu
Hefei University of Technology
12 September, 2019

Contents

Introduction

Since the times of Felix Klein and his work on the icosahedron [18], there has existed a deep dictionary between invariant polynomials of finite groups and the modular forms of number theory. More recently, since the seminal talk of Gleason at the ICM 1970 in Nice [17], invariant theory has been found to have important applications in combinatorics in connection with the study of weight enumerators of self-dual codes over finite fields [19]. In 1972, these two trends fused when Broué and Enguehard [3] established a ring isomorphism between, on the one hand, the space of invariant polynomials for the finite group that leaves weight enumerators of Type II binary codes invariant and, on the other hand, the ring of modular forms of weight a multiple of four. In the 1990's, this topic was met with renewed interest for two main reasons:

- new alphabets (finite rings) appeared;
- new weight enumerators were discovered.

As an example of the latter, we can cite polynomial analogues for modular forms of a number of different types: Jacobi [2,8,10,11] (a generalized split weight enumerator, a complete weight enumerator); Hilbert [5,7,16] (a Lee weight enumerator, a complete weight enumerator); Siegel [5,7,15,30] (a higher joint weight enumerator); and Jacobi forms of higher genus [9] (a higher complete and symmetrized weight enumerator). As for the former argument, one can mention the fact that the complete weight enumerator of Type II codes over the ring $\mathbb{Z}_{2m}$ determines a Jacobi form for the full Jacobi group [8]. The Broué-Enguehard map itself was extended by Ozeki [2,21] and independently by Runge [23] to the ring of modular forms of even weight. A similar connection between the ring of modular forms of integral weight and self-dual binary codes was established by Rains and Sloane [22, p.282]. In another related direction, shadows and generalized shadows of self-dual codes are studied to construct self-dual codes of higher lengths, and Jacobi forms of even weights are constructed by establishing a similar connection [1, 4, 6, 13, 14, 24–26].

In this book, we study Broué and Enguehard type maps from the algebra of invariants, where the symmetric weight enumerators of self-dual codes over some finite ring live, to some infinite ring of modular forms of some type (one variable, Siegel, Jacobi, Hilbert). To make the book self-contained, we have included basic material on codes and lattices, that can be found, with greater detail, in [12] for instance. We have been, however, minimalists in the treatement of invariants, which have received a very thorough exposition in the book [20].

This research is supported by the National Natural Science Foundation of China (61672036), the Excellent Youth Foundation of Natural Science Foundation of Anhui Province (1808085J20), the Academic Fund for Outstanding Talents in Universities (gxbjZD03). YoungJu Choie is partially supported by NRF2018R1A4A1023590 and NRF2017R1A2B2001807.

References

[1] E. Bannai, S. T. Dougherty, M. Harada, M. Oura, Type II codes, even unimodular lattices, and invariant rings, *IEEE Trans. Inform. Theory* 45(4), pp. 1194–1205 (1999).

[2] E. Bannai, M. Ozeki, Construction of Jacobi forms from certain combinatorial polynomials, Proceedings of the Japan Ac., Vol. 72, Ser. A, No 1 (1996) 12–15.

[3] M. Broué, M. Enguehard, Polynômes des poids de certains codes et fonctions theta de certains réseaux, Ann. Sc. ENS 5 (1972) 157–181.

[4] K. Betsumiya, Y. Choie, Jacobi forms over totally real fields and type II codes over Galois rings $GR(2^m, f)$, European J. Combin. 25 (2004), no. 4, 475–486.

[5] K. Betsumiya, Y. Choie, Codes over $\mathbb{F}_4$, Jacobi forms and Hilbert-Siegel modular forms over $\mathbb{Q}(\sqrt{5})$, European J. Combin. 26 (2005), no. 5, 629–650.

[6] Y. Choie, S. T. Dougherty, Codes over σ_{2m} and Jacobi forms over the quaternions, Appl. Algebra Engrg. Comm. Comput. 15 (2004), no. 2, 129–147.

[7] Y. Choie, S. T. Dougherty, H. Liu, Jacobi forms and Hilbert-Siegel modular forms over totally real fields and self-dual codes over polynomial rings $\mathbb{Z}_{2m}[x]/g(x)$, Ars Combin. 107 (2012), 141–160.

[8] Y. Choie, N. Kim, The complete weight enumerator of Type II codes over $\mathbb{Z}_{2m}$ and Jacobi Forms, IEEE Trans. on Information Theory, Vol. IT-47 (2001), 396–399.

[9] Y. Choie, H. Kim, Codes over Z_{2m} and Jacobi forms of genus n, J. Combin. Theory Ser. A 95 (2001), no. 2, 335–348.

[10] Y. Choie, E. Jeong, Jacobi forms over totally real fields and codes over $\mathbb{F}_p$, Illinois J. Math. 46 (2002), no. 2, 627–643.

[11] Y. Choie, P, Solé, Ternary codes and Jacobi forms, Discrete Math. 282 (2004), no. 1-3, 81–87.

[12] J. H. Conway, N. J. A. Sloane, *Sphere Packings, Lattices and Groups*, Springer (1998) 3rd edition.

[13] S. T. Dougherty, M. Harada, P. Solé, Shadow lattices and shadow codes, *Discrete Mathematics* 219, pp. 49–64 (2000).

[14] S. T. Dougherty, M. Harada, P. Solé, Shadow codes over $\mathbb{Z}_4$, *Finite Fields and Their Applications* 7(4), pp. 507–529 (2001).

[15] W. Duke, Codes and Siegel Modular Forms, International Research Notices in Math., 3, p. 125 in Duke Math. J. 70 (1993).

[16] W. Ebeling, *Lattices and Codes*, Vieweg (1994). Birkhäuser (1985).

[17] A. M. Gleason, Weight enumerators of self-dual codes and MacWilliams identities, Actes du Congrés International de mathématiques, vol. 3, Gauthier-Villars (1970) 211–215.

[18] F. Klein, Vorlesungen uber das Ikosaeder und die Auflosung der Gleichungen vom funften Grade. (German) [Lectures on the icosahedron and the solution of equations of fifth degree] Reprint of the 1884 original. Edited, with an introduction and commentary by Peter Slodowy. Birkhauser Verlag, Basel; B. G. Teubner, Stuttgart, (1993).

[19] F. J. MacWilliams, N. J. A. Sloane, *The theory of Error Correcting Codes*, North Holland second edition (1981).

[20] G. Nebe, E. M. Rains, N. J. A. Sloane, *Self-dual codes and invariant theory*, Springer, ACM vol. 17 (2006).

[21] M. Ozeki, On the notion of Jacobi polynomials for codes, Math. Proc. Cambridge Phil. Soc.(1997).

[22] E. M. Rains, N. J. A. Sloane, Self-dual codes, in *The Handbook of Coding Theory I*, V. S. Pless and W. C. Huffman, eds, North-Holland (1998).

[23] B. Runge, On Siegel Modular forms I, Crelle 436 (1993) 57–85.

[24] A. Sharma, A. K. Sharma, Construction of self-dual codes over $\mathbb{Z}_{2^m}$, *Cryptography and Communications* 8(1), pp. 83–101 (2016).

[25] K. Suzuki, Complete m-spotty weight enumerators of binary codes, Jacobi forms, and partial Epstein zeta functions, Discrete Math. 312 (2012), no. 2, 265–278.

[26] Han-Ping Tsai, Existence of some extremal self-dual codes, *IEEE Trans. Inform. Theory* 38(6), pp. 1829–1833 (1992).

Chapter 1

Codes and Lattices: A Dictionary

1.1 Lattices

A *lattice* in $\mathbb{R}^n$ is defined as a discrete additive subgroup: imagine a grid infinite in all directions like $\mathbb{Z}^n$ for instance. They bear no connection, but their name, with the special posets of order theory. They occur in many areas of modern science from pure mathematics (algebraic geometry [16], Lie groups [23]), to physics (crystallography [4]), to computer science (cryptography [20]). They are used in electrical engineering as group codes for the Gaussian channel and as codebooks for vector quantization [13]. Every lattice can be represented as the row span over $\mathbb{Z}$ of an $m \times n$ generator matrix M, say with $m \leq n$. That is, we have

$$L = \{uM : u \in \mathbb{Z}^m\}.$$

The rows of M form an integral basis of the lattice. For instance, a version of the hexagonal lattice sitting in the hyperplane $x + y + z = 0$ is obtained for $m = 2$ and $n = 3$, upon letting

$$M = \begin{pmatrix} 1 & -1 & 0 \\ 0 & 1 & -1 \end{pmatrix}.$$

A square generator matrix for that lattice is given in the next subsection.

1.1.1 *Sphere Packing Problem*

By a *sphere* in $\mathbb{R}^n$ of center x and radius r, we shall mean the following open ball in the sense of topology:

$$B(x, r) := \{y \in \mathbb{R}^n : d_E(x, y) < r\},$$

where $d_E(.,.)$ is the standard Euclidean distance in $\mathbb{R}^n$. A sphere packing S of $\mathbb{R}^n$ is defined as an infinite collection of such spheres of the same radius but with different centers. The *density* $\delta(S)$ of S is, roughly speaking, the proportion of $\mathbb{R}^n$ covered by S (See §1.2 or [13, p.7] for a definition). The sphere packing problem is to maximize the density $\delta(S)$ over S. This optimal density is known for $n = 1, 2, 3$. A sphere packing S meeting the optimal density is not always unique.

- For $n = 1$, the spheres are the intervals of constant width. For example, one may take $S = \bigcup_{k \in \mathbb{Z}} [k, k+1]$.
- For $n = 2$, the optimal sphere packing S is, up to rotations and translations, the so-called hexagonal packing well-known from honeycombs, obtained by taking the centers of a hexagonal tiling. This lattice is spanned by the rows of
$$M = \begin{pmatrix} 1 & 0 \\ 1/2 & \sqrt{3}/2 \end{pmatrix}.$$
- For $n = 3$, there are infinitely many optimal sphere packings. The centers of the simplest optimal sphere packing S are obtained by the points of the lattice D_3, as conjectured by Kepler and proved by Hales [16].
$$D_3 = \{(x, y, z) \in \mathbb{Z}^3 \mid x + y + z \equiv 0 \pmod{2}\}.$$
This packing is known in chemistry books as *the face centered cubic packing*. A truncated version of it can be seen sometimes at fruit stands, and cannon balls stacks.
- For many n's, the densest sphere packing in $\mathbb{R}^n$ is conjectured or known (for $n = 1, 2, 3$) to be a lattice packing. Very recently, the lattices E_8 and L_{24} to be defined below, were shown to be optimal by using modular forms [12, 33].

In general, the sphere packing with the highest density obtained from a lattice is obtained as $\bigcup_{x \in L} B(x, r)$ for any r such that the balls of radius r about the lattice points are disjoint. Define the *norm* $N(L)$ of a lattice as the smallest non zero squared norm of one of its vectors. Then the largest radius in such a packing is
$$\rho(L) = \frac{\sqrt{N(L)}}{2}.$$

The quantity $\rho(L)$ is usually called the *packing radius* of L. Thus the packing $\bigcup_{x \in L} B(x, \rho(L))$ is called the lattice packing attached to L.

1.1.2 *Determinant*

Let L be a lattice having generator matrix M, i.e., L is spanned by the rows of M. The *Gram matrix* associated with the lattice L is defined as

$$G = MM^t.$$

The determinant of the Gram matrix of L is called *the determinant* of the lattice L, and is denoted henceforth by $\det(L)$. Geometrically, the determinant $\det(L)$ of the lattice L equals the square of the fundamental volume of L, which is the volume of the convex body spanned by the rows of M. In particular, it controls the density of the lattice packing $S(L)$ attached to L, in the sense that

$$\delta(S(L)) = \frac{V_n \rho(L)^n}{\sqrt{\det(L)}},$$

where

$$V_n = \frac{\pi^{n/2}}{\Gamma(n/2+1)}$$

is the volume of the unit ball in $\mathbb{R}^n$ [13, p.9]. Here Γ is the classic Γ function, which is defined by

$$\Gamma(x) = \int_0^\infty t^{x-1} e^{-t} dt \quad \text{for} \quad x > 0.$$

1.1.3 *The Theta Series*

The theta series of a lattice L can be defined as a formal power series in the variable q by the formula

$$\theta_L(q) = \sum_{x \in L} q^{\langle x, x \rangle},$$

where $\langle x, y \rangle = \sum_{i=1}^{n} x_i y_i$ denotes the Euclidean inner product of the vectors $x = (x_1, x_2, \ldots, x_n)$ and $y = (y_1, y_2, \ldots, y_n)$. Intuitively, it is a generating function counting the points of the lattice at a given distance from the origin. Thus

$$\theta_L = \sum_{m \geq 0} N_m(L) q^m,$$

where $N_m(L) = |\{x \in L \mid \langle x, x\rangle = m\}|$. A trivial, but fundamental example is

$$\theta_{\mathbb{Z}}(q) = \sum_{n\in\mathbb{Z}} q^{n^2} = 1 + 2\sum_{n=1}^{\infty} q^{n^2}.$$

Thinking now of q as a complex variable, it can be shown that $\theta_L(q)$ converges in the open unit disk. For an integral lattice, a substitution of the type $q = e^{ci\tau}$, with c a positive real number and i a square root of -1, yields an analytic function of τ, when τ belongs to

$$\mathcal{H} = \{z \in \mathbb{C} : \Im(z) > 0\},$$

the so-called Poincaré upper half-plane.

1.1.4 *Duality*

There is a natural notion of dual L^* of a lattice L, which is fundamental for physicists studying crystal diffraction and for number theorists involved with modular forms. The *dual* L^* of the lattice L is defined as

$$L^* = \{y \in \mathbb{R}^n \mid \langle x, y\rangle \in \mathbb{Z}\ \forall\ x \in L\}.$$

The lattice L is called an *integral lattice* if it satisfies $L \subseteq L^*$, while the lattice L is called a *unimodular lattice* if it satisfies $\det(L) = 1$. The lattice L is said to be *self-dual* if $L = L^*$. There is a simple formula relating the theta series of a lattice with that of its dual, called the *Poisson-Jacobi* formula:

$$\theta_L(e^{-\pi\tau}) = \frac{\tau^{-\frac{n}{2}}}{\sqrt{\det(L)}}\theta_{L^*}(e^{-\frac{\pi}{\tau}}).$$

The formula is especially nice and is useful when L is a self-dual lattice. It is well-known that a lattice is self-dual if and only if L is both integral and unimodular. Equivalently, if M is a generator matrix of the lattice L, then the lattice L is self-dual if and only if the following two conditions are met:

- MM^t has integral entries and
- $\det(M) = \pm 1$.

It is classical to consider the function $\vartheta_L(\tau) = \theta_L(e^{\pi i\tau})$. If L is an integral lattice, then the function ϑ_L is analytic in the open unit disk. When L is a self-dual lattice, the Poisson-Jacobi formula relates the functions $\vartheta_L(\tau)$ and $\vartheta_L(-1/\tau)$ by the relation

$$\vartheta_L(-1/\tau) = (\frac{\tau}{i})^{\frac{n}{2}}\vartheta_L(\tau).$$

The integrality conditions entail a periodicity in τ, namely $\vartheta_L(\tau) = \vartheta_L(\tau+2)$. Thus it makes sense to consider the group Γ defined by the following two generators:

- $S: \ z \mapsto -1/z$, and
- $T^2: \ z \mapsto z+2$.

In the matrix form, we can write

$$\Gamma = \left\langle \begin{pmatrix} 0 & -1 \\ 1 & 0 \end{pmatrix}, \begin{pmatrix} 1 & 2 \\ 0 & 1 \end{pmatrix} \right\rangle.$$

We give the transformation law for an arbitrary $A = \begin{pmatrix} a & b \\ c & d \end{pmatrix} \in \Gamma$. Consider the character χ of the group Γ defined on its generators S and T^2 by

- $\chi(S) = i$, and
- $\chi(T^2) = 1$.

With this notation, we have

$$\vartheta_L\left(\frac{a\tau+b}{c\tau+d}\right) - \left(\frac{c\tau+d}{\chi(A)}\right)^{\frac{n}{2}} \vartheta_L(\tau).$$

More generally, any function f satisfying the preceding functional equation, analytic on $\mathcal{H}$, and "holomorphic at the cusp" [35] of Γ is called a *modular form* attached to Γ of weight $n/2$. When n varies, the modular forms constitute a ring $\mathcal{M}$ under addition and multiplication. A deep theorem due to Hecke gives the algebraic structure of the ring $\mathcal{M}$, which is also a complex vector space. Denote by $\mathcal{M}_n$, the subspace of $\mathcal{M}$ consisting of modular forms of weight $n/2$. This notation gives the structure of a graded ring to $\mathcal{M}$, since $\mathcal{M}_k\mathcal{M}_\ell \subseteq \mathcal{M}_{k+\ell}$.

Theorem 1. *([13, Chap. 7, Th. 7])*

- *For all* $n \geq 1$, *we have* $\dim_{\mathbb{C}}(\mathcal{M}_n) = 1 + \lfloor \frac{n}{8} \rfloor$.

- *The ring $\mathcal{M}$ is generated by $\vartheta_{\mathbb{Z}}(e^{\pi i\tau})$ and*
$$\Delta_8(\tau) = q\prod_{m=1}^{\infty}(1-q^{2m-1})^8(1-q^{4m})^8$$
with $q = e^{\pi i\tau}$.

In this theorem, the form $\Delta_8(\tau)$ is a so-called *cusp form*, that is, it vanishes when $q = 0$, or equivalently, when $\tau \to i\infty$. A trivial example of an element of $\mathcal{M}_n$ is the theta series of the unimodular lattice $\mathbb{Z}^n$, which is none other than $\vartheta_{\mathbb{Z}^n}$. Now the larger group generated by

- $S: z \mapsto -1/z$,
- $T: z \mapsto z+1$,

is $SL_2(\mathbb{Z}) = \{A \in GL(2,\mathbb{Z}) : \det(A) = 1\}$.

Is there a natural notion of modular forms for $SL_2(\mathbb{Z})$ for which the theta series of a special class of lattices is a modular form? Invariance under T will happen if we assume that L is a self-dual lattice yielding integral inner products $\langle x, x\rangle$ of lattice vectors x and, furthermore, that $2|\langle x, x\rangle\ \forall x \in L$. Such a lattice, by a coding analogy, is called a Type II lattice (or even lattice) in [13]. A self-dual lattice L is called a Type I lattice if there exists $x \in L$ satisfying $\langle x, x\rangle \not\equiv 0 \pmod 2$. Denote by χ', the character of $SL_2(\mathbb{Z})$ satisfying

- $\chi'(S) = i$, and
- $\chi'(T) = 1$.

Then we see that the theta series of a Type II lattice is a modular form of weight $n/2$ for $SL_2(\mathbb{Z})$ and the character χ' in the sense that it satisfies the transformation law
$$\vartheta_L\left(\frac{a\tau+b}{c\tau+d}\right) = \left(\frac{c\tau+d}{\chi'(A)}\right)^{\frac{n}{2}}\vartheta_L(\tau)$$
for an arbitrary $A = \begin{pmatrix} a & b \\ c & d \end{pmatrix} \in SL_2(\mathbb{Z})$.

To describe the first example of a Type II lattice, we consider the lattice in 8 dimensions, usually called E_8, a notation coming from Lie group theory, defined by
$$E_8 = \{x \in \mathbb{Z}^8 \cup (1/2+\mathbb{Z})^8 : \sum_{i=1}^{8} x_i \in 2\mathbb{Z}\}.$$

To see that the norms of the lattice vectors are even, just check the property on a basis, and use the parallelogram identity, combined with self-duality. Denote by $\mathcal{M}'$, the ring of modular forms for $SL_2(\mathbb{Z})$ and the character χ'. Denote by $\mathcal{M}'_n$, the subspace of $\mathcal{M}'$ of weight $n/2$. Now the structure theorem for Type II lattices is stated as follows.

Theorem 2. *(Hecke, [13, Th.17 of Chap. 7])*

- *For all $n \geq 1$, we have $\dim_{\mathbb{C}}(\mathcal{M}'_n) = 1 + \lfloor \frac{n}{24} \rfloor$.*
- *The ring $\mathcal{M}'$ is generated by the two forms $\vartheta_{E_8}(q)$ and*

$$\Delta_{24}(\tau) = q^2 \prod_{m=1}^{\infty} (1 - q^{2m})^{24}$$

with $q = e^{\pi i \tau}$.

The form Δ_{24} is a cusp form of weight 12, which was considered by Ramanujan in his work on the function $\tau(n)$, the coefficient of q^n in the theta series expansion of $\Delta_{24}(\tau)$ [31]. By a theorem of Conway [13, Chap. 12], there is a unique Type II lattice in dimension 24 of norm 4, namely the Leech lattice L_{24}. Its theta series can be computed by using the above theorem as follows:

$$\vartheta_{L_{24}} = \vartheta_{E_8}^3 - 720\Delta_{24}.$$

1.1.5 *Modular Lattices*

A generalization of unimodular lattices is the concept of *modular lattices* [25, 26, 30]. A lattice L is said to be *ℓ-modular* for some prime number $\ell > 0$ if L is a Type II lattice and L is isometric to $\sqrt{\ell}L^*$ [25]. A more complicated definition (strongly modular lattices) holds for ℓ, a product of two primes [26]. The ring of theta series of such lattices admits a "nice" description similar to classical modular forms (two generators, one theta series, one cusp form) for ℓ being, remarkably, the square-free order of an element of the Mathieu group M_{23}, or, more explicitly, ℓ being one of the ten integers such that $\sigma_1(\ell)$, the sum of their divisors divide 24. Specifically, we have that

$$\ell \in \{1, 2, 3, 5, 6, 7, 11, 14, 15, 23\}.$$

The role of the modular group is played by $\Gamma_0(\ell)^+$, an extension of $\Gamma_0(\ell)$ by the so-called Atkin-Lehner involutions [25, 26] (see the Appendix for a formal definition). Here $\Gamma_0(\ell)$ denotes the Hecke congruence subgroup of level ℓ, consisting of those matrices in $SL_2(\mathbb{Z})$ with lower left entry multiple of ℓ.

A cusp form (see Appendix for a definition) of weight d_ℓ for $\Gamma_0(\ell)^+$ is then given by

$$\Delta_\ell(\tau) := \prod_{d|\ell} \eta(d\tau)^{24/\sigma_1(\ell)},$$

where η denotes the Dedekind η-function, which is a modular form of weight $\frac{1}{2}$, defined by

$$\eta(\tau) = q^{\frac{1}{12}} \prod_{n=1}^{\infty} (1 - q^{2n})$$

for $q = e^{\pi i \tau}$ and d_ℓ as in Table 1.

Let Λ_ℓ^0 denote the even ℓ-modular lattice of the smallest dimension. Let Θ_ℓ denote the elliptic theta series of Λ_ℓ^0, and let k_ℓ denote the weight of Θ_ℓ. Let Λ_ℓ denote the even strongly ℓ-modular lattice of the highest norm in the so-called critical dimension (E^ℓ in the notation of [22]). For lattice names and notations, we follow the Nebe–Sloane catalogue of lattices [27].

Table 1 Levels and weights.

ℓ	1	2	3	5	6	7	11	14	15	23
k_ℓ	4	2	1	2	2	1	1	2	4	1
Λ_ℓ^0	E_8	D_4	J_3	$QQF.4.a$	$J_3 \oplus \sqrt{2}J_3$	H_7	J_{11}	$H_7 + 2H_7$	$E(15)$	H_{23}
Λ_ℓ	Λ_{24}	BW_{16}	K_{12}	$Q_8(1)$	$G_2 \otimes F_4$	$A_6^{(2)}$	Λ_{11}	Λ_{14}	$E(15)$	$[4, 1; 1, 6]$
d_ℓ	12	8	6	4	4	3	2	2	2	1

The following result is a combination of [25] for ℓ prime and [26] for ℓ a product of two primes. The proof is omitted.

Theorem 3. *(Quebbemann) Assume* $\ell \in \{1, 2, 3, 5, 6, 7, 11, 14, 15, 23\}$. *If* f *is a modular form for* $\Gamma_0(\ell)^+$, *then, with the above notation,* f *is a so-called* isobaric *polynomial in* Θ_ℓ *and* Δ_ℓ. *This means that for modular form* f *of weight* k, *we have*

$$f = \sum f_{i,j} \Theta_\ell^i \Delta_\ell^j,$$

where the summation is over the indices such that $ik_\ell + jd_\ell = k$.

This result is valid in particular if f is the theta series of an ℓ-modular lattice for the ten special ℓ's.

1.2 Codes

Historically, codes are much younger than lattices as they appeared officially in 1948 as a way to detect and correct errors in transmissions [17]. Let A be a finite ring with identity. In most classical works, A was a finite field, but this has changed in the 1990's as explained in the introduction. Out of tradition, we let $q = |A|$, not to be confused with the complex variable in a theta series.

1.2.1 *Balls*

The *Hamming distance* on A^n is defined between vectors $x, y \in A^n$ as

$$d_H(x, y) = |\{i \in [n] : x_i \neq y_i\}|,$$

where $[n] = \{1, 2, \cdots, n\}$. A *code* of length n over A is just a non empty subset of A^n. Its elements are called *codewords.* The *minimum distance* d of a code $C \subseteq A^n$ is

$$d = \min\{d_H(x, y) : x, y \in C \text{ and } x \neq y\}.$$

This parameter controls the Hamming space analogue of Euclidean sphere packing. Define a ball of center x and radius r as

$$B(x, r) := \{y \in A^n : d_H(y, x) \leq r\}.$$

The largest r such that the balls of radius r centered on the codewords of C are pairwise disjoint, is called the *packing radius* or *error-correcting capacity* of C, and is usually denoted by e. Then the triangle inequality for the Hamming distance easily shows that $e = \lfloor \frac{d-1}{2} \rfloor$.

1.2.2 *Dimension*

When A is a finite field, the analogue of the notion of lattice is the notion of linear codes. A *linear code* C of length n over A is defined as the row span of a matrix G over A, which is called a *generator matrix* of C. This matrix is supposed to be of full-rank and size $k \times n$ over A. That is, we have

$$C = \{uG|\ u \in A^k\}.$$

The integer k is called the *dimension* of the code. It is the dimension of C regarded as an A-vector space. It controls the cardinality of the code, i.e., we have

$$|C| = |A|^k.$$

An important example of a linear code is the extended Hamming code of length 8 with generator matrix G as

$$G = \begin{pmatrix} 1\,0\,0\,0\,1\,1\,1\,0 \\ 0\,1\,0\,1\,0\,1\,1\,0 \\ 0\,0\,1\,1\,1\,0\,1\,0 \\ 1\,1\,1\,1\,1\,1\,1\,1 \end{pmatrix}.$$

It is customary to sum up the parameters of a code by the string $[n, k, d]$.

1.2.3 *Weight Enumerators*

Since A is equipped with the structure of an Abelian group, it is handy to call the quantity $d(0, x)$ as the *Hamming weight* $w(x)$ of the vector $x \in A^n$ in the Hamming space. Thus we have

$$d_H(x, y) = w(x - y) \text{ for all } x, y \in A^n.$$

The Hamming weight enumerator of a linear code is a generating function counting codewords by their Hamming weight:

$$W_C(x, y) = \sum_{c \in C} x^{n-w(c)} y^{w(c)} = \sum_{i=1}^{n} A_i x^{n-i} y^i,$$

where A_i is the number of codewords of weight i. Some easy examples are the *universe code* A^n for which

$$W_{A^n}(x, y) = (x + (q - 1)y)^n,$$

the *repetition code* $R_n = \langle (1, \ldots, 1) \rangle$ of length n over A for which

$$W_{R_n}(x, y) = x^n + (q - 1)y^n,$$

and the binary extended Hamming code E_8 of length 8 for which

$$W_{E_8}(x, y) = x^8 + 14x^4y^4 + y^8.$$

The reason for having two variables instead of one is to be understood in terms of invariant theory to appear in the next subsection. A generalization of the Hamming weight enumerator that we will use is the *complete weight enumerator* in $q = |A|$ variables x_a with $a \in A$

$$cwe_C((x_a)_a) = \sum_{c\in C} \prod_{a\in A} x_a^{n_a(c)}$$

where

$$n_a(c) = |\{i \in [n] : c_i = a\}|.$$

For instance, the complete weight enumerator of the universe code A^n is

$$cwe_{A^n}((x_a)_a) = \left(\sum_{a\in A} x_a\right)^n.$$

The Hamming weight enumerator can be obtained from the complete weight enumerator by letting $x_0 = x$ and $x_a = y$ for $a(\neq 0) \in A$. Of course, they coincide in the two variable case when $A = \mathbb{F}_2$.

When $A = \mathbb{Z}_m$ for some integer m, an important weight enumerator in the context of lattices is the Euclidean weight enumerator. To define this weight enumerator, let us represent the elements of $\mathbb{Z}_m$ by m integers in the $[-m/2, m/2]$ real range. The *Euclidean weight* of $x \in \mathbb{Z}_m$ is then defined as $w_E(x) = |x|^2$. The Euclidean weight of $x \in \mathbb{Z}_m^n$ is further defined as the sum of the Euclidean weights of its coordinates. The Euclidean weight enumerator of a code C of length n over A is defined as

$$EW_C((y_i)) = \sum_{c\in C} \prod_{i=0}^{\lfloor m/2 \rfloor} y_i^{N_i(c)},$$

where $N_i(c) = |\{j \in [n] : w_E(c_j) = i^2\}|$ for $0 \leq i \leq \lfloor m/2 \rfloor$. It can be obtained from the complete weight enumerator by letting $x_a = y_i$ whenever $w_E(a) = i^2$ for each $a \in \mathbb{Z}_m$. For future use, we denote by d_E, the minimum (non zero) Euclidean weight of such a code.

1.2.4 *Self-dual Codes*

There is a natural notion of the dual of a code induced by the standard inner product on A^n, which is defined by $\langle x, y \rangle = \sum_{i=1}^{n} x_i y_i$ for all $x =$

$(x_1, x_2, \cdots, x_n), y = (y_1, y_2, \cdots, y_n) \in A^n$. Given a code C of length n over A, the *dual* code $C^\perp$ of C is defined as

$$C^\perp = \{y \in A^n \mid \langle x, y\rangle = 0 \text{ for all } x \in C\}.$$

The code C is said to be *self-dual* if $C = C^\perp$. An application of the Fourier transform is the following formula due to MacWilliams:

$$W_{C^\perp}(x, y) = \frac{1}{|C|} W_C(x + (q-1)y, x - y).$$

Assume that C is a self-dual code of length n over A. Then the dimension of C is $n/2$ and $|C| = q^{n/2}$. Since W_C is a homogeneous polynomial of degree n, the above formula becomes

$$W_C(x, y) = W_C\left(\frac{x + (q-1)y}{\sqrt{q}}, \frac{x - y}{\sqrt{q}}\right)$$

when C is a self-dual code. In the special case when C is a binary self-dual code, we see that the weight of every codeword of C is even, which gives rise to the extra symmetry $W_C(x, -y) = W_C(x, y)$. Thus the weight enumerator of a binary self-dual code is a polynomial invariant under the group $\mathcal{G}_I$ of 2×2 matrices acting by linear substitutions on the variables x and y, where

$$\mathcal{G}_I = \left\langle \frac{1}{\sqrt{2}}\begin{pmatrix} 1 & 1 \\ 1 & -1 \end{pmatrix}, \begin{pmatrix} 1 & 0 \\ 0 & -1 \end{pmatrix} \right\rangle.$$

In general, the polynomials invariant under a matrix group of substitutions form a ring called the *ring of invariants*, graded by the degree of polynomials. A trivial example of such a code is the binary repetition code $R_2 = \{00, 11\}$ of length 2, whose weight enumerator is given by $\psi_2 = x^2 + y^2$. For a gentle introduction to invariant theory in the context of Coding Theory, we refer to [9]. Gleason proved the following theorem [15].

Theorem 4. *(Gleason) The ring of invariants of $\mathcal{G}_I$ is free on two generators $\psi_2 = x^2 + y^2$ and $\xi_8 = x^2y^2(x^2 - y^2)^2$.*

There are similar theorems for self-dual codes that are *divisible*, that is, the codewords weight of which are multiples of some integral constant κ. Next, a *formally self-dual code* is a code whose weight enumerator is invariant under MacWilliams transform. In a celebrated theorem [9,19], Gleason and Pierce classified divisible formally self-dual codes into the following 5 types:

I. $q = 2$ and $\kappa = 2$,
II. $q = 2$ and $\kappa = 4$,
III. $q = 3$ and $\kappa = 3$,
IV. $q = 4$ and $\kappa = 2$,
V. q arbitrary and the weight enumerator is $(x^2 + y^2)^{n/2}$.

The hypothesis of formal self-duality was further weakened to having parameters $[n, n/2]$ by Ward [33].

In Table 2, the two generators of the ring of invariants are denoted by p_1 and p_2. For Types I and IV, the polynomial p_1 is the weight enumerator of the repetition code of length 2. For Type III, the generator p_1 is the weight enumerator of the tetracode, a $[4, 2, 3]$ code with generator matrix $\begin{pmatrix} 1 & 0 & 1 & 1 \\ 0 & 1 & 1 & -1 \end{pmatrix}$. We see the generator $p_2(x, y)$ that satisfies $p_2(x, 0) = 0$ is an analogue of a cusp form. We believe, but cannot prove, that Gleason was inspired by modular form theory!

Table 2 Generators p_1 and p_2 of the ring of invariants under the group G_T.

Type T	Group G_T	p_1	p_2
I	$\left\langle \frac{1}{\sqrt{2}} \begin{pmatrix} 1 & 1 \\ 1 & -1 \end{pmatrix}, \begin{pmatrix} 1 & 0 \\ 0 & -1 \end{pmatrix} \right\rangle$	ψ_2	η_8
II	$\left\langle \frac{1}{\sqrt{2}} \begin{pmatrix} 1 & 1 \\ 1 & -1 \end{pmatrix}, \begin{pmatrix} 1 & 0 \\ 0 & i \end{pmatrix} \right\rangle$	$x^8 + 14x^4y^4 + y^8$	$x^4y^4(x^4 - y^4)^4$
III	$\left\langle \frac{1}{\sqrt{3}} \begin{pmatrix} 1 & 2 \\ 1 & -1 \end{pmatrix}, \begin{pmatrix} 1 & 0 \\ 0 & j \end{pmatrix} \right\rangle$	$x^4 + 8xy^3$	$y^3(x^3 - y^3)^3$
IV	$\left\langle \frac{1}{2} \begin{pmatrix} 1 & 3 \\ 1 & -1 \end{pmatrix}, \begin{pmatrix} 1 & 0 \\ 0 & -1 \end{pmatrix} \right\rangle$	$x^2 + 3y^2$	$y^2(x^2 - y^2)^2$

1.3 Dictionary

Since the 1970's and as illustrated in the previous sections, there has been a dictionary between codes and lattices as shown in the following table.

$\mathbb{F}_2^n$	$\mathbb{R}^n$
Hamming distance	Euclidean distance
linear codes	lattices
minimum distance	norm
dimension	determinant
Weight enumerator	Theta function
MacWilliams relations	Poisson-Jacobi formula
Self-dual codes	Unimodular lattices
Gleason formulas	Hecke theorems

This analogy is materialized by the so-called *Construction A* in the terminology of [13], which, in its simplest form, associates to a binary code C a lattice $A(C)$ through the following symbolic formula:

$$\sqrt{2}A(C) = C + 2\mathbb{Z}^n,$$

which really means

$$\sqrt{2}A(C) = \bigcup_{c\in C}(c + 2\mathbb{Z}^n),$$

where $\mathbb{F}_2 = \mathbb{Z}_2$ is identified with $\{0, 1\} \subseteq \mathbb{Z}$. The following features of Construction A [13, Th. 2, p.183] support the said analogy.

- If $C = \langle (I_k, U)\rangle$, then the matrix $M = \left\langle \begin{pmatrix} I_k & U \\ 0 & 2I_{n-k}\end{pmatrix} \right\rangle$ is a generator matrix of the lattice $A(C)$.
- If C is self-dual, so is $A(C)$.
- $\det(A(C)) = 2^{n-k}$.
- The norm of the lattice $A(C)$ is $\min\{2, \frac{d}{2}\}$.
- $\theta_{A(C)} = W_C(\theta_{\mathbb{Z}}, \theta_{\mathbb{Z}+\frac{1}{2}})$.

This is very consistent, but from a lattice density view point, Construction A limits the norm of the lattice to 2. For dimensions > 8, a more sophisticated construction is needed to reach norms > 2: Construction B in the terminology of [13] (cf. §3.1 for the definition). For instance, to construct the very dense and very symmetric Leech lattice (no less than three finite sporadic simple groups are built from it), John Leech in 1965 used the extended Golay code, Construction B, and sliding a coset to "fill-in the holes" [13].

1.3.1 *Construction A_4*

To simplify Construction B, and more importantly, to reach norms > 2 for self-dual lattices, Bonnecaze, Solé, and Calderbank [3] introduced construction A_4 which associates the lattice $A_4(C)$ to a quaternary code C by the formula

$$2A_4(C) = C + 4\mathbb{Z}^n.$$

The notion of a Type II $\mathbb{Z}_4$-code was introduced in [3]. A $\mathbb{Z}_4$-code C is said to be a Type II code if it is self-dual and the Euclidean weight of each of its codewords is a multiple of 8. The following facts [19, Th. 12.5.12] show that the analogy between $\mathbb{Z}_4$-codes and lattices might be even closer than the analogy between binary codes and lattices. Assume a quaternary code of type $4^{k_1}2^{k_2}$ as an Abelian group and generator matrix of the type

$$\begin{pmatrix} I_{k_1} & U & V \\ 0 & 2I_{k_2} & 2W \end{pmatrix},$$

where the matrices U, W have binary entries and the matrix V has quaternary entries.

- $A_4(C)$ is the row integer span of the matrix

$$\begin{pmatrix} I_{k_1} & U & V \\ 0 & 2I_{k_2} & 2W \\ 0 & 0 & 4I_{n-k_1-k_2} \end{pmatrix}.$$

- C is a self-dual code if and only if $A_4(C)$ is a self-dual lattice.
- C is a Type II code if and only if $A_4(C)$ is a Type II code.
- $\det(A_4(C)) = 4^{n-2k_1-k_2}$.
- The norm of $A_4(C)$ is $\min\{4, \frac{d_E}{4}\}$.
- $\theta_{A_4(C)} = W_C(\theta_{\mathbb{Z}}, \theta_{\mathbb{Z}+\frac{1}{2}})$.

For instance, the lattice E_8 can be obtained by this construction applied to the Klemm code $K_8 = R_8 + 2P_8^{\perp}$ of length 8. The first benefit is a simple interpretation of construction B for a binary code C_2 as construction A_4 applied to $C_2 + 2P_n$, where P_n is the single parity check code of length n. This approach yields a simple proof for many theta series identities, old and new [32]. The above table becomes

$\mathbb{Z}_4^n$	R^n
Euclidean distance	Euclidean distance
Lee weight enumerator	Theta function
MacWilliams relations	Poisson-Jacobi formula
Self-dual codes	Unimodular Lattices
Type II codes	Unimodular Even Lattices

where the Euclidean weight of $0, \pm 1, 2$ is 0,1,4, respectively. The so-called Lee weight enumerator or symmetrized weight enumerator is the trivariate generating function associated to this weight. Type II codes are self-dual codes containing the all-one vector and with Euclidean weights multiple of 8 [3]. This class contains half of the quadratic residue codes introduced in [2]. These latter codes give the Gosset lattice in dimension 8 (case of the octacode), the Leech lattice in dimension 24 [2], and an extremal lattice of the same sound but with a different shape than the Barnes-Wall lattice in dimension 32 [3]. The quaternary Reed-Muller code of length 32 yields the Barnes-Wall lattice [2]. In the same spirit, lifting a doubly circulant code of length 40 yields an extremal Type II lattice in [6], where it is conjectured that there should be nine quaternary constructions of the Leech lattice like there are four quaternary constructions of the Gosset lattice E_8 [2].

1.3.2 *Generalized Constructions A*

What if we want norm 8? Well, use codes over $\mathbb{Z}_8$! Very recently, the first Type II lattice of norm 8 was constructed by Nebe in dimension 72 [20]. A coding construction modulo 8 was found thereafter by Harada et al. [18]. In general, the construction A_{2^a} is defined in [5] as

$$2^{a/2} A_{2^a}(C_a) = C_a + 2^a \mathbb{Z}^n,$$

where C_a denotes a code of length n over the integers mod 2^a. Again this construction sends self-dual codes into unimodular lattices, and, more surprisingly, the lift of extended binary self-dual cyclic codes of length $\equiv -1 \pmod{2^{a+1}}$ into Type II lattices [5, Corollary 2.6].

In general, any ring R that can be construed as the quotient of the ring of integers in a number field, or of a maximal order in a quaternion algebra, can lead to alternative Constructions A, and often to constructions of modular lattices [1, 7–11]. For instance, the Construction A_c in [13, Chap. 7] leads to 2-modular and 3-modular lattices by taking two different quotients from the ring of integers of $\mathbb{Q}(\sqrt{-3})$, the so-called ring of Eisenstein integers.

1.4 Appendix: Modular Forms and Atkin-Lehner Involutions

1.4.1 *Modular Forms on* $\Gamma = SL_2(\mathbb{Z})$

For a complex-valued function f defined on $\mathcal{H} = \{\tau \in \mathbb{C} \,|\, \Im(\tau) > 0\}$, a matrix $A = \begin{pmatrix} a & b \\ c & d \end{pmatrix} \in \Gamma$ and a fixed non-negative integer k, define the slash operator on f as

$$f|_A(\tau) = \left(\frac{c\tau + d}{\chi(A)}\right)^{-k} f\left(\frac{a\tau + b}{c\tau + d}\right),$$

where χ is the character of Γ defined by

$$\chi\begin{pmatrix} 0 & -1 \\ 1 & 0 \end{pmatrix} = i \text{ and } \chi\begin{pmatrix} 1 & 1 \\ 0 & 1 \end{pmatrix} = 1.$$

Note that

$$f|_{A_1 A_2} = (f|A_1)\,|_{A_2}$$

and

$$f|_{\alpha_1 A_1 + \alpha_2 A_2} = \alpha_1 (f|_{A_1}) + \alpha_2 (f|_{A_2})$$

for all $A_1, A_2 \in SL_2(\mathbb{Z})$ and $\alpha_1, \alpha_2 \in \mathbb{C}$.

A complex-valued function f defined on $\mathcal{H}$ is called a modular form attached to a subgroup G of finite index in Γ if it satisfies the following two conditions for some integer $k \geq 0$:

- $f(\tau) = f|_A(\tau) = \left(\frac{c\tau+d}{\chi_G(A)}\right)^{-k} f\left(\frac{a\tau+b}{c\tau+d}\right)$ for all $A \in G$, where χ_G is a character of G.
- $f(\tau) = \sum_{n=0}^{\infty} \alpha_n q^n$, where $q = e^{2\pi i \tau}$ with $\tau \in \mathcal{H}$ and $\alpha_n \in \mathbb{C}$ for all $n \geq 0$.

The integer $k \geq 0$ is called the weight of the modular form f. Further, the modular form f on $\mathcal{H}$ is said to be holomorphic at the cusp if f is holomorphic as $\tau \to i\infty$.

1.4.2 *Cusp Form of Weight k*

For a positive integer N, consider the congruence subgroup $\Gamma_0(N)$ of the modular group Γ, defined as

$$\Gamma_0(N) = \left\{ \begin{pmatrix} a & b \\ c & d \end{pmatrix} \in \Gamma : c \equiv 0 \pmod{N} \right\}.$$

Note that $\Gamma_0(1) = \Gamma$. Further, if M is a positive divisor of N, then $\Gamma_0(N)$ is a subgroup of $\Gamma_0(M)$.

Note that any element of Γ maps $\mathcal{H}$ into itself. The points of compactification of the connected fundamental domain of $\Gamma_0(N)$ on the boundary of $\mathcal{H}$ are called cusps of $\Gamma_0(N)$.

Now a modular form f attached to $\Gamma_0(N)$ is called a cusp form of weight k on $\Gamma_0(N)$ if it satisfies the following three conditions:

- f is holomorphic on $\mathcal{H}$.
- f is zero at each cusp of $\Gamma_0(N)$.
- $f(\tau) = f|_A(\tau) = \left(\frac{c\tau+d}{\chi(A)}\right)^{-k} f\left(\frac{a\tau+b}{c\tau+d}\right)$ for each $A \in \Gamma_0(N)$.

1.4.3 *Atkin-Lehner Involutions*

Now for a positive divisor e of N satisfying $\gcd(e, N/e) = 1$, define an integral matrix W_e as follows:

$$W_e = \begin{pmatrix} ae & b \\ cN & de \end{pmatrix}$$

with $\det(W_e) = e$. If $A \in \Gamma_0(N)$, then $W_e A W_e^{-1} \in \Gamma_0(N)$. The matrix W_e^2 has determinant e^2 and can be expressed as $W_e^2 = eD$ for some $D \in \Gamma_0(N)$. As $f|_{W_e^2} = f$ for all cusp forms of weight k on $\Gamma_0(N)$, the matrix operator W_e is an involution on the space of all cusp forms of weight k on $\Gamma_0(N)$ and is called *the Atkin-Lehner involution* of $\Gamma_0(N)$.

References

[1] C. Bachoc, Applications of Coding Theory to the Construction of Modular Lattices, *Jour. Comb. Th. (A)*, **(78)**, (1997), 92–119.

[2] A. Bonnecaze, P. Solé, A. R. Calderbank, Quaternary Construction of Unimodular Lattices IEEE Trans. Inform. Theory, **41**, (1995) 366–376.

[3] A. Bonnecaze, P. Solé, C. Bachoc, B. Mourrain, Type II quaternary codes, IEEE Trans. Inform. Theory **43**, (1997), 969–976.

[4] H. Brown, R. Bülow, J. Neubüser, Wondratschek, Hans; Zassenhaus, Hans (1978), *Crystallographic groups of four-dimensional space*, New York: Wiley-Interscience.

[5] A. R. Calderbank, W.-C. W. Li, B. Poonen, A 2-adic approach to the analysis of cyclic codes, IEEE Trans. Inform. Theory, **43**, (1997), 977–986.

[6] A. R. Calderbank, N. J. A. Sloane, Double circulant codes over $\mathbb{Z}_4$ and even unimodular lattices, Journal of Algebraic Combinatorics, **6**, (1997), 119–131.

[7] H. H. Chan, K. S. Chua, P. Solé, Quadratic iterations to π associated with elliptic functions to the cubic and septic base, Trans. of the AMS **355**, (2003) 1505–1520.

[8] H. H. Chan, K. S. Chua, P. Solé, 7-modular lattices and septic base Jacobi identity, J. Number Theory, **99**, (2003), 361–372.

[9] Y. Choie, S. T. Dougherty, Codes over Σ_{2m} and Jacobi forms over the quaternions, Appl. Algebra Engrg. Comm. Comput. 15 (2004), no. 2, 129–147.

[10] Y. Choie, S. T. Dougherty, Codes over rings, complex lattices and Hermitian modular forms. European J. Combin. 26 (2005), no. 2, 145–165.

[11] Y. Choie, P. Solé, Self-dual codes over $\mathbb{Z}_4$, and half-integral weight modular forms. Proc. Amer. Math. Soc. 130 (2002), no. 11, 3125–3131.

[12] H. Cohn, A. Kumar, S. D. Miller, D. Radchenko, M. Viazovska, The sphere packing problem in dimension 24, Annals of Math **185**, (2017), 1017–1033.

[13] J. H. Conway, N. J. A. Sloane, *Sphere Packings, Lattices and Groups*, Springer (1992).

[14] N. D. Elkies, Lattices, Linear Codes, and Invariants (2-part expository article), Notices of the American Math. Society **47** (2000), 1238–1245 and 1382–1391.

[15] M. Gleason, Weight polynomials of self-dual codes and the MacWilliams identities. Actes du Congrés International des Mathématiciens (Nice, 1970), Tome 3, pp. 211–215. Gauthier-Villars, Paris, 1971.

[16] T. C. Hales, A proof of the Kepler conjecture, Annals of Math. (2) **162** (2005), no. 3, 1065–1185.

[17] R. W. Hamming, Error detecting and error correcting codes. Bell System Tech. J. **29**, (1950). 147–160.

[18] M. Harada, M. Tsuyoshi, On the existence of extremal Type II $\mathbb{Z}_{2k}$-codes. Math. Comp. **83** (2014), no. 287, 1427–1446.

[19] W. C. Hufman, V. Pless, *Fundamentals of Error Correcting Codes*, Cambridge (2003).

[20] D. Micianccio, S. Goldwasser, *Complexity of Lattice Problems: a cryptographic perspective.* Kluwer Academic Publishers, Boston (2002).

[21] F. J. MacWilliams, N. J. A. Sloane, *The theory of error correcting codes*, North-Holland (1977).

[22] D. Mumford ,*Abelian varieties*, Tata Institute of Fundamental Research Studies in Mathematics, 5, [1970], Providence, R.I.: American Mathematical Society (2008).

[23] G. Nebe, An even unimodular 72-dimensional lattice of minimum 8. J. Reine Angew. Math. **673** (2012), 237–247.

[24] V. Pless, W. C. Huffman, *Handbook of Coding Theory*, North Holland (1998), vol. I.

[25] H. G. Quebbemann, Modular lattices in euclidean spaces, J. Number Th. **54** (1995) 190–202.

[26] H. G. Quebbemann, Atkin-Lehner eigenforms and strongly modular lattices, Ens. Math. **43** (1997) 55–65.

[27] `http://www.research.att.com/ njas/lattices/index.html`

[28] M. S. Raghunathan, *Discrete subgroups of Lie groups.* Ergebnisse de Mathematik und ihrer Grenzgebiete. Springer-Verlag (1972).

[29] E. M. Rains, N. J. A. Sloane, Self-dual codes, chapter I of [24].

[30] E. M. Rains, N. J. A. Sloane, The shadow theory of modular lattices, J. Number Theory, 73 (1998), pp. 359–389.

[31] S. Ramanujan, On certain arithmetical functions, Trans. Cambr. Phil. Soc. **22**(9) (1916) 159–184.

[32] P. Solé, P. Loyer, U_n lattices, construction B and AGM iterations, European Journal of Combinatorics, **19**(2), (1998) 227–236.

[33] M. S. Viazovska, The sphere packing problem in dimension 8, Annals of Math **185**, (2017), 991–1015.

[34] H. Ward, Divisible codes, Archiv. Math Basel, **36**, (1981), 485–494.

[35] D. Zagier, Elliptic modular forms and their applications, in J. H. Bruinier, G. van der Geer, G. Harder, D. Zagier, *The* $1-2-3$ *of modular forms*, Universitext, Springer (2008).

Chapter 2

Modular Forms in One Variable

2.1 Broué-Enguehard Map

As explained in Chapter I, Construction A is a partial explanation of the dictionary between codes and lattices. In analytic terms, this leads to the equation

$$\theta_{A(C)} = W_C\left(\theta_{\mathbb{Z}}, \theta_{\mathbb{Z}+\frac{1}{2}}\right),$$

which connects the theta series of the lattice $A(C)$ to the weight enumerator of the code C. Now, all invariants are not weight enumerators, and all modular forms are not theta series. Can the preceding formula be extended to these more general objects? Let us define a Broué-Enguehard map BE as follows:

$$f \in \mathbb{C}[x, y] \mapsto f(\theta_{\mathbb{Z}}, \theta_{\mathbb{Z}+\frac{1}{2}}).$$

Let $\mathbb{C}[x, y]^{\mathcal{G}}$ denote the ring of homogeneous polynomials in $\mathbb{C}[x, y]$ that are invariant under the matrix group $\mathcal{G}$. Recall that $\mathcal{G}_I$ and $\mathcal{G}_{II}$ denote the matrix groups acting on the weight enumerators of codes of Type I and Type II, respectively. Likewise, the rings of modular forms containing the theta series of lattices of Type I and Type II are denoted by $\mathcal{M}$ and $\mathcal{M}'$, respectively.

Theorem 5. *([2, Theorem 7]) The map BE is a ring isomorphism from $\mathbb{C}[x, y]^{\mathcal{G}_I}$ onto $\mathcal{M}$ and from $\mathbb{C}[x, y]^{\mathcal{G}_{II}}$ onto $\mathcal{M}'$.*

Proof. (sketch) The map BE is a ring morphism. The theorems of Gleason and Hecke of Chapter I give some generators for the rings in the statement.

The result then follows by checking that BE maps some system of generators of $\mathbb{C}[x,y]^{\mathcal{G}_I}$ (resp. $\mathbb{C}[x,y]^{\mathcal{G}_{II}}$) onto a system of generators of $\mathcal{M}$ (resp. $\mathcal{M}'$). For instance, it is easy to check in the Type I case that

$$BE(\psi_2) = \theta_{\mathbb{Z}} \quad \text{and} \quad BE(\xi_8) = \Delta_8.$$

The Type II case is dealt with in a similar manner, but requires the knowledge of the binary extended Golay code [7]. □

2.2 Ozeki Map

Ozeki [10] defined formal weight enumerators as relative invariants for the group $\mathcal{G}_{II}$ and the character Ψ, defined by

$$\Psi\left(\begin{pmatrix}1 & 0\\ 0 & i\end{pmatrix}\right) = 1,$$

$$\Psi\left(\frac{1}{\sqrt{2}}\begin{pmatrix}1 & 1\\ 1 & -1\end{pmatrix}\right) = -1.$$

In other words, they are homogeneous polynomials in x, y with exponents of y multiples of 4 and transformed under the MacWilliams transform of §2.4 into their opposite. The elements g of $\mathcal{G}_{II}$ that satisfy $\Psi(g) = 1$ form a subgroup $\mathcal{H}_{II}$ (say) of $\mathcal{G}_{II}$. Denote by $\mathcal{M}''$, the ring of modular forms of even weight. Thus we have $\mathcal{M}'' \subseteq \mathcal{M}'$.

Theorem 6. *([10, Proposition 4]) The map BE is a ring isomorphism from $\mathbb{C}[x,y]^{\mathcal{H}_{II}}$ onto $\mathcal{M}''$.*

Proof. (sketch) The map BE is a ring morphism. The matrix equations combined with the transformation laws for $\theta_{\mathbb{Z}}$, $\theta_{\mathbb{Z}+\frac{1}{2}}$ show that the image of an invariant of the group $\mathcal{H}_{II}$ under the map BE transforms like a modular form of $\mathcal{M}''$. □

It is well-known [7] that $\mathcal{M}''$ is generated by the Eisenstein series $E_4 = \theta_{E_8}$ and E_6, which is defined by its q-series expansion

$$E_6 = 1 - 504\sum_{n=1}^{\infty} \sigma_5(n)q^n.$$

On the other hand, Ozeki shows that $\mathbb{C}[x,y]^{\mathcal{H}_{II}}$ is generated by ψ_8 and κ_{12}, a polynomial encountered by Klein [15, p. 54] in his research on the icosahedron:

$$\kappa_{12}(x,y) = x^{12} - 33x^8y^4 - 33x^4y^8 + y^{12}.$$

As noticed by several authors [2, 11], there is an analogous result in relation with Type III codes. Denote by $\mathcal{G}_{III}$, the group that preserves the weight enumerator of these codes. Define the map BE_3 by the rule $f(x,y) \mapsto f(\theta_{3\mathbb{Z}}, \theta_{3\mathbb{Z}+1})$.

Theorem 7. *([2, 11]) The map BE_3 is a ring isomorphism from $\mathbb{C}[x,y]^{\mathcal{H}_{II}}$ onto $\mathcal{M}''$.*

Proof. The same argument as in the preceding theorem can be made. One can also look at the image of the generators of $\mathbb{C}[x,y]^{\mathcal{H}_{II}}$ as given in Gleason formula (Theorem 4 of Chapter I) under the BE map. □

2.3 Rankin-Cohen Brackets

The main content in this section follows [6].

2.3.1 *Background*

Let $f \in \mathcal{M}'_k$ and $g \in \mathcal{M}'_\ell$ be modular forms of weights k and ℓ, respectively. Consider the differential operator $D = q\frac{d}{dq}$. The Rankin-Cohen (RC) bracket of order ν, hereby denoted by $[f,g]_\nu$, associates a modular form of weight $k+\ell$ to the pair (f,g), as follows:

$$[f,g]_\nu = H_\nu(k,\ell; D_{\tau_1} D_{\tau_2}) fg|_{\tau_1=\tau_2=\tau},$$

where $H_\nu(k,\ell;X,Y)$ is the bivariate polynomial, defined as

$$H_\nu(k,\ell;X,Y) = \sum_{r+s=\nu} \binom{\nu+k-1}{s}\binom{\nu+\ell-1}{s} X^r Y^s.$$

These differential operators were introduced by Cohen [8] in 1975 and studied by Zagier [21] and Connes et al. [9] in the 1990s. Since then, the references have been too many to be collected here.

2.3.2 *RC vs. BE*

In this paragraph, we use *BE* maps of the previous sections to construct a RC bracket for some spaces of polynomial invariants. For simplicity, denote by ϕ_2 (resp. ϕ_3), the BE map attached to binary (resp. ternary) codes. It is natural to define the RC bracket of invariants in $\mathbb{C}[x,y]^{\mathcal{G}_{II}}$ or $\mathbb{C}[x,y]^{\mathcal{H}_{II}}$ as the RC bracket of their images by ϕ_2 or ϕ_3 in (in both cases) the ring $\mathcal{M}''$. It is well-known that the derivation D acts on $\mathcal{M}''_k$ as

$$Df = \frac{k}{12} + \delta(f),$$

where δ is a derivation of order 2 on the ring $\mathcal{M}''$ (see [21, eq. (32)]). To define the analogue $\mathcal{D}$ of δ, write Q, R for the two generators of either $\mathbb{C}[x,y]^{\mathcal{G}_{II}}$ and $\mathbb{C}[x,y]^{\mathcal{H}_{II}}$. Then we define $\mathcal{D}$ as the conjugate by the BE map of δ, or more explicitly

$$\mathcal{D} = -\frac{R}{3}\frac{\partial}{\partial Q} - \frac{Q^2}{2}\frac{\partial}{\partial R}.$$

Some expressions of $\mathcal{D}$ in terms in variables x, y can be found in [6, Cor. 1 and 2]. The main result of this section is as follows:

Theorem 8. *The RC bracket for invariants K, L in either $\mathbb{C}[x,y]^{\mathcal{G}_{II}}$ of degree $2k, 2\ell$ or $\mathbb{C}[x,y]^{\mathcal{H}_{II}}$ of degree k, ℓ is given by*

$$\sum_{r+s=\nu,\, 1\le r,s\le \nu-1} \binom{\nu+k-1}{s}\binom{\nu+\ell-1}{s} f_r g_s,$$

where f_r and g_s are defined by the recurrences

$$f_{r+1} = \mathcal{D}(f_r) - \frac{Q}{144} r(r+k-1) f_{r-1}, \text{ and}$$

$$g_{s+1} = \mathcal{D}(g_s) - \frac{Q}{144} s(s+\ell-1) f_{r-1},$$

of initial conditions $f_0 = K$ and $g_0 = L$, respectively.

2.4 Half-integral Weight

2.4.1 *Background*

We need a refinement of Hecke theorem of the previous chapter. Let Γ be an arbitrary subgroup of finite index in $SL_2(\mathbb{Z})$, and let χ be a character of Γ. A complex valued function $f : \mathcal{H} \to \mathbb{C}$ is called a modular form of weight ω *for* Γ *with respect to* χ if

(1) $f(\tau)$ is holomorphic for $Im(\tau) > 0$;

(2) $f\left(\frac{a\tau+b}{c\tau+d}\right) = \left(\frac{c\tau+d}{\chi(M\tau)}\right)^{\omega} f(\tau)$ for all $M = \begin{pmatrix} a & b \\ c & d \end{pmatrix} \in \Gamma$;

(3) $f(\tau)$ is "holomorphic" at every cusp of Γ.

Note that the definition of modular form in chapter one is the special case of $\Gamma = SL_2(\mathbb{Z})$ and $\chi = 1$. Now let us recall the following classical result studied by Hecke (see [7, p.187] for a short account).

Theorem 9.

Let $\mathcal{M}_{\frac{n}{2}}(\Gamma_0(4), id)$ be the space of modular forms of weight $\frac{n}{2}$ on $\Gamma_0(4)$ with trivial character $\chi = id$. Then

$$\mathcal{M}(\Gamma_0(4), id) = \bigoplus_{n=0}^{\infty} \mathcal{M}_{\frac{n}{2}}(\Gamma_0(4), id) = \mathbb{C}[\theta_3(2\tau), g(\tau)].$$

Here, $g(\tau)$, defined as

$$g(\tau) := \theta_2^4(2\tau) - 2\theta_2^2(2\tau)\theta_4^2(2\tau), \tag{2.1}$$

is a modular form of weight 2 in $\Gamma_0(4)$.

Remark 10.

(1) In general, it is known that the ring $\mathcal{M}_{\frac{n}{2}}(\Gamma_0(4), id)$ is isomorphic to $\mathbb{C}[\theta_3(\tau), F_2(\tau)]$, where F_2 is any modular form of weight 2 on $\Gamma_0(4)$, which is linearly independent from $\theta_3(\tau)^4$.
(2) One can check, from Proposition 31, that $g(\tau) := \theta_2^4(\tau) - 2\theta_2^2(2\tau)\theta_4^2(2\tau)$ is a modular form of weight 2 on $\Gamma_0(4)$.

2.4.2 *Some Special Theta Series*

We also recall the following theta series $\theta_{2,\mu}(\tau, z)$ which was introduced to find the correspondence between Jacobi forms and modular forms of half integral weight (see [8]):

$$\theta_{2,\mu}(\tau, z) := \sum_{r\in\mathbb{Z},\ r\equiv\mu \pmod{4}} e^{\frac{\pi i r^2}{4}\tau}\xi^r, \quad \text{where} \quad \xi = e^{2\pi i z}.$$

In particular, when $m = 2$, we get the following relations among the theta series:

$$\theta_{2,0}(\tau, 0) = \frac{\theta_3(\tau) + \theta_4(\tau)}{2},$$

$$\theta_{2,1}(\tau, 0) = \theta_{2,3}(\tau, 0) = \frac{\theta_2(\tau)}{2},$$

$$\theta_{2,2}(\tau, 0) = \frac{\theta_3(\tau) - \theta_4(\tau)}{2},$$

where the three Jacobi "theta-nulls" θ_3, θ_2, θ_4 are defined by

$$\theta_2(\tau) = \sum_{\ell \in \mathbb{Z}} q^{(\ell+\frac{1}{2})^2}, \theta_3(\tau) = \sum_{\ell \in \mathbb{Z}} q^{\ell^2},$$

$$\theta_4(\tau) = \theta_3(\tau + 1) = \sum_{\ell \in \mathbb{Z}} (-q)^{\ell^2} \text{ with } q = e^{\pi i \tau}.$$

The following relations are well-known identities among theta series.

Proposition 11. *For all τ in the upper half-plane, we have*

$$\theta_2(\tau + 1) = \sqrt{i}\theta_2(\tau),$$

$$\theta_3(\tau + 1) = \theta_4(\tau),$$

$$\theta_4(\tau + 1) = \theta_3(\tau),$$

$$\theta_2(-\frac{1}{\tau}) = (\frac{\tau}{i})^{\frac{1}{2}}\theta_4(\tau),$$

$$\theta_3(-\frac{1}{\tau}) = (\frac{\tau}{i})^{\frac{1}{2}}\theta_3(\tau),$$

$$\theta_4(-\frac{1}{\tau}) = (\frac{\tau}{i})^{\frac{1}{2}}\theta_2(\tau)$$

$$\theta_3(\tau)\theta_4(\tau) = \theta_4(2\tau)^2,$$

$$\theta_2(\tau)^4 + \theta_4(\tau)^4 = \theta_3(\tau)^4,$$

$$\theta_3(\tau)^2 - \theta_4(\tau)^2 = 2\theta_2(2\tau)^2.$$

Proof. See, for instance, [7]. For a coding-theoretic proof of the Jacobi identity on fourth powers, see [20]. □

2.4.3 *Quaternary Codes*

This paragraph follows [6]. A linear code C of length n over $\mathbb{Z}_4$ is an additive subgroup of $\mathbb{Z}_4^n$ and an element of C is called a codeword. We denote by $|C|$ the number of codewords in C. The inner product of x and y in $\mathbb{Z}_4^n$ is given by

$$\langle x, y\rangle = x_1y_1 + \cdots + x_ny_n \pmod 4.$$

The dual code $C^\perp$ of C is defined as $C^\perp = \{y \in \mathbb{Z}_4^n \mid \langle x, y\rangle = 0 \text{ for all } x \in C\}$. If $C = C^\perp$, then the code C is called a self-dual code. The complete weight enumerator cwe_C of C is defined as

$$cwe_C(W, X, Y, Z) = \sum_{u\in C} W^{n_0(u)}X^{n_1(u)}Y^{n_2(u)}Z^{n_3(u)},$$

where $n_k(u)$ denotes the number of components of u that are equal to k modulo 4. The symmetrized weight enumerator swe_C of C over $\mathbb{Z}_4$ is given by

$$swe_C(W, X, Y) = cwe_C(W, X, Y, X).$$

The MacWilliams identities for the complete and symmetrized weight enumerators can be written as follows.

Theorem 12. *For all linear codes C, we have:*

(1) $cwe_{C^\perp}(W, X, Y, Z) = \frac{1}{|C|}cwe_C(W + X + Y + Z, W + iX - Y - iZ, W - X + Y - Z, W - iX - Y + iZ)$,
(2) $swe_{C^\perp}(W, X, Y) = \frac{1}{|C|}swe_C(W + 2X + Y, W - Y, W - 2X + Y)$.

Proof. See [5] or [7], for instance. □

Corollary 13. *Let C be a self-dual code of length n over $\mathbb{Z}_4$.*

(1) Let G_4 be the group generated by

$$M_4 = \frac{1}{2}\begin{pmatrix} 1 & 1 & 1 & 1 \\ 1 & i & -1 & -i \\ 1 & -1 & 1 & -1 \\ 1 & -i & -1 & i \end{pmatrix}$$

and

$$N_4 = \begin{pmatrix} 1 & 0 & 0 & 0 \\ 0 & i & 0 & 0 \\ 0 & 0 & 1 & 0 \\ 0 & 0 & 0 & i \end{pmatrix}.$$

Then $cwe_C(W, X, Y, Z) \in \mathbb{C}[W, X, Y, Z]^{G_4}$.

(2) Let G_3 *be the matrix group generated by the matrices*

$$N_3 = \begin{pmatrix} 1 & 0 & 0 \\ 0 & i & 0 \\ 0 & 0 & 1 \end{pmatrix}$$

and

$$M_3 = \frac{1}{2} \begin{pmatrix} 1 & 2 & 1 \\ 1 & 0 & -1 \\ 1 & -2 & 1 \end{pmatrix}.$$

Then $swe_C(W, X, Y)$ *is in* $\mathbb{C}[W, X, Y]^{G_3}$.

Proof. The results follows directly from the functional equations of $\theta_{2,\mu}(\tau, z)$ and $\theta_{\Lambda(C),2}(\tau, z)$ and by applying Theorem 12. So, we omit the detailed proof. □

The following result is proved in [6]. It can be thought out as a BE map for symmetrized weight enumerators.

Theorem 14. *The map,*

$$\tilde{\phi} : \mathbb{C}[W, X, Y]^{G_3} \to \mathbb{C}[\theta_3(2\tau), g(\tau)],$$

defined as

$$\tilde{\phi}(f(W, X, Y)) = f\left(\frac{\theta_3(2\tau) + \theta_4(2\tau)}{2}, \frac{\theta_2(2\tau)}{2}, \frac{\theta_3(2\tau) - \theta_4(2\tau)}{2}\right),$$

is an algebra homomorphism. Here $g(\tau)$ *is a modular form of weight* 2 *defined by (2.1).*

2.5 Higher Levels

2.5.1 *Background*

The aim of the present section, based on [5], is to give a unified Broué-Enguehard map for each of the ten special levels given in Chapter I, subsection 1.5. However we were only able to compute it when the alphabet of the codes occurring in Construction A was one of the three rings of order four and characteristic two: $\mathbb{F}_4$ (levels 3, 6, 11), $\mathbb{F}_2 \times \mathbb{F}_2$ (levels 1, 2, 5), $\mathbb{F}_2 + u\mathbb{F}_2$ (levels 7, 14, 23). So level 15 is left as an open problem. The expert reader will note that the maps in levels 1 and 3 are different from the original maps of [2], which were obtained from binary and ternary codes, respectively. Also an Ozeki type enlargment of the space of invariants was needed for some levels.

2.5.2 *Generalized Construction A*

Let Z_ℓ denote an ℓ-modular lattice of dimension 2 (ℓ prime) or 4 (ℓ composite). Assume a ring structure on Z_ℓ (ring of integers of a number field). Let R_ℓ denote the quotient ring of Z_ℓ by some ideal I_ℓ that serves as the alphabet for our codes. Let $w_\ell := \frac{-1+\sqrt{-\ell}}{2}$. With this datum, define Construction A_ℓ as

$$A_\ell(C) = \{x \in Z_\ell^n \,|\, \exists c \in C,\, x \equiv c \pmod{I_\ell}\}.$$

For the desired ℓ's, it is known by computing the different of a suitable number field [1] that $A_\ell(C)$ is ℓ-modular if C is a self-dual code over R_ℓ. Define *formally* the Broué Enguehard substitution BE_ℓ, for instance, as

$$BE_\ell(x) = \theta_{I_\ell}(q),$$

and so on; the variable attached to the element a in the quotient ring R_ℓ being replaced by the theta series of the inverse image of a by reduction modulo I_ℓ. It is also a folklore theorem that $\theta_{A_\ell(C)}$ is obtained by doing the Broué Enguehard substitution in the suitable weight enumerator of C (either W_C or swe_C according to the number of variables).

Proposition 15. *If $\ell = 3, 6$ and C is a code over $\mathbb{F}_4$, then we have*

$$\theta_{A_\ell(C)} = W_C(BE_\ell(x), BE_\ell(y)).$$

If $\ell \in \{1, 2, 5, 7, 11, 14, 23\}$ *and* C *is a code over* R_ℓ*, then we have*

$$\theta_{A_\ell(C)} = swe_C(BE_\ell(x), BE_\ell(y), BE_\ell(z)).$$

ℓ	1	2	3	5	6	7	11	14	23
Z_ℓ	$\mathbb{Z}_{[\sqrt{-1}]}$	$\mathbb{Z}_{[\sqrt{-2}]}$	$\mathbb{Z}_{[w_3]}$	$\mathbb{Z}_{[\sqrt{-5}]}$	$\mathbb{Z}_{[\sqrt{2},\, w_3]}$	$\mathbb{Z}_{[w_7]}$	$\mathbb{Z}_{[w_{11}]}$	$\mathbb{Z}_{[\sqrt{2},\, w_7]}$	$\mathbb{Z}_{[w_{23}]}$
I_ℓ	(2)	(2)	(2)	(2)	$(\sqrt{2})$	(2)	(2)	$(\sqrt{2})$	(2)
R_ℓ	$\mathbb{F}_2 + u\mathbb{F}_2$	$\mathbb{F}_2 + u\mathbb{F}_2$	$\mathbb{F}_4$	$\mathbb{F}_2 + u\mathbb{F}_2$	$\mathbb{F}_4$	$\mathbb{F}_2 \times \mathbb{F}_2$	$\mathbb{F}_4$	$\mathbb{F}_2 \times \mathbb{F}_2$	$\mathbb{F}_2 \times \mathbb{F}_2$

Table 2: Lattices and alphabets

For nine of the ten special levels ℓ, we *explicitly* define the Broué Enguehard map BE_ℓ by its substitutions on x, y (two variable case) or x, y, z (three variable case). We specify the inverse images of Θ_ℓ and Δ_ℓ. See [5] for details.

The polynomial invariants are named as per the appendix. For $d = 1, 2, 3, 6$, we define the quadratic $Q_d(x, y) := x^2 + xy + dy^2$, and from there

$$A_d(\tau) = \sum_{m,n\in\mathbb{Z}} q^{2Q_d(m,n)},$$
$$C_d(\tau) = \sum_{m,n\in\mathbb{Z}} q^{2Q_d(m+1/2,n)},$$
$$G_d(\tau) = \sum_{m,n\in\mathbb{Z}} q^{2Q_d(m,n+1/2)},$$
$$H_d(\tau) = \sum_{m,n\in\mathbb{Z}} q^{2Q_d(m+1/2,n+1/2)},$$

which can be evaluated [7] as

$$A_d(\tau) = \theta_3(q^2)\theta_3(q^{8d-2}) + \theta_2(q^2)\theta_2(q^{8d-2}),$$
$$C_d(\tau) = \theta_2(q^2)\theta_3(q^{8d-2}) + \theta_3(q^2)\theta_2(q^{8d-2}),$$
$$2G_d(2\tau) = \theta_2(q)\theta_2(q^{4d-1}),$$
$$2H_d(2\tau) = \theta_2(q)\theta_2(q^{4d-1}).$$

2.5.2.1 $\ell = 1$

The BE_1 map is defined by

$$x = \theta_3^2(q),\ y = \theta_2^2(q),\ z = \theta_3(q)\theta_2(q).$$

With this notation

$$BE_1(Q_1^4+8Q_2^2-4Q_1^2Q_2-4Q_4)=\Theta_1=E_4,\ BE_1((Q_2^2-Q_4)^2(Q_1^2-2Q_2)^2)=\frac{16}{1728}\Delta_1,$$

yielding the isomorphism

$$SL_2(\mathbb{Z})\cong\mathbb{C}[Q_1^4+8Q_2^2-4Q_1^2Q_2-4Q_4,(Q_2^2-Q_4)^2(Q_1^2-2Q_2)^2]<\mathbb{C}[x,y,z]^{G_6'}\cong\mathbb{C}[Q_1,Q_2,Q_4].$$

The expression for $\Delta_1(\tau)$ is immediate from the definition of BE_1 and from the expression for Δ in [11, p. 69].

2.5.2.2 $\ell = 2$

The BE_2 map is defined by

$$x=\theta_3(q)\theta_3(q^2),\ y=\theta_2(q)\theta_3(q^2),\ z=\theta_3(q)\theta_2(q^2).$$

With this notation

$$BE_2(Q_1^2-2Q_2)=\Theta_2,\ BE_2(Q_1^8-(Q_1^2-2Q_2)^4)=16\Delta_2$$

yielding the isomorphism

$$\mathcal{M}(\Gamma_0(2),1)\cong\mathbb{C}[Q_1^2-2Q_2,\ (Q_1^2-2Q_2)^4-Q_1^8]<\mathbb{C}[x,y,z]^{G_6'}.$$

The expression for Θ_2 comes from the construction of the unique 4-dimensional 2-modular lattice $D_4=F_4$ from the repetition code of length 2, whose swe is $x^2+y^2+2z^2=Q_1^2-2Q_2$. The expression for the cusp form Δ_2 comes from subtracting the theta series of Z_2^8 from that of $(\Lambda_2^0)^4$.

2.5.2.3 $\ell = 3$

The BE_3 map is defined by

$$x=A_1(\tau),\ y=C_1(\tau).$$

With this notation

$$BE_3(R_1)=\Theta_3,\ BE_3(D_6)=4\Delta_3$$

yielding the isomorphism

$$\mathcal{M}(\Gamma_0(3),1)\cong\mathbb{C}[R_1,\ D_6]<\mathbb{C}[x,y]^{H_3'}.$$

Note that it can be shown that

$$64D_6=R_1^6-2R_1^4D_2+R_1^2D_2^2,$$

and that

$$\mathbb{C}[x,y]^{H_3}\cong\mathbb{C}[R_2,D_6]\cong\mathbb{C}[\Theta_3^2,\Delta_3].$$

2.5.2.4 $\ell = 5$

The BE_5 map is defined by

$$x = \theta_3(q)\theta_3(q^5),\ y = \theta_2(q)\theta_3(q^5),\ z = \theta_3(q)\theta_2(q^5).$$

With this notation

$$BE_5(Q_1^2 - 2Q_2) = \Theta_5,\ BE_5(Q_2(Q_1^2 - Q_2)) = 2\Delta_5,$$

yielding the isomorphism

$$\mathcal{M}(\Gamma_0(5), 1) \cong \mathbb{C}[Q_1^2 - 2Q_2, Q_2(Q_2 - Q_1^2)].$$

The expression for Θ_2 comes from the construction of the unique (see [19]) 4-dimensional 5-modular lattice $QQF4.a$ from the repetition code of length 2, whose swe is $x^2 + y^2 + 2z^2 = Q_1^2 - 2Q_2$.

2.5.2.5 $\ell = 6$

The BE_6 map is defined (notation of $\ell = 3$ section) by

$$x = A_1(\tau)A_1(2\tau),\ y = A_1(\tau)C_1(2\tau).$$

With this notation

$$BE_6(R_1) = \Theta_6,\ BE_6(R_1^2 - D_2) = 2\Delta_6,$$

yielding the isomorphism

$$\mathcal{M}(\Gamma_0(6), 1) \cong \mathbb{C}[R_1, D_2 - R_1^2] < \mathbb{C}[x, y]^{H_3'}.$$

The invariant of the cusp form is computed from subtracting the invariant for theta series of Λ_6^0 from that of A_6 on the length 2 repetition code.

2.5.2.6 $\ell = 7$

The BE_7 map is defined by

$$x = A_2(\tau),\ y = C_2(\tau),\ z = G_2(\tau).$$

With this notation

$$BE_7(P_1) = \Theta_7,\ BE_7(P_3) = 2\Delta_7$$

yielding the isomorphism

$$\mathcal{M}(\Gamma_0(7), 1) \cong \mathbb{C}[P_1, P_3] < \mathbb{C}[x, y, z]^{G_6}.$$

The motivation for the inverse image of Δ_7 is the fact that $A_6^{(2)}$ can be constructed from a code C_3 in the notation of [1] with an explicit swe S, say. Taking $S - P_1^3$ yields a multiple of P_3.

2.5.2.7 $\ell = 11$

The BE_{11} map is defined by

$$x = A_3(\tau),\ y = C_3(\tau),\ z = G_3(\tau).$$

With this notation

$$BE_{11}(Q_1) = \Theta_{11},\ BE_{11}(r_2 - P_1^2) = 4\Delta_{11},$$

yielding the isomorphism

$$\mathcal{M}(\Gamma_0(11), 1) \cong \mathbb{C}[Q_1,\ r_2 - P_1^2] < \mathbb{C}[x, y, z]^{G_2}.$$

The invariant of the cusp form is computed from subtracting the invariant for theta series of $A_{11}(\{0,\ 1\})^2$ from that of A_{11} on the repetition code of length 2.

2.5.2.8 $\ell = 14$

The BE_{14} map is defined by

$$x = A_2(\tau)A_2(2\tau),\ y = A_2(\tau)C_2(2\tau),\ z = A_2(\tau)G_2(2\tau).$$

With this notation

$$BE_{14}(P_1) = \Theta_{14},\ BE_{14}(Q_1 - P_1) = 2\Delta_{14},$$

yielding the isomorphism

$$\mathcal{M}(\Gamma_0(14), 1) \cong \mathbb{C}[P_1,\ Q_1] \leq \mathbb{C}[x, y, z]^{G_2}.$$

The invariant of the cusp form is computed from subtracting the invariant for theta series of Λ_{14}^0 from that of Z_{14}.

2.5.2.9 $\ell = 23$

Here, we can only give a map for $\mathcal{M}(\Gamma_0(23), 1)^{(2)}$, which is the even part of $\mathcal{M}(\Gamma_0(23), 1)$, namely

$$\mathcal{M}(\Gamma_0(2), 1)^{(2)} = \bigoplus_{j=1}^{\infty} \mathcal{M}(\Gamma_0(2), 1)_{2j}(23).$$

This BE_{23} map is defined by

$$x = A_6(\tau),\ y = C_6(\tau),\ z = G_6(\tau).$$

With this notation

$$BE_{23}(r_2) = \theta_{A_{23}(C_2)},\ BE_{23}(P_2) = \Theta_{23}\Delta_{23},$$

yielding the isomorphism

$$\mathcal{M}(\Gamma_0(23), 1)^{(2)} \cong \mathbb{C}[P_1,\ P_2] \leq \mathbb{C}[x, y, z]^{G_2}.$$

2.5.3 *Summary of Results*

We summarize the results of the above subsections in the following tables.

ℓ	1	2
$BE^{-1}(\theta_\ell)$	$Q_1^4 + 8Q_2^2 - 4Q_1^2Q_2 - 4Q_4$	$Q_1^2 - 2Q_2$
$BE^{-1}(\Delta_\ell)$	$(1728/16)(Q_2^2 - Q_4)^2(Q_1^2 - 2Q_2)^2$	$(Q_1^8 - (Q_1^2 - 2Q_2)^4)/16$
$\mathcal{M}(\Gamma_0(\ell), 1) \cong$	$\mathbb{C}[Q_1^2 - 2Q_2, Q_1^8 - (Q_1^2 - 2Q_2)^4]$	$\mathbb{C}[R_2,\, D_6]$

ℓ	3	5	6
$BE^{-1}(\theta_\ell)$	R_1	$Q_1^2 - 2Q_2$	R_1
$BE^{-1}(\Delta_\ell)$	$D_6/4$	$Q_2(Q_1^2 - Q_2)/2$	P_2
$\mathcal{M}(\Gamma_0(\ell), 1) \cong$	$\mathbb{C}[Q_1^2 - 2Q_2, Q_2(Q_1^2 - Q_2)]$	$\mathbb{C}[Q_1^2 - 2Q_2, Q_2(Q_1^2 - Q_2)]$	$\mathbb{C}[R_1,\, R_1^2 - D_2]$

ℓ	7	11	14	23
$BE^{-1}(\theta_\ell)$	P_1	Q_1	P_1	r_2
$BE^{-1}(\Delta_\ell)$	$R_1^2 - D_2$	$P_3/2$	$(r_2 - P_1^2)/4$	$(Q_1 - P_1)/2$
$\mathcal{M}(\Gamma_0(\ell), 1) \cong$	$\mathbb{C}[P_1,\, P_3]$	$\mathbb{C}[Q_1,\, r_2 - P_1^2]$	$\mathbb{C}[P_1,\, P_2]$	$\mathcal{M}(\Gamma_0(23), 1)^{(2)} \cong \mathbb{C}[P_1,\, P_2]$

2.6 Appendix: Invariants and Codes

2.6.1 *Two Variables*

This section is related to $\mathbb{F}_4$-codes, following [18, p. 226]. We consider the matrix group H_3 of order 6 and Weyl type G_2 generated by M_2, M_3, where

$$M_2 = \frac{1}{2}\begin{pmatrix} 1 & 3 \\ 1 & -1 \end{pmatrix},$$

$$M_3 = \begin{pmatrix} 1 & 0 \\ 0 & -1 \end{pmatrix}.$$

The Molien series of H_3 is

$$\frac{1}{(1-t^2)(1-t^6)}.$$

The ring of invariants is generated by

$$R_2 := x^2 + 3y^2,\ D_6 := y^2(x^2 - y^2)^2.$$

Consider the subgroup H_3' generated by M_2 alone. Its ring of invariants contains the subring generated by

$$R_1 := x + y,\ D_2 := (x - 3y)^2.$$

2.6.2 *Three Variables*

2.6.2.1 $\mathbb{F}_4$-*codes*

Following [18, p. 227], we consider the matrix group G_{48} of order 48 and Weyl type B_3 generated by matrices M_1, M_4 and M_5, where

$$M_1 = \frac{1}{2}\begin{pmatrix} 1 & 1 & 2 \\ 1 & 1 & -2 \\ 1 & -1 & 0 \end{pmatrix},$$

$$M_4 = \begin{pmatrix} 0 & 1 & 0 \\ 1 & 0 & 0 \\ 0 & 0 & 1 \end{pmatrix} \text{ and } M_5 = \begin{pmatrix} 1 & 0 & 0 \\ 0 & -1 & 0 \\ 0 & 0 & -1 \end{pmatrix}.$$

The Molien series of G_{48} is

$$\frac{1}{(1-t^2)(1-t^4)(1-t^6)} = 1 + t^2 + 2t^4 + 3t^6 + 4t^8 + 5t^{10} + \cdots.$$

The ring of invariants is generated by r_2, d_4, h_6, where

$$\begin{aligned} r_2 &= x^2 + y^2 + 2z^2, \\ d_4 &= (x^2 - z^2)(y^2 - z^2), \\ h_6 &= x^6 + y^6 + 2z^6 + 15(2x^2y^2z^2 + x^2z^4 + y^2z^4). \end{aligned}$$

2.6.2.2 $\mathbb{F}_2 \times \mathbb{F}_2$-*codes*

On the other hand, following [1, Th. 4.4 (1)], let G_6 be the group of order 6 generated by M_1 and

$$M_6 = \begin{pmatrix} 1 & 0 & 0 \\ 0 & -1 & 0 \\ 0 & 0 & 1 \end{pmatrix}.$$

The Molien series of G_6 is

$$\frac{1}{(1-t)(1-t^2)(1-t^3)}.$$

The ring of invariants is generated by P_1, P_2, P_3, where

$$\begin{aligned} P_1 &= x + z, \\ P_2 &= 2xz - y^2 - z^2, \\ P_3 &= z(x^2 - y^2). \end{aligned}$$

2.6.2.3 $\mathbb{F}_2 + u\mathbb{F}_2$-*codes*

Now, following [1, Th. 4.4 (2)], let

$$M_6' = \begin{pmatrix} 1 & 0 & 0 \\ 0 & 1 & 0 \\ 0 & 0 & -1 \end{pmatrix}$$

and define the group of order 6, say $G_6' = \langle M_1, M_6' \rangle$. The ring of invariants of G_6' is generated by Q_1, Q_2, Q_4, where

$$Q_1 = x + y,$$
$$Q_2 = xy - z^2,$$
$$Q_4 = z^2(x - y)^2.$$

Let G_2 be the group of order 2 generated by M_1. It has two invariants of degree one, namely $Q_1 = x + y$ and $P_1 = x + z$.

References

[1] C. Bachoc, Applications of Coding Theory to the Construction of Modular Lattices, *Jour. Comb. Th. (A)*, **(78)** (1997) 92–119.

[2] M. Broué, M. Enguehard, Polynômes des poids de certains codes et fonctions theta de certains réseaux, Ann. Sc. ENS **5** (1972) 157–181.

[3] Y. Choie and N. Kim, The complete weight enumerator of Type II codes over $\mathbb{Z}_{2m}$ and Jacobi Forms, IEEE Transactions on Information Theory, Vol. **47**, (2001) 396–399.

[4] Y. Choie, P. Solé, Self-Dual Codes over $\mathbb{Z}_4$, and Half-Integral Weight Modular Forms Proceedings of the American Mathematical Society, **130**(11) (2002) 3125–3131.

[5] Y. Choie, P. Solé, Broué-Enguehard maps and Atkin-Lehner involutions. European J. Combin. 29(1) (2008) 24–34.

[6] Y. Choie, P. Solé, Rankin-Cohen brackets and Invariant theory, J. of Algebraic Combinatorics **13** (2001) 5–13.

[7] K.-S. Chua, Extremal modular lattices, McKay Thompson series, quadratic iterations and new series for π, Experimental Math. **14** (2005) 343–357.

[8] H. Cohen, Sums involving the values at negative integers of L-functions of quadratic characters, Math. Annal. **217** (1975) 271–285.

[9] A. Connes, H. Moscovici, Rankin-Cohen Brackets and the Hopf Algebra of Transverse Geometry, Mosc. Math. J., **4** (2004) 111–130.

[10] J. H. Conway, N. J. A. Sloane, *Sphere Packings, Lattices and Groups* Springer (1998) 3rd edition.

[11] W. Ebeling, *Codes and lattices, a course partially based on lectures by F. Hirzebruch*, Viehweg (1994).

[12] M. Eichler, D. Zagier, *The theory of Jacobi forms*, Birkhäuser (1985).

[13] F. Klein, Vorlesungen uber das Ikosaeder und die Auflosung der Gleichungen vom funften Grade. (German) [Lectures on the icosahedron and the solution of equations of fifth degree] Reprint of the 1884 original. Edited, with an introduction and commentary by Peter Slodowy. Birkhauser Verlag, Basel; B. G. Teubner, Stuttgart (1993).

[14] M. Ozeki, On the notion of Jacobi polynomials for codes, Math. Proc. Cambridge Phil. Soc. (1997).

[15] V. S. Pless, C. W. Huffman, *the handbook of Coding Theory*, North-Holland (1998).

[16] H. G. Quebbemann, Modular lattices in euclidean spaces, J. Number Th. 54 (1995) 190–202.

[17] H. G. Quebbemann, Atkin-Lehner eigenforms and strongly modular lattices, Ens. Math. 43 (1997) 55–65.

[18] E. M. Rains, N. J. A. Sloane, Self-dual codes, chapter I of [15].

[19] E. M. Rains, N. J. A. Sloane, The shadow theory of modular lattices, J. Number Theory, 73 (1998) pp. 359–389.

[20] P. Solé, D_4, E_6, E_8 and the AGM, Springer Lecture Notes in Computer Science **948** (1995) 448–455.

[21] D. Zagier, Modular forms an differential operators, Proc. Indian Acad. of Sc. 104 (1994) 57–75.

Chapter 3

Siegel Modular Forms

3.1 Introduction

We follow the nice survey [22] in this section and in the next three sections. Siegel modular forms are a multivariate generalization of classical modular forms. Let $g > 1$ be an integer, hereby called, for geometric reasons, the *genus.* Let $\mathcal{H}_g$ be the *Siegel upper half plane of genus* g, defined as

$$\mathcal{H}_g = \{Z \in M_{g \times g}(\mathbb{C}) \mid Z^t = Z,\ Im(Z) > 0\}.$$

Here, the symbol $>$ is to be understood in the sense of quadratic forms.

The Siegel modular group $\Gamma_g = Sp_{2g}(\mathbb{Z})$ consists of matrices of size $2g$ and in block form $M = \begin{pmatrix} A & B \\ C & D \end{pmatrix}$, where A, B, C, D are invertible matrices of size g with integral entries satisfying $M^t J M = J$, where $J = \begin{pmatrix} 0 & I_g \\ -I_g & 0 \end{pmatrix}$.

If n is an even positive integer, called *the level*, a subgroup $\Gamma_g(n)$ of Γ_g is obtained by keeping only matrices congruent to I_{2g} modulo n. This is called the *principal congruence subgroup of level* n of Γ_g. For an even positive integer n, we shall define a subgroup $\Gamma_g(n, 2n)$ of $\Gamma_g(n)$ by the condition $(B)_0 \equiv (D)_0 \equiv 0 \pmod{2n}$, where for a matrix X of size g, $(X)_0$ denotes the vector of the diagonal coefficients of X. Let Γ denote a discrete subgroup of Γ_g.

A holomorphic function f with domain $\mathcal{H}_g$ is a *Siegel modular form* of weight k and genus g for the group Γ if it transforms as

$$f((AZ + B)(CZ + D)^{-1}) = \det(CZ + D)^k f(Z),$$

for all $\begin{pmatrix} A & B \\ C & D \end{pmatrix} \in \Gamma$. The set of such functions forms a ring $A(\Gamma)_k$, which is also a complex vector space. The graded ring of modular forms of integral weights is denoted by

$$A(\Gamma) = \bigoplus_{k=0}^{\infty} A(\Gamma)_k,$$

where $A(\Gamma)_0 = \mathbb{C}$. For a graded ring $S = \bigoplus_{k=0}^{\infty} S_k$ and an integer d, we let $S^{(d)} = \bigoplus_{k=0}^{\infty} S_{dk}$. We recall the notations of Riemann theta functions

$$\theta \begin{bmatrix} \alpha \\ \beta \end{bmatrix} (\tau) := \sum_{x \in \mathbb{Z}^g} \exp 2\pi\sqrt{-1}\left(\frac{1}{2}\tau\left[x + \frac{\alpha}{2}\right] + \langle x + \frac{\alpha}{2}, \frac{\beta}{2}\rangle\right), \ \alpha, \beta \in \mathbb{F}_2^g, \ \tau \in \mathcal{H}_g.$$

In the sequel, we will use the simplified notation $\theta_m(Z)$ with $m' = \alpha$ and $m'' = \beta$. At the time of Igusa, it was only known that

$$\dim A(\Gamma_2)_2 = 0, \dim A(\Gamma_2)_4 = \dim A(\Gamma_2)_6 = \dim A(\Gamma_2)_8 = 1,$$

which are proved by Maass [16] and partly by Witt [33]. Applying his theory of moduli of curves in genus two, Igusa showed that the ring $A(\Gamma_2)^{(2)}$ can be generated by Eisenstein series of weights four, six, ten, and twelve. These four Eisenstein series are algebraically independent over $\mathbb{C}$, and the dimension formula of $A(\Gamma_2)^{(2)}$ also follows. In [10], he showed that the graded ring $A(\Gamma_g(4,8))$ is the normalization of the graded ring $\mathbb{C}[\theta_m\theta_n]$ generated over $\mathbb{C}$ by the products of theta-constants with $m, n \in \mathbb{Z}^{2g}$, i.e.,

$$A(\Gamma_g(4,8)) = (\mathbb{C}[\theta_m\theta_n])^N,$$

in which the exponent N denotes the integral closure of the ring in its field of fractions.

This is called by him a fundamental lemma. In [11], it is shown that the ring $\mathbb{C}[\theta_m\theta_n]$ is normal for $g = 2$, when it coincides with $A(\Gamma_2(4,8))$ by the fundamental lemma. Finally the going-down process, i.e., taking successively the invariant subrings of finite groups, provides $A(\Gamma_2)$. As a consequence, we know that this ring can be generated by five elements and that there is essentially a unique relation among the generators. The explicit formula of this relation can be found in [13]. The fundamental lemma is generalized in [12] as

$$A(\Gamma_g(r^2, 2r^2)) = (\mathbb{C}[\theta_m\theta_n])^N,$$

with $\frac{rm}{2}, \frac{rn}{2} \in \mathbb{Z}^{2g}$ for any even positive integer r. This generalization, again with going-down process, enables us to investigate $A(\Gamma_g(\ell))$ for any level ℓ. In [13], he defined the ρ-homomorphism from a subring of $A(\Gamma_g)$ to the ring of projective invariants of a binary form of degree $2g+2$, denoted by $S(2, 2g+2)$ (see [22] for a precise definition). This ρ-homomorphism gives a bijection from $A(\Gamma_1)$ to $S(2,4)$, and also an injection from $A(\Gamma_2)$ to $S(2,6)$. In this way, he obtained the structure theorems of $A(\Gamma_g)$ for $g = 1, 2$. In the case $g = 3$, he determines the kernel of ρ, which is a principal ideal, and the vector spaces of low weights. The latter leads to an affirmative answer of a problem of Witt [33], which asks whether the two theta series of even unimodular lattices E_8^2 and D_{16}^+ coincide in genus 3. This problem is independently settled by Kneser [15].

3.2 Genus 2

We review the theory of the classical joint weight enumerator in the case of binary codes. See [17] for more details. Let u, v denote binary vectors of length n. We define $i(u,v)$, $j(u,v)$, $k(u,v)$ and $l(u,v)$ to be the number of indices $i \in [n]$ with $(u_i, v_i) = (0,0)$, $(0,1)$, $(1,0)$ and $(1,1)$, respectively.

The *joint weight enumerator* $J(A,B)$ of two binary linear codes A, B is the four-variable polynomial defined by the formula

$$J(A,B)(a,b,c,d) = \sum_{u\in A,\, v\in B} a^{i(u,v)} b^{j(u,v)} c^{k(u,v)} d^{l(u,v)}. \tag{3.1}$$

Regrouping the terms on the basis of monomials, we get

$$J(A,B)(a,b,c,d) = \sum_{i+j+k+l=n} M(i,j,k,l) a^i b^j c^k d^l. \tag{3.2}$$

Note that the number of (i,j,k,l)'s is $\binom{n-1}{3}$, the number of compositions (ordered partitions) of the integer n into four parts. Thus, the joint weight enumerator of two codes generalizes the classical weight enumerator of a single code in the same way as the joint probability distribution function of two random variables generalizes the probability distribution function of one variable. There are several MacWilliams relations for this weight enumerator.

For instance, in [19], we can find

$$J(A, B^{\perp})(a, b, c, d) = \frac{1}{|B|} J(A, B)(a + b, a - b, c + d, c - d).$$

When the two codes A and B are equal, the joint weight enumerator $B_A = J(A, A)$ is called the *biweight* enumerator of the code A. If A is self-dual, then B_A is an invariant of degree n [9, Theorem 2.3] of the group

(1) $G_I = \langle T, \Delta_I \rangle$ if C is Type I,
(2) $G_{II} = \langle \Delta_{II} \rangle$ if C is Type II,

where $T = diag(H, H)$ with $H = \frac{1}{\sqrt{2}} \begin{bmatrix} 1 & 1 \\ 1 & -1 \end{bmatrix}$ as the normalized matrix of the MacWilliams transform, and

(1) Δ_I is the group of diagonal matrices $diag(\pm 1, \pm 1, \pm 1, \pm 1)$;
(2) $\Delta_{II} = \langle D, \Delta_I \rangle$ where $D = diag(1, i, 1, i)$.

There is a Gleason-like formula for biweight enumerators [7, Th. 2]. To state it, we borrow the following notation from [7].

(1) r_n is the binary repetition code of length n;
(2) d_n is a maximal self-orthogonal code generated by tetrads [7];
(3) $d_n^+ = d_n \bigcup (a + d_n)$, where $a = (1, 0, 1, 0, \dots)$;
(4) g_{24} is the extended Golay code of parameters $[24, 12, 8]$.

We can now state this important result.

Theorem 16. *([7])*

(1) The space of invariants $B(II)$ containing the biweight enumerators of Type II codes is freely generated by the biweight enumerators of r_4, r_8, r_{12}, d_8. It has Molien series

$$\frac{1}{(1 - t^4)(1 - t^8)^2(1 - t^{12})}.$$

(2) *The space of invariants $B(I)$ containing the biweight enumerators of Type I codes affords the Molien series*

$$\frac{1+t^{32}}{(1-t^8)(1-t^{24})^2(1-t^{40})}.$$

It is of form $\mathcal{W} \oplus B_{d_{32}^+}\mathcal{W}$, where $\mathcal{W}$ is freely generated by the biweight enumerators of $d_8^+, d_{24}^+, g_{24}, d_{40}^+$.

By combining this information with powerful results of Igusa on Siegel modular forms of genus 2, it is shown in [7, Theorem 1] that the natural BE map obtained by substitution of Riemann theta constants into the variables of an invariant of the group that leaves $B(II)$ (resp. $B(I)$) invariant is an isomorphism from that space to the suitable space of modular forms.

3.3 Genus 3

We need to introduce multivariate analogues of the biweight enumerator. The weight enumerator of a binary code C in genus g is the homogeneous polynomial in 2^g variables, defined by

$$W_C^{(g)}(x_a \mid a \in \mathbb{F}_2^g) = \sum_{x_1,\cdots,x_g \in C} \prod_{a \in \mathbb{F}_2^g} x_a^{n_a(x_1,\cdots,x_g)},$$

where $n_a(x_1, \cdots, x_g) = |\{i|a = (x_{1i}, x_{2i}, \cdots, x_{gi})\}|$. Here, as we promised, we consider the problem of Witt in coding theory. The general method to attack this problem is given by Nebe [20]. See also [7, 23, 29]. There exist 1, 2, 9, 85 classes of Type II codes of length 8, 16, 24, 32, respectively. For length 16, two weight enumerators are linearly independent if and only if $g \geq 3$. For length 24, nine weight enumerators are linearly independent if and only if $g \geq 6$. For length 32, eighty-five weight enumerators are linearly independent if and only if $g \geq 10$. The Gleason type theorem for weight enumerators of higher genera also holds, i.e., the ring over $\mathbb{C}$ generated by the weight enumerators $W_C^{(g)}$ of all Type II codes C coincides with the invariant ring of a certain finite group G_g for any positive integer g whose precise definition we omit. If we consider an index 2 subgroup H_g of G_g, the invariant ring of H_g is not generated by the weight enumerators any more, but Runge [29] showed $A(\Gamma_g)^{(2)}$ is isomorphic to the normalized of the quotient of the invariant ring of H_g by the theta relations. For a

Type II code C of length n, the function on the Siegel upper half plane $W_C^{(g)}(\theta_{a0}(2Z))$ is a modular form of weight $n/2$ for Γ_g.

Runge ([29]) obtained the following isomorphism:

$$A(\Gamma_3) \simeq \mathbb{C}[xa]^{H_3}/(W^{(3)}_{d_{16}^+} - W^{(3)}_{e_8^2}).$$

From this, the dimension formula of $A(\Gamma_3)$ reduces to the computation of the dimension formula of the invariant ring of H_3, i.e.,

$$\sum_{d=0}^{\infty}(\dim_{\mathbb{C}} A(\Gamma_3)_d)t^{2d} = (1-t^{16})\sum_{j=0}^{\infty}(\dim_{\mathbb{C}}(\mathbb{C}[xa]^{H_3})_j)t^j.$$

Note that an exponent of t in the left-hand side is $2d$. In [30], the dimension formula of the invariant ring of H_3 is computed and the dimension formula of $A(\Gamma_3)$ in [32] is reproved. Tsuyumine used Igusa "ρ-homomorphism" to determine the dimension formula and the generators of $A(\Gamma_3)$. The relations among the generators are not given in [32].

3.4 Genus 4

The degree 24 part of the invariant ring of H_4 is spanned by the weight enumerators of Type II codes of length 24. There are 9 Type II codes of this length and the dimension of the complex vector space spanned by the weight enumerators of those codes is 7, see [10]. On the other hand, the dimension of $A(\Gamma_4)_{12}$ is 6 by geometric arguments, see [25]. In [8], the authors analysed the BE-like map

$$\mathbb{C}[x_a]_{24}^{H_4} = \langle W_{C_1}, W_{C_2}, \cdots, W_{C_9}\rangle \to A(\Gamma_4)_{12},$$

given by $x_a \to \theta_{a0}(2Z)$. This linear map is surjective and the kernel is explicitly described. The obtained relation gives a non-trivial cusp form in genus 5. It is an open problem to decide if this relation comes from Riemann's theta relations. The same problem in the case of length 32 is treated in [23]. The dimension of the degree 32 part of the invariant ring $\mathbb{C}[x_a]^{H_4}$ is 19, while the dimension of the corresponding vector space of modular forms is 14. See [10, 27], respectively. Finally we get 5 relations among theta series of weight 16 in genus 4. In the course of these computations, Oura uses the "Restriction technique" of ([26, 27]) to overcome a difficulty which does not appear in [8].

If we only care for Type II lattices, which only exist in dimensions multiples of 8, we can focus on the subring $A(\Gamma_2)^{(4)}$ of $A(\Gamma_2)$. This ring is the normalization of the ring generated over C by theta series of even unimodular lattices, i.e., $A(\Gamma_2)^{(4)} = \mathbb{C}[\theta_*]^N$, see [24]. No normalization is necessary if $g \leq 4$, see [18]. Explicit ring structures are known by Ozeki [24] to be as follows. The lattice notation is as per [7]

$$A(\Gamma_2)^{(4)} = \mathbb{C}[\theta_{E_8}, \theta_{A_1^{24}}, \theta_{D_{24}^+}, \theta_{D_{32}^+}, \theta_{D_{40}^+}].$$

3.5 Higher Genera and Larger Alphabets

In the previous sections all codes were binary. In this section, following [1, 3, 4], we introduce several types of weight enumerators of codes over $\mathbb{Z}_{2k}$ for k an integer ≥ 1. For these weight enumerators, we establish the MacWilliams identities and study invariants. From now on R denotes the ring $\mathbb{Z}_{2k}$.

3.5.1 *Weight Enumerators and MacWilliams Identities*

First let us fix the notations. We denote the primitive m-th root $e^{2\pi i/m}$ of unity by η_m. Define $A\,[B] := A^t BA$ for matrices A and B, where A^t denotes the matrix transpose of A.

Definition 17 (Complete Weight Enumerators in Genus g). For a code C over R, we define the complete weight enumerator in genus g by

$$\mathfrak{C}_{C,g}(z_a \mid a \in R^g) = \sum_{c_1,\ldots,c_g \in C} \prod_{a \in R^g} z_a^{n_a(c_1,\ldots,c_g)},$$

where $n_a(c_1, \ldots, c_g)$ denotes the number of i satisfying $a = (c_{1i}, \ldots, c_{gi})^t$.

Remark 3.

(1) For the case $g = 1$, these weight enumerators are the same as ordinary complete weight enumerators defined in Chapter I.
(2) For the case $k = 1$, these weight enumerators were introduced in [7] and [23].

We define a relation $\sim$ in R^g by

$$a \sim b \Longleftrightarrow a = b \text{ or } a = -b,$$

where $a, b \in R^g$. Then the relation $\sim$ becomes an equivalence relation in R^g and we denote the natural projection using the conventions $\overline{a}$. Note that $wt_E(a) = wt_E(-a)$ and $wt_L(a) = wt_L(-a)$.

Definition 18 (Symmetrized Weight Enumerators in Genus g). For a code C over R, we define the symmetrized weight enumerator in genus g by

$$\mathfrak{S}_{C,g}(z_{\overline{a}} \,|\, \overline{a} \in \overline{R^g}) = \sum_{c_1,\ldots,c_g \in C_g} \prod_{\overline{a} \in \overline{R^g}} z_{\overline{a}}^{n_{\overline{a}}(c_1,\ldots,c_g)},$$

where $n_{\overline{a}}(c_1, \ldots, c_g)$ denotes the number of i satisfying $\overline{a} = \overline{(c_{1i}, \ldots, c_{gi})}^t$.

From now on, we often write complete and symmetrized weight enumerators in genus g by $\mathfrak{C}_{C,g}(z_a)$, $\mathfrak{S}_{C,g}(z_{\overline{a}})$, respectively, for simplicity.

We have the *MacWilliams identity* for the complete weight enumerators. Here we consider that an n by n matrix M acts on the polynomial ring $\mathbb{C}[x_1, x_2, \ldots, x_n]$ naturally, that is,

$$M \cdot f(x_1, x_2, \ldots, x_n) = f\Big(\sum_{1 \le j \le n} a_{1j}x_j, \ldots, \sum_{1 \le j \le n} a_{nj}x_j\Big),$$

where $f \in \mathbb{C}[x_1, x_2, \ldots, x_n]$ and $A = (a_{ij})$.

Theorem 19 (MacWilliams Identity). *For a code C over R, we have*

$$\mathfrak{C}_{C^\perp,g}(z_a) = \frac{1}{|C|^g} T \cdot \mathfrak{C}_{C,\,g}(z_a),$$

where $T = \left(\eta_{2k}^{\langle a,b \rangle}\right)_{a,b \in R^g}$.

Similarly, we have the *MacWilliams identity* for the symmetrized weight enumerators.

Corollary 20 (MacWilliams Identity). *For a code C, we have*

$$\mathfrak{S}_{C^\perp,g}(z_{\overline{a}}) = \frac{1}{|C|^g} \overline{T} \cdot \mathfrak{S}_{C,g}(z_{\overline{a}}),$$

where $\overline{T} = \left(t(\overline{a}, \overline{b})\right)_{\overline{a},\overline{b} \in \overline{R^g}}$ *and* $t(\overline{a}, \overline{b}) = \sum_{d \in R^g \text{ with } \overline{d} = \overline{b}} \eta_{2k}^{\langle a,d \rangle}$.

3.5.2 *Invariant Rings*

In this subsection, we study the invariance properties of complete and symmetrized weight enumerators.

We define a subgroup $G^8_{g,k}$ of $GL((2k)^g, \mathbb{C})$ as

$$G^8_{g,k} = \langle T_g, D_S, \eta_8 |\ S \text{ runs over all integral symmetric matrices}\rangle,$$

where

$$T_g = \left(\frac{\eta_8}{\sqrt{2k}}\right)^g T, D_S = \text{ diag } (\eta_{4k}^{S[a]} \,|\, a \in R^g).$$

Theorem 21. *For any Type II code C over R, the complete weight enumerator in genus g is invariant under the action of the group $G^8_{g,k}$.*

Proof. We have only to check three types of generators, T_g, D_S, and η_8. The invariance property of T_g, η_8 is derived in [1]. We shall show that $D_S \cdot \mathfrak{C}_{C,g}(z_a) = \mathfrak{C}_{C,g}(z_a)$. We have

$$\begin{aligned} D_S \cdot \mathfrak{C}_{C,\,g}(z_a) &= \sum_{c_1,\ldots,c_g \in C} \prod_{a \in R} (\eta_{4k}^{S[a]} z_a)^{n_a(c_1,\ldots,c_g)} \\ &= \sum_{c_1,\ldots,c_g \in C} \prod_{a \in R} \eta_{4k}^{S[a]\cdot n_a(c_1,\ldots,c_g)} z_a^{n_a(c_1,\ldots,c_g)}. \end{aligned}$$

In order to prove the theorem, we have to show $\sum_{a\in R} S[a] \cdot n_a(c_1, \ldots, c_g) \equiv 0$ (mod $4k$).

$$\begin{aligned} \sum_{a\in R} S[a] \cdot n_a(c_1, \ldots, c_g) &= \sum_{1\le i\le n} S[{}^t(c_{1i}, \ldots, c_{gi})] \\ &= \sum_{1\le i\le n} \left\{ \sum_{1\le k\le g} s_{kk}(c_{ki})^2 + 2 \sum_{1\le l<m\le g} s_{lm} c_{li}\, c_{mi} \right\} \\ &= \sum_{1\le k\le g} s_{kk} \sum_{1\le i\le n} (c_{ki})^2 + 2 \sum_{1\le l<m\le g} s_{lm} \sum_{1\le i\le n} c_{li}\, c_{mi}. \end{aligned}$$

For $\forall\ c_k$, we have $wt_E(c_k) = \sum_{1\le i\le n} (c_{ki})^2 \equiv 0 \pmod{4k}$. $\sum_{1\le i\le n} c_{li}c_{ki} \equiv 0 \pmod{2k}$ follows from the calculation $d_E(c_l, c_k) \equiv 0 \pmod{4k}$. Therefore it turns out that $\sum_{a\in R} S[a] \cdot n_a(c_1, \ldots, c_g) \equiv 0 \pmod{4k}$. This completes the proof of the theorem. □

Remark. (1) $G_{g,k}$ is (up to ±1) the homomorphic image of the modular group Γ_g under the theta representation of index k (see [31]).
(2) Theorem 21 says that the ring generated by complete weight enumerators for Type II codes is contained in the invariant ring of the group $G^8_{g,k}$. For $k = 1$, the two rings coincide (see Theorem 3.6 in [23]).

We now define a subgroup $H^8_{g,k}$ of $GL(2^{g-1}(k^g+1), \mathbb{C})$ as

$$H^8_{g,k} = \langle \overline{T_g}, \overline{D_S}, \eta_8 |\ S \text{ runs over all integral symmetric matrices } \rangle,$$

where

$$\overline{T_g} = \left(\frac{\eta_8}{\sqrt{2k}}\right)^g \overline{T} \text{ and } \overline{D_S} = \text{ diag } (\eta_{4k}^{S[a]} \text{ with } \overline{a} \in \varphi(R^g)).$$

Similarly to complete weight enumerators, we have the following MacWilliams identity for symmetrized weight enumerators in genus g.

Corollary 22. *For any Type II code C over R, the symmetrized weight enumerator in genus g is invariant under the action of the group $H^8_{g,k}$.*

In concluding this subsection, we would like to emphasize that the groups $G_{g,k}$, as well the groups $H_{g,k}$, $G^8_{g,k}$ and $H^8_{g,k}$, are all finite groups. This is explained as follows. Here we assume that the reader is familiar with some of the basic concepts of theta functions, such as given in Runge [31].

The group $H_{g,k} = \langle \overline{T_g}, \overline{D_S} \,|\, S$ runs over all integral symmetric matrices$\rangle$ acts linearly on the space spanned by the theta constants $f_a^{(k)}$ of index k, where

$$f_a^{(k)}(\tau) = \sum_{x \in \mathbb{Z}^g} \exp\left(2\pi i k \tau[x + \frac{a}{2k}]\right).$$

Note that here $k \in \mathbb{N}, a \in (\mathbb{Z}_{2k})^g$.

It is known that the group $H_{g,k}/(\pm 1)$ is a homomorphic image of the Siegel modular group $\Gamma_g = Sp(2g, \mathbb{Z})$ under the theta representation of index k:

$$\rho_{\theta,k} : \Gamma_g \longrightarrow Aut(\mathcal{TH}^{(k)}_{g,(2)})$$

in the notation of [31]. The kernel of this representation is completely described in Runge [31, Theorem 2.4]. In particular, this kernel contains the subgroup $\Gamma_g(4k)$. Since $\Gamma_g/\Gamma_g(4k) \cong Sp(2g, \mathbb{Z}_{4k})$ is a finite group, the finiteness of the group $H_{g,k}$ follows immediately.

Similarly, the group $G_{g,k} = \langle T_g, D_S \mid S \text{ as above}\rangle$ acts linearly on the space spanned by the theta functions $f_a^{(k)}(\tau, z)$ of index k, where

$$f_a^{(k)}(\tau, z) = \sum_{x\in\mathbb{Z}^g} \exp\Big(2\pi i(k\tau[x + \frac{a}{2k}] + \langle x + \frac{a}{2k}, 2kz\rangle\Big).$$

Again, $G_{g,k}/(\pm 1)$ is a homomorphic image of $\Gamma_g = Sp(2g, \mathbb{Z})$ under the theta representation:

$$\rho_{\theta,k} : \Gamma_g \longrightarrow Aut(\mathcal{THET}^{(\leq k)}_{g,(2)})$$

in the notation of [31]. From the relation

$$\begin{pmatrix} 1 & S \\ 0 & 1 \end{pmatrix} \cdot f_a^{(k)}(\tau, z) = \exp 2\pi i(\frac{S[a]}{4k}) \cdot f_a^{(k)}(\tau, z),$$

it is again proved that $\Gamma(4k)$ is in the kernel of the theta representation, see e.g., Runge [31] or Kac [14, Theorem 13.5, p. 169]. Since the group $\Gamma_g/\Gamma_g(4k)$ is finite and since $|G_{g,k}| \leq 2 \cdot |\Gamma_g/\Gamma_g(4k)|$, we have the finiteness of the group $G_{g,k}$. The finiteness of the groups $G^8_{g,k}$ and $H^8_{g,k}$ are immediately obtained as $|G^8_{g,k}| \leq 8 \cdot |G_{g,k}|$ and $|H^8_{g,k}| \leq 8 \cdot |H_{g,k}|$.

Although we will not discuss the details here, it is possible to determine the orders and the structures of the groups $G_{g,k}$, $H_{g,k}$, $G^8_{g,k}$ and $H^8_{g,k}$ more explicitly, by using the known explicit determinations of the kernels of the theta representations $\rho_{\theta,k} : \Gamma_g \longrightarrow Aut(\mathcal{TH}^{(k)}_{g,(2)})$ given in Runge [31].

We give in Table 3.1 the orders of the groups $G_{g,k}$, $H_{g,k}$, $G^8_{g,k}$ and $H^8_{g,k}$ for $g = 1$ and $k \leq 8$. It can be shown, for example, that

$$|G^8_{1,2^m}| = 192 \cdot 2^{m-1}.$$

Table 3.1 Orders of the Groups $G_{g,k}$, $H_{g,k}$, $G^8_{g,k}$ and $H^8_{g,k}$.

k	1	2	3	4	5	6	7	8
$\lvert G_{1,k}\rvert$	96	384	2304	3072	11520	9216	32256	24576
$\lvert H_{1,k}\rvert$	96	384	1152	3072	5760	9216	16128	24576
$\lvert G^8_{1,k}\rvert$	192	1536	4608	12288	23040	368648	64512	98304
$\lvert H^8_{1,k}\rvert$	192	768	2304	6144	11520	18432	32256	49152

3.5.3 *Shadows and Weight Enumerators*

We first prove that the complete (resp. symmetrized) weight enumerator of the shadow of a Type I code C over $\mathbb{Z}_{2k}$ is uniquely determined from the complete (resp. symmetrized) weight enumerator of C.

Lemma 23. *If C is a Type I code over $\mathbb{Z}_{2k}$, then*

$$cwe_{C_0}(x_0, x_1, \ldots, x_{2k-1}) = \frac{1}{2}(swe_C(x_0, x_1, \ldots, x_{2k-1}) + swe_C(\eta_{4k}^{0^2} x_0, \eta_{4k}^{1^2} x_1, \ldots, \eta_{4k}^{(2k-1)^2} x_{2k-1})),$$

$$swe_{C_0}(x_0, x_1, \ldots, x_k) = \frac{1}{2}(swe_C(x_0, x_1, \ldots, x_k) + swe_C(\eta_{4k}^{0^2} x_0, \eta_{4k}^{1^2} x_0, \ldots, \eta_{4k}^{k^2} x_k)),$$

where η_{4k} denotes the primitive $4k$-th root of unity.

Proof. Let $c = (c_1, c_2, \ldots, c_n)$ be a codeword in C then

$$\prod_{i=1}^{n} (\eta_{4k}^{i^2} x_i)^{n_i(c)} = \prod_{i=1}^{n} (\eta_{4k})^{i^2 n_i(c)} \prod_{i=1}^{n} (x_i)^{n_i(c)} = (\eta_{4k})^{\sum_{i=1}^{n} i^2 n_i(c)} \prod_{i=1}^{n} (x_i)^{n_i(c)}.$$

Since C is self-dual, c has Euclidean weight $\equiv 0 \pmod{2k}$. Since $wt_E(c) \equiv \sum_{i=1}^{n} i^2 n_i(c) \pmod{4k}$,

$$\prod_{i=1}^{n} (\eta_{4k}^{i^2} x_i)^{n_i(c)} = \begin{cases} -\prod_{i=1}^{n} x_i^{n_i(c)} & \text{if } wt_E(c) \equiv 2k \pmod{4k}, \\ \prod_{i=1}^{n} x_i^{n_i(c)} & \text{if } wt_E(c) \equiv 0 \pmod{4k}. \end{cases}$$

This proves the lemma. The *swe* is computed from the *cwe*. □

Theorem 24. *Let C be a Type I code over $\mathbb{Z}_{2k}$ and let S be its shadow. Then the cwe and swe of S is related to the cwe and swe of C by the relation*

$$cwe_S(x_0, x_1, \ldots, x_{2k-1}) = cwe_C(A(x_0, x_1, \ldots, x_{2k-1})),$$
$$swe_S(x_0, x_1, \ldots, x_k) = swe_C(B(x_0, x_1, \ldots, x_k))$$

where $A = (a_{ij})$ is the $2k$ by $2k$ matrix with $a_{ij} = \frac{1}{\sqrt{2k}} \eta_{4k}^{i^2+2ij}$, and $B = (b_{ij})$ is the $(k+1)$ by $(k+1)$ matrix with $b_{ij} = \sum_{i' \equiv i} a_{i'j}$ where $i' \equiv i$ if $i' = i$ or $i' = -i$.

Proof. We proceed as in [5, p. 1323] by computing first by the MacWilliams identity

$$cwe_{C^\perp}(x_0, x_1, \ldots, x_{2k-1}) = \frac{1}{|C|} cwe_C(M(x_0, x_1, \ldots, x_{2k-1}),$$

where $M = (m_{ij})$ is the $2k$ by $2k$ matrix with $m_{ij} = \eta_{2k}^{ij}$, the cwe of $C^{\perp}$, then the cwe of its $4k$-weight subcode, the cwe of the dual of the latter, and finally the cwe of the shadow by the difference of the cwe of $C_0^{\perp}$ and the cwe of C. The swe follows similarly. □

Definition 25 (Complete Joint Weight Enumerators). The complete joint weight enumerator for codes C and K of length n over R is defined as

$$\mathfrak{J}_{C,K}(X_{\mathbf{a}} \text{ with } \mathbf{a} \in R \times R) = \sum_{(c,k)\in C\times K} \prod_{\mathbf{a}\in R\times R} X_{\mathbf{a}}^{n_{\mathbf{a}}(c,k)},$$

where $n_{\mathbf{a}}(c,k) = |\{j|(c_j,k_j) = \mathbf{a}\}|$, $c = (c_1,\ldots,c_n)$ and $k = (k_1,\ldots,k_n)$. Similarly to complete weight enumerators, we often simply denote the weight enumerators by $\mathfrak{J}_{C,K}(X_{\mathbf{a}})$.

In a similar argument to Theorem 19, we have the MacWilliams identity for complete joint weight enumerators.

Theorem 26 (MacWilliams Identity). *Let $\tilde{A}$ denote either A or $A^{\perp}$. Then*

$$\mathfrak{J}_{\tilde{C},\tilde{K}}(X_{\mathbf{a}}) = \frac{1}{|C|^{\delta_{\tilde{C},C^{\perp}}}|K|^{\delta_{\tilde{K},K^{\perp}}}}(T^{\delta_{\tilde{C},C^{\perp}}} \otimes T^{\delta_{\tilde{K},K^{\perp}}})\mathfrak{J}_{C,K}(X_{\mathbf{a}}),$$

where

$$T = \left(\eta_{2k}^{\langle a,b\rangle}\right)_{a,b\in R} \quad \text{and } \delta_{\tilde{A},A^{\perp}} = \begin{cases} 0 & \text{if } \tilde{A} = A, \\ 1 & \text{if } \tilde{A} = A^{\perp}. \end{cases}$$

Proof. Similar to that of Theorem 19. □

We give relationships between a Type I code and its shadow using the weight enumerators.

Given the complete joint weight enumerator for $\mathfrak{J}_{C,C}$ we can find $\mathfrak{J}_{C,C_0}$, $\mathfrak{J}_{C_0,C}$, and $\mathfrak{J}_{C_0,C_0}$.

Proposition 27. *Let C be a Type I code over R and let C_0 be the $4k$-weight subcode of C. Then*

$$\mathfrak{J}_{C,C_0}(X_{\mathbf{a}}) = \frac{1}{2}(\mathfrak{J}_{C,C}(X_{\mathbf{a}}) + \mathfrak{J}_{C,C}(X_{\phi(\mathbf{a})})),$$

$$\mathfrak{J}_{C_0,C}(X_{\mathbf{a}}) = \frac{1}{2}(\mathfrak{J}_{C,C}(X_{\mathbf{a}}) + \mathfrak{J}_{C,C}(X_{\psi(\mathbf{a})})),$$

$$\mathfrak{J}_{C_0,C_0}(X_{\mathbf{a}}) = \frac{1}{4}(\mathfrak{J}_{C,C}(X_{\mathbf{a}}) + \mathfrak{J}_{C,C}(X_{\phi(\mathbf{a})}) + \mathfrak{J}_{C,C}(X_{\psi(\mathbf{a})}) + \mathfrak{J}_{C,C}(X_{\theta(\mathbf{a})})),$$

where $\phi(\mathbf{a}) = \eta_{4k}^{b^2}(a,b)$, $\psi(\mathbf{a}) = \eta_{4k}^{a^2}(a,b)$ *and* $\theta(\mathbf{a}) = \eta_{4k}^{a^2+b^2}(a,b)$ *for* $\mathbf{a} = (a,b) \in R \times R$.

Proof. Notice that the substitution $X_{\phi(\mathbf{a})}$ fixes each monomial representing codewords with Euclidean weight divisible by $4k$ and negates each monomial representing codewords whose Euclidean weight $\equiv 2k \pmod{4k}$, which gives the result. The remaining two cases are similar. □

We can apply the MacWilliams identity to find all the joint weight enumerators involving $C, C_0, C_0^{\perp}$, and S. In particular we have the following:

Proposition 28. *Let* C *be a Type I code over* R *and let* S *be its shadow then*

$$\mathfrak{J}_{S,C}(X_{\mathbf{a}}) = (T \otimes I)\,\mathfrak{J}_{C,C}(X_{\phi(\mathbf{a})}),$$
$$\mathfrak{J}_{C,S}(X_{\mathbf{a}}) = (I \otimes T)\,\mathfrak{J}_{C,C}(X_{\psi(\mathbf{a})}),$$
$$\mathfrak{J}_{S,S}(X_{\mathbf{a}}) = (T \otimes T)\,\mathfrak{J}_{C,C}((X_{\theta(\mathbf{a})}).$$

Proof. We compute $\mathfrak{J}_{C,C_0}$, $\mathfrak{J}_{C_0,C}$ and $\mathfrak{J}_{C_0,C_0}$ by the above proposition, apply the MacWilliams identity and then compute the desired weight enumerators from these weight enumerators. □

Lemma 23, Theorem 24, Propositions 27 and 28 determine complete, symmetrized and joint weight enumerators for C_0 and S from ones of C. For the code to exist all of these weight enumerators must have non-negative integral coefficients. Our results seem to be useful for proving the non-existence of a certain Type I code over $\mathbb{Z}_{2k}$. In fact, for the case $k = 1$ the non-existence of some Type I codes with high minimum weight was proved in [5] using their shadows.

3.5.4 *Construction of Siegel Modular Forms*

This section follows from [3, 4, 31]. We first recall the notations of theta functions (for more detail, e.g. see [31]):

$$\theta\begin{bmatrix}\alpha\\ \beta\end{bmatrix}(\tau) := \sum_{x\in\mathbb{Z}^g} \exp 2\pi\sqrt{-1}\left(\frac{1}{2}\tau\left[x+\frac{\alpha}{2}\right] + \langle x+\frac{\alpha}{2}, \frac{\beta}{2}\rangle\right),$$
$$\alpha, \beta \in \mathbb{F}_2^g,\ \tau \in \mathcal{H}_g,$$

where $\mathcal{H}_g$ denotes the Siegel upper half-space $\mathcal{H}_g = \{Z = X + iY \in GL(g, \mathbb{C}) | Z = {}^tZ, Y > 0\}$.

We define for any positive integer k the following theta functions:

$$f_a^{(k)}(\tau) := \theta \begin{bmatrix} a/k \\ 0 \end{bmatrix} (2k\tau).$$

It is well known that the modular group $\Gamma_g = Sp(2g, \mathbb{Z})$ is generated by the elements $J = \begin{pmatrix} 0 & I \\ -I & 0 \end{pmatrix}$ and $D_S = \begin{pmatrix} I & S \\ 0 & I \end{pmatrix}$, where S runs over the symmetric g by g matrices. They act on the theta functions as follows:

$$D_S(f_a^{(k)})(\tau) = \exp 2\pi i \left(\frac{S[a]}{4k} \right) f_a^{(k)}(\tau), \quad \frac{J(f_a^{(k)})(\tau)}{\sqrt{\det(-\tau)}} = \sum_{b \in (\mathbb{Z}_{2k})^g} (T_g)_{a,b} f_b^{(k)}(\tau).$$

Moreover, the theta functions for a lattice L are defined by

$$\theta_{L,g}(\tau) := \sum_{x_1, \ldots, x_g \in L} \prod_{1 \le i,j \le g} q_{ij}^{[x_i, x_j]},$$

where $q_{ij} = \exp \pi \sqrt{-1} \tau_{ij}$.

A Siegel modular form of weight k for $\Gamma_g = Sp(2g, \mathbb{Z})$ is a holomorphic function f on the Siegel upper half-space such that for all $\begin{pmatrix} A & B \\ C & D \end{pmatrix} \in \Gamma_g$, we have

$$f((A\tau + B)(C\tau + D)^{-1}) = \det(C\tau + D)^k f(\tau).$$

We need more conditions for the case $g = 1$.

Theorem 29. *[3] Let C be a Type II code of length n over R and let $\Lambda(C)$ be the even unimodular lattice constructed from C by Construction A. Then*

$$\mathfrak{C}_{C,g}(f_a^{(k)}(\tau)) = \mathfrak{S}_{C,g}(f_{\overline{a}}^{(k)}(\tau)) = \theta_{\Lambda(C),g}(\tau)$$

and these functions give Siegel modular forms of weight $n/2$ for Γ_g.

3.5.5 *Molien Series for Small Cases*

The weight enumerator of a self-dual code belongs to the ring of polynomials fixed by the group of substitutions. In this subsection, we give the Molien series for the invariant rings of the groups of small k and g.

First, let us recall the general invariant theory of finite groups. Let G be a finite subgroup of $GL(n,\mathbb{C})$. Then G acts on the polynomial ring $\mathbb{C}[x_1,\dots,x_n]$ ($\mathbb{C}[x_k]$ for short) naturally, i.e.,

$$A\cdot f(x_1,\dots,x_n)=f\Bigg(\sum_{1\le j\le n}A_{1j}x_j,\dots,\sum_{1\le j\le n}A_{nj}x_j\Bigg),$$

where $f\in\mathbb{C}[x_k]$ and $A=(A_{ij})_{1\le i,j\le n}$. There exists a *homogeneous system of parameters* $\{\theta_1,\dots,\theta_n\}$ such that the invariant ring $\mathbb{C}[x_k]^G$ is finitely generated free $\mathbb{C}[\theta_1,\dots,\theta_n]$-module. The invariant ring has the *Hironaka decomposition*

$$\mathbb{C}[x_k]^G=\bigoplus_{1\le m\le s}g_m\mathbb{C}[\theta_1,\dots,\theta_n],\text{ where } g_1=1.$$

The invariant ring is a graded ring and the dimension formula is defined by

$$\Phi_G(t)=\sum_{d\ge 1}\dim\mathbb{C}[x_k]_d^G t^d,$$

where $\mathbb{C}[x_k]_d^G$ is the d-th homogeneous part of $\mathbb{C}[x_k]^G$. The dimension formula for the *Hironaka decomposition* given in the above form is

$$\Phi_G(t)=\frac{1+t^{\deg(g_2)}+\dots+t^{\deg(g_s)}}{(1-t^{\deg(\theta_1)})\cdots(1-t^{\deg(\theta_n)})}.$$

In general, the converse is not true. It is known that we have the identity

$$\Phi_G(t)=\sum_{A\in G}\frac{1}{\det(1-tA)}.$$

This was shown by Molien and is called *Molien series.*

We recall the notations:

$$\begin{aligned}
R &:= \mathbb{Z}_{2k},\\
G_{g,k} &:= \left\langle \left(\frac{\eta_8}{\sqrt{2k}}\right)^g \left(\eta_{2k}^{\langle a,b\rangle}\right)_{a,b\in R^g},\ \operatorname{diag}\,(\eta_{4k}^{S[a]}\text{ with } a\in R^g)\right\rangle,\\
G_{g,k}^8 &:= \langle G_{g,k},\eta_8\rangle,\\
H_{g,k} &:= \left\langle \left(\frac{\eta_8}{\sqrt{2k}}\right)^g (t(\overline{a},\overline{b}))_{\overline{a},\overline{b}\in\overline{R^g}},\ \operatorname{diag}\,(\eta_{4k}^{S[a]}\text{ with } \overline{a}\in\overline{R^g})\right\rangle,\\
H_{g,k}^8 &:= \langle H_{g,k},\eta_8\rangle,
\end{aligned}$$

where $t(\overline{a},\overline{b})=\sum_{d\in R^g\text{ with }\overline{d}=\overline{b}}\eta_{2k}^{\langle a,d\rangle}$.

In the following, we give the Molien series in the form

$$\begin{aligned}\Phi_G(t) &= \text{the expansion}\\ &= \text{the Hironaka decomposition}\\ &= \text{the Hironaka decomposition with factored numerators.}\end{aligned}$$

If the numerator is irreducible, we omit the third line for each case.

$|G_{1,1}| = 96$ and

$$\begin{aligned}\Phi_{G_{1,1}}(t) &= 1 + t^8 + t^{12} + t^{16} + t^{20} + 2t^{24} + t^{28} + 2t^{32} + \cdots\\ &= 1/(1-t^8)(1-t^{12}).\end{aligned}$$

$|G_{1,2}| = 384$ and

$$\begin{aligned}\Phi_{G_{1,2}}(t) &= 1 + 4t^8 + 2t^{10} + 3t^{12} + 2t^{14} + 11t^{16} + 7t^{18} +\\ &\quad 11t^{20} + 9t^{22} + 25t^{24} + 18t^{26} + 27t^{28} + 23t^{30} + 48t^{32} + \cdots\\ &= (1 + t^8 + 2t^{10} + 2t^{12} + 2t^{14} + 2t^{16} + t^{18} + t^{20} + t^{22} +\\ &\quad t^{26} + t^{28} + t^{30})/(1-t^8)^3(1-t^{12})\\ &= \left(1+t^2\right)\left(1+t^4\right)\\ &\quad \left(1 - t^2 + 2t^8 + 2t^{16} - t^{18} + t^{24}\right)/(1-t^8)^3(1-t^{12}).\end{aligned}$$

$|G_{1,3}| = 2304$ and

$$\begin{aligned}\Phi_{G_{1,3}}(t) &= 1+t^8+15t^{12}+37t^{16}+78t^{20}+229t^{24}+419t^{28}+721t^{32}+\cdots\\ &= (1+12t^{12}+36t^{16}+63t^{20}+148t^{24}+233t^{28}+303t^{32}+366t^{36}\\ &\quad +444t^{40}+460t^{44}+427t^{48}+338t^{52}+272t^{56}+174t^{60}+96t^{64}\\ &\quad +53t^{68}+24t^{72}+5t^{76}+t^{80})/(1-t^8)(1-t^{12})^3(1-t^{24})^2\\ &= (1-t+t^2)(1+t+t^2)(1-t^2+t^4)(1-t^4+13t^{12}+23t^{16}+27t^{20}\\ &\quad +98t^{24}+108t^{28}+97t^{32}+161t^{36}+186t^{40}+113t^{44}+128t^{48}+97t^{52}\\ &\quad +47t^{56}+30t^{60}+19t^{64}+4t^{68}+t^{72})/(1-t^8)(1-t^{12})^3(1-t^{24})^2.\end{aligned}$$

$|G^8_{1,1}| = 192$ and

$$\begin{aligned}\Phi_{G^8_{1,1}}(t) &= 1 + t^8 + t^{16} + 2t^{24} + 2t^{32} + \cdots\\ &= 1/(1-t^8)(1-t^{24}).\end{aligned}$$

$|G^8_{1,2}| = 1536$ and

$$\begin{aligned}\Phi_{G^8_{1,2}}(t) &= 1 + 4t^8 + 11t^{16} + 25t^{24} + 48t^{32} + \cdots\\ &= (1 + t^8 + 2t^{16} + 2t^{24} + t^{32} + t^{40})/(1-t^8)^3(1-t^{24})\\ &= (1+t^8)(1+t^{16})^2/(1-t^8)^3(1-t^{24}).\end{aligned}$$

$|G^8_{1,3}| = 4608$ and

$$\begin{aligned}\Phi_{G^8_{1,3}}(t) &= 1 + t^8 + 37t^{16} + 229t^{24} + 721t^{32} + \cdots \\ &= (1 + 35t^{16} + 188t^{24} + 456t^{32} + 1099t^{40} + 1677t^{48} \\ &\quad + 1829t^{56} + 1793t^{64} + 1246t^{72} + 590t^{80} + 241t^{88} + 56t^{96} \\ &\quad + 5t^{104})/(1 - t^8)(1 - t^{16})(1 - t^{24})^4 \\ &= (1 + t^8)(1 - t^8 + 36t^{16} + 152t^{24} + 304t^{32} + 795t^{40} \\ &\quad + 882t^{48} + 947t^{56} + 846t^{64} + 400t^{72} + 190t^{80} + 51t^{88} \\ &\quad + 5t^{96})/(1 - t^8)(1 - t^{16})(1 - t^{24})^4.\end{aligned}$$

$|H_{1,2}| = 384$ and

$$\begin{aligned}\Phi_{H_{1,2}}(t) &= 1 + 2t^8 + t^{12} + 4t^{16} + 2t^{20} + 7t^{24} + 4t^{28} + 10t^{32} + \cdots \\ &= (1 + t^{16})/(1 - t^8)^2(1 - t^{12}).\end{aligned}$$

$|H_{1,3}| = 1152$ and

$$\begin{aligned}\Phi_{H_{1,3}}(t) &= 1 + t^8 + 3t^{12} + 4t^{16} + 5t^{20} + 15t^{24} \\ &\quad + 14t^{28} + 24t^{32} + \cdots \\ &= (1 + 2t^{12} + 3t^{16} + 2t^{20} + 6t^{24} + 6t^{28} + 7t^{32} + 6t^{36} \\ &\quad + 5t^{40} + 6t^{44} + t^{48} + 2t^{52} + t^{56})/(1 - t^8)(1 - t^{12})(1 - t^{24})^2 \\ &= (1 - t + t^2)(1 + t + t^2)(1 + t^4)(1 - t^2 + t^4)(1 - t^4 + t^8) \\ &\quad (1 - t^4 + 2t^{12} + t^{16} - t^{20} + 5t^{24} - t^{28} + t^{32} + t^{36}) \\ &\quad /(1 - t^8)(1 - t^{12})(1 - t^{24})^2.\end{aligned}$$

$|H^8_{1,2}| = 768$ and

$$\begin{aligned}\Phi_{H^8_{1,2}}(t) &= 1 + 2t^8 + 4t^{16} + 7t^{24} + 10t^{32} + \cdots \\ &= (1 + t^{16})/(1 - t^8)^2(1 - t^{24}).\end{aligned}$$

$|H^8_{1,3}| = 2304$ and

$$\begin{aligned}\Phi_{H^8_{1,3}}(t) &= 1 + t^8 + 4t^{16} + 15t^{24} + 24t^{32} + \cdots \\ &= (1 + 2t^{16} + 9t^{24} + 6t^{32} + 5t^{40} + 7t^{48} + 2t^{56}) \\ &\quad /(1 - t^8)(1 - t^{16})(1 - t^{24})^2 \\ &= (1 + t^8)(1 - t^8 + 3t^{16} + 6t^{24} + 5t^{40} + 2t^{48}) \\ &\quad /(1 - t^8)(1 - t^{16})(1 - t^{24})^2.\end{aligned}$$

Remark. The Molien series $\Phi_{G_{1,3}}(t)$ and $\Phi_{G_{1,4}}(t)$ were determined by Runge [30] and Oura [10], respectively.

Finally, we describe the invariant rings for these Molien series. We first consider the Hamming weight enumerators of binary Type II codes. In this case, the invariant ring for $G^8_{1,1}$ is generated by the weight enumerators of the extended Hamming $[8,4,4]$ code and the extended Golay $[24,12,8]$ code. Now let us consider the complete and symmetrized weight enumerators of Type II codes over $\mathbb{Z}_4$. In [2], the invariant ring for $H^8_{1,2}$ was investigated under the condition that Type II codes contain all-one vector, that is, they investigated the invariant ring for the group K generated by $H^8_{1,2}$ and the matrix

$$\begin{pmatrix} 0 & 0 & 1 \\ 0 & 1 & 0 \\ 1 & 0 & 0 \end{pmatrix}.$$

The group K has the same order as $H^8_{1,2}$. Thus the invariant ring for $H^8_{1,2}$ is

$$\mathbb{C}[\phi_8, \phi'_8, \phi_{24}] \oplus \phi_{16}\mathbb{C}[\phi_8, \phi'_8, \phi_{24}]$$

where $\phi_8, \phi'_8, \phi_{16}$ and ϕ_{24} are the symmetrized weight enumerators of Type II codes $O_8, Q_8, RM(1,4) + 2RM(2,4)$ and the lifted Golay code G_{24} over $\mathbb{Z}_4$, respectively. For the complete weight enumerators, a Magma computation shows that the invariant ring of $G^8_{1,2}$ has the homogenous system of parameters of degrees $8, 8, 8$ and 24. This means that the invariant ring has exactly the Molien series of the form

$$\frac{1 + t^8 + 2t^{16} + 2t^{24} + t^{32} + t^{40}}{(1-t^8)^3(1-t^{24})}.$$

Let $\mathcal{W}(n)$ be the ring generated by the g-th complete weight enumerators of Type II codes of length n. We have verified by computer that $\dim \mathcal{W}(8) = 4$ and $\dim \mathcal{W}(16) = 11$, however we have checked only $\dim \mathcal{W}(24) \geq 23$. Thus it is not known if the invariant ring for $G^8_{1,2}$ is generated by the complete weight enumerators of Type II codes over $\mathbb{Z}_4$.

References

[1] E. Bannai, S. Dougherty, M. Harada, M. Oura, Type II Codes, Even Unimodular Lattices, and Invariant Rings, IEEE Trans. on Information Theory **45**, (1999), 1194–1205.

[2] A. Bonnecaze, P. Solé, C. Bachoc, B. Mourrain, Type II codes over $|Z_4$, IEEE Trans. Inform. Theory **43**, (1997), 969–976.

[3] Y. Choie, S. T. Dougherty, H. Kim, Complete Joint Weight Enumerators and Self-Dual Codes, IEEE Trans. on Information Theory, **49**, (2003), 1275–1282.

[4] Y. Choie, S. T. Dougherty, H. Liu, Jacobi forms and Hilbert-Siegel modular forms over totally real fields and self-dual codes over polynomial rings $\mathbb{Z}_{2m}[x]/ < g(x) >$, Ars Combin. 107 (2012), 141–160.

[5] J. H. Conway and N. J. A. Sloane, "A new upper bound on the minimal distance of self-dual codes," *IEEE Trans. Inform. Theory*, **36**, (1990), 1319–1333.

[6] J. H. Conway, N. J. A. Sloane, *Sphere Packings, Lattices and Groups* Springer (1998) 3rd edition.

[7] W. Duke, Codes and Siegel Modular Forms, International Research Notices in Math., 3, p. 125 in Duke Math. J. **70** (1993).

[8] E. Freitag, M. Oura, A theta relation in genus 4, Nagoya Math. J. **161**, 69–83 (2001).

[9] W. Huffman, Cary The biweight enumerator of self-orthogonal binary codes. Discrete Math. **26** (1979), 129–143.

[10] J. Igusa, On the graded ring of theta-constants, Am. J. Math. **86**, (1964), 219–246.

[11] J. Igusa, On Siegel modular forms of genus two II, Am. J. Math. **86**, (1964), 392–412.

[12] J. Igusa, On the graded ring of theta-constants II, Am. J. Math. **88**, (1966), 221–236.

[13] J. Igusa, Modular forms and projective invariants, Am. J. Math. **89**, (1967), 817–855.

[14] V. G. Kac, *Infinite Dimensional Lie Algebras*, Prog. in Math. Vol. 44: Birkhäuser, 1983.

[15] M. Kneser, Lineare Relationen zwischen Darstellungsanzahlen quadratischer Formen, Math. Ann. **168**, (1967), 31–39.

[16] H. Maass, Uber die Darstellung der Modulformen n-ten Grades durch Poincarésche Reihen, Math. Ann. **123**, (1951), 125–151.

[17] F. J. MacWilliams, N. J. A. Sloane, *The theory of error correcting codes*, North Holland (1977).

[18] R. S. Manni, Modular forms of the fourth degree (Remark on a paper of Harris and Morrison), Classification of irregular varieties, minimal models

and Abelian varieties, Proc. Conf., Trento/Italy 1990, Lect. Notes Math. 1515, (1992), 106–111.

[19] F. J. MacWilliams, C. L. Mallows, N. J. A. Sloane Generalizations of Gleason's theorem on weight enumerators of self-dual codes IEEE Trans. Information Theory, **18**, (1972), 794–805.

[20] G. Nebe, Kneser-Hecke-operators in coding theory, Abh. Math. Sem. Univ. Hamburg **76**, (2006), 79–90.

[21] M. Oura, The dimension formula for the ring of code polynomials in genus 4, Osaka J. Math. **34**, (1997), 53–72.

[22] M. Oura, Invariant theoretical approaches to the ring of Siegel modular forms and the related topics (a survey), The 4-th Spring Conference on Modular Forms and Related Topics "Siegel Modular Forms and Abelian Varieties", at Hotel Curreac, Hamana-Lake, organized by T. Ibukiyama (Osaka Univ.), February 5–9, 2007.

[23] M. Oura, C. Poor, D. S. Yuen, Towards the Siegel ring in genus four. Int. J. Number Theory, **4**, (2008), 563–586.

[24] M. Ozeki, On basis problem for Siegel modular forms of degree 2, Acta Arith. **31**, (1976), 17–30.

[25] C. Poor, D. S. Yuen, Dimensions of spaces of Siegel modular forms of low weight in degree four, Bull. Aust. Math. Soc. **54**, 309–315. (1996).

[26] C. Poor, D. S. Yuen, Restriction of Siegel modular forms to modular curves, Bull. Aust. Math. Soc. **65**, 239–252 (2002).

[27] C. Poor, D. S. Yuen, Computations of spaces of Siegel modular cusp forms, J. Math. Soc. Japan **9**, 185–222 (2007).

[28] B. Runge, Codes and Siegel modular forms, Discrete Math **148**, (1996), 175–204.

[29] B. Runge, On Siegel modular forms I, J. Reine Angew. Math. **436**, (1993), 57–85.

[30] B. Runge, On Siegel modular forms II, Nagoya Math. J. **138**, (1995), 179–197.

[31] B. Runge, "Theta functions and Siegel-Jacobi forms," *Acta Math.*, **175**, (1995), 165–196.

[32] S. Tsuyumine, On Siegel modular forms of degree three, Am. J. Math. 108, 755–862; Addendum 1001–1003 (1986).

[33] E. Witt, Eine Identität zwischen Modulformen zweiten Grades, Abh. Math. Semin. Hansische Univ. **14**, (1941), 323–337.

Chapter 4

Jacobi Forms

4.1 Background

A Jacobi form is a modular form in two variables, say τ and z. The precise transformation law is in the next section. Intuitively, the first variable lives in the Poincaré upper half-plane and the second one in the complex plane. A Jacobi form is thus an hybrid object between a one variable theta series of Chapter II and an elliptic function. Concrete examples include Taylor coefficients of genus 2 Siegel modular forms of Chapter III, and Jacobi theta functions $\theta_2, \theta_3, \theta_4$ of the next section. More recently, Jacobi modular forms occurred in the theory of mock modular forms introduced by Ramanujan and developped by Zagier [15] and Zwegers [16]. The foundations of the theory are described in the monograph [8].

The analogue object on the coding side is the Jacobi polynomial (in Ozeki terms) or Ozeki polynomial [7] (our terms). This polynomial is essentially the joint weight enumerator of a code and a nonzero vector in ambient space. The two main applications are the study of coset weight distributions of self-dual codes, as done by Ozeki [11–14]; and the study of combinatorial designs held by the codewords of given weight of self dual codes [2].

4.2 Jacobi Forms and Theta Series

Let Γ be an arbitrary subgroup of finite index in $SL_2(\mathbb{Z})$ and let χ be a character of Γ. A function $\phi : \mathcal{H} \times \mathbb{C} \to \mathbb{C}$ is said to be a Jacobi form of a weight ω and an index m for Γ with respect to χ (see [8]) if

(1) $\phi(\frac{a\tau+b}{c\tau+d}, \frac{z}{c\tau+d}) = \left(\frac{c\tau+d}{\chi(M)}\right)^{\omega} e^{2\pi i m \frac{z^2}{c\tau+d}} \phi(\tau, z) \ \forall \ M = \left(\begin{smallmatrix} a & b \\ c & d \end{smallmatrix}\right) \in \Gamma$;

(2) $\phi(\tau, z + \lambda\tau + \mu) = e^{2\pi i m(\lambda^2\tau + 2\lambda z)} \phi(\tau, z) \ \forall \ (\lambda, \mu) \in \mathbb{Z}^2$;

(3) $\phi(\tau, z)$ has a Fourier expansion of the following type

$$\phi(\tau, z) = \sum_{n, r \in \mathbb{Z}, r^2 \leq 4mn} c(n, r) e^{2\pi i n\tau} e^{2\pi i r z}.$$

We also recall the following theta series $\theta_{2,\mu}(\tau, z)$ which was introduced to find the correspondence between Jacobi forms and modular forms of half integral weight(see [8]):

$$\theta_{2,\mu}(\tau, z) := \sum_{r \in \mathbb{Z}, r \equiv \mu \pmod 4} e^{\frac{\pi i r^2}{4}\tau} \xi^r \quad \text{with} \quad \xi = e^{2\pi i z}.$$

Remark 30. The theta series $\theta_{2,\mu}(\tau, z) = \sum_{r \in \mathbb{Z}, r \equiv \mu \pmod 4} e^{\frac{\pi i r^2}{4}\tau} \xi^r$ satisfies the following relations:

(i) $\theta_{2,\mu}(\tau, z)$ is well-defined and holomorphic on $\mathbb{H} \times \mathbb{C}$;

(ii) $\theta_{2,\mu}(\tau + 1, z) = e^{\frac{\pi i}{4}\mu^2} \theta_{2,\mu}(\tau, z), \mu \in \{0, 1, 2, 3\}$;

(iii) $\theta_{2,\mu}\left(-\frac{1}{\tau}, \frac{z}{\tau}\right) = \sqrt{\frac{\tau}{4i}} \ e^{4\pi i \frac{z^2}{\tau}} \sum_{\nu (mod 4)} e^{(-\frac{\pi i}{4}\mu\nu)} \theta_{2,\nu}(\tau, z)$;

(iv) $\theta_{2,\mu}(\tau, z + \lambda\tau + \nu) = e^{-4\pi i(\lambda^2\tau + 2\lambda z)} \theta_{2,\mu}(\tau, z)$ for all $(\lambda, \nu) \in \mathbb{Z}^2$.

The more general type of theta series $\theta_{m,\mu}(\tau, z)$ has been used to study the connection between Jacobi forms and codes over the ring $\mathbb{Z}_{2m}$ (see [5]). In particular, when $m = 2$, we get the following relations among the theta series.

$$\theta_{2,0}(\tau, 0) = \frac{\theta_3(\tau) + \theta_4(\tau)}{2}, \theta_{2,1}(\tau, 0) = \theta_{2,3}(\tau, 0) = \frac{\theta_2(\tau)}{2},$$
$$\theta_{2,2}(\tau, 0) = \frac{\theta_3(\tau) - \theta_4(\tau)}{2},$$

where

$$\theta_2(\tau) = \sum_{\ell \in \mathbb{Z}} q^{(\ell + \frac{1}{2})^2}, \theta_3(\tau) = \sum_{\ell \in \mathbb{Z}} q^{\ell^2},$$
$$\theta_4(\tau) = \theta_3(\tau + 1) = \sum_{\ell \in \mathbb{Z}} (-q)^{\ell^2} \ \text{with} \ q = e^{\pi i \tau}.$$

The following relations are well-known identities among theta series.

Proposition 31. *For all τ in the Poicaré upper half-plane, we have*

$$\theta_2(\tau+1) = \sqrt{i}\theta_2(\tau), \theta_3(\tau+1) = \theta_4(\tau), \theta_4(\tau+1) = \theta_3(\tau),$$
$$\theta_2(-\frac{1}{\tau}) = (\frac{\tau}{i})^{\frac{1}{2}}\theta_4(\tau), \theta_3(-\frac{1}{\tau}) = (\frac{\tau}{i})^{\frac{1}{2}}\theta_3(\tau), \theta_4(-\frac{1}{\tau}) = (\frac{\tau}{i})^{\frac{1}{2}}\theta_2(\tau),$$
$$\theta_3(\tau)\theta_4(\tau) = \theta_4(2\tau)^2, \theta_2(\tau)^4 + \theta_4(\tau)^4 = \theta_3(\tau)^4, \theta_3(\tau)^2 - \theta_4(\tau)^2 = 2\theta_2(2\tau)^2.$$

Proof. See, for instance [6]. □

4.3 Jacobi Forms

Let $\mathcal{M}'_k$ denote the complex vector space of modular forms of weight k. Recall its dimension [1, p. 119]:

$$\begin{cases} [\frac{k}{12}] & \text{if } k \equiv 2 \pmod{12}, \\ [\frac{k}{12}] + 1 & \text{otherwise.} \end{cases}$$

A standard result for classical modular forms is that the ring $\mathcal{M}'$ of modular forms of even weight for the full modular group is a free ring in two generators E_4, E_6, the Eisenstein series of weight 4 and 6. The analogous result for Jacobi forms is Theorem 8.3 in [8].

Theorem 32. *(Eichler-Zagier) Let $J_{2*,*}$ denote the ring of Jacobi forms of even weight and integral index, and let $E_{4,1}$ and $E_{6,1}$ denote the Jacobi Eisenstein series of weight 4 and 6, respectively, with index 1. Let $\Delta = (E_4^3 - E_6^2)/1728$. Then $J_{2*,*}$ is contained in $\mathcal{M}'[\frac{1}{\Delta}][E_{4,1}, E_{6,1}]$.*

For application to codes, we shall require the following generalization.

Theorem 33. *Let $J_{4*,*}$ denote the ring of Jacobi forms of weight a multiple of 4 and integral index, and let $\phi_{12,1}$ (in the notations of [8, p.39]) denote the Jacobi cusp form of weight 12 and index 1. Then $J_{4*,*}$ is contained in $\mathcal{M}'[\frac{1}{\Delta}][E_{4,1}, \phi_{12,1}]$.*

Proof. First, we recall that Jacobi cusp forms $\phi_{10,1}$ and $\phi_{12,1}$ are algebraically independent (see the proof of Theorem 8.1 in [8]). We claim that

$E_{4,1}(\tau, z)$ and $\phi_{12,1}(\tau, z)$ are also algebraically independent over $\mathcal{M}'$. But, from the following well-known relations [8], given by

$$\phi_{10,1}(\tau, z) = \frac{1}{144}(E_6(\tau)E_{4,1}(\tau, z) - E_4(\tau)E_{6,1}(\tau, z)), \tag{4.1}$$

$$\phi_{12,1}(\tau, z) = \frac{1}{144}(E_4^2(\tau)E_{4,1}(\tau, z) - E_6(\tau)E_{6,1}(\tau, z)), \tag{4.2}$$

we have

$$E_{4,1}(\tau, z) = \frac{144}{E_4^3(\tau) - E_6^2(\tau)}(E_4(\tau)\phi_{12,1}(\tau, z) - E_6(\tau)\phi_{10,1}(\tau, z)). \tag{4.3}$$

If $E_{4,1}(\tau, z)$ and $\phi_{12,1}(\tau, z)$ are algebraically dependent over $\mathcal{M}'$, relation (3) shows that $\phi_{10,1}(\tau, z)$ and $\phi_{12,1}$ are also algebraically dependent, which is a contradiction.

Now, since the space $J_{4*,\ell}$ is a module of rank $\ell + 1$ over $\mathcal{M}'$, we can write, for any $f(\tau, z) \in J_{4*,\ell}$,

$$f(\tau, z) = \sum_{j=0}^{\ell} g_j(\tau)E_{4,1}^j(\tau, z)E_{6,1}^{\ell-j}(\tau, z),$$

where $g_j(\tau)$ is a meromorphic modular form. From Theorem 1 and from relation (2), we have the coefficient $g_j(\tau)$ in $\mathcal{M}'[\frac{1}{\Delta}]$. This means that $J_{4*,*} \subset \mathcal{M}'[\frac{1}{\Delta}][E_{4,1}, \phi_{12,1}(\tau, z)]$. □

We shall require the theory of weak Jacobi forms to control the exponent of Δ in these theorems. Let $\tilde{J}_{2*,*}$ denote the ring of so-called weak Jacobi forms of even weight.

Theorem 34. *(Eichler-Zagier) The ring $\tilde{J}_{2*,*}$ is a polynomial algebra over $\mathcal{M}'$ on two generators*

$$\tilde{\phi}_{0,1} = \frac{\phi_{10,1}}{\Delta} \in \tilde{J}_{-2,1},$$

$$\tilde{\phi}_{-2,1} = \frac{\phi_{12,1}}{\Delta} \in \tilde{J}_{0,1},$$

where $\phi_{10,1}$, $\phi_{12,1}$ are like in (1) and (2).

The following results improve on Theorem 3 for low m by removing the denominator Δ from the formulas. The proofs of the four theorems are all modelled on the proof of [8, Theorem 3.5]. First, a map from modular to Jacobi forms is injective by algebraic independence reasons. Next, the dimension of image and the relevant space of Jacobi forms coincide by the dimension formula of [8, p. 121]. Surjectivity of the said map follows.

First, we deal with forms of weight multiple of 4.

Theorem 35. *Let $l \geq 3$ be an integer. Every Jacobi form of $J_{4l,1}$ can be written as*

$$f_{4l-4}E_{4,1} + f_{4l-12}\phi_{12,1},$$

where $f_i \in \mathcal{M}'_i$ if $i > 0$ and zero otherwise.

Proof. Observe that, by the argument in the proof of Theorem 4 the forms $E_{4,1}, \phi_{12,1}$ are algebraically independent. It is known (see [8, p. 121]) that, for $k \geq m, k \equiv 0 \pmod 2$,

$$\dim(J_{k,m}) = \sum_{\nu=0}^{m}(\dim(\mathcal{M}'_{k+2\nu}) - \lceil \frac{\nu^2}{4m} \rceil).$$

Here, $\lceil x \rceil$ is the nearest integer $\geq x$. A computation combining the dimension formulas for the space of modular forms M_j yields

$$\dim(J_{4l,1}) = \dim(\mathcal{M}'_{4l}) + \dim(\mathcal{M}'_{4l+2}) - 1 = \dim(\mathcal{M}'_{4l-4}) + \dim(\mathcal{M}'_{4l-12}),$$

so the result follows. □

Theorem 36. *Let $l \geq 7$ be an integer. Every Jacobi form of $J_{4l,2}$ can be written as*

$$f_{4l-8}E_{4,1}^2 + f_{4l-16}E_{4,1}\phi_{12,1} + f_{4l-24}\phi_{12,1}^2 + \lambda E_4^{l-1}E_{4,2} + \mu f_{4l-20}\phi_{10,1}^2,$$

where λ, μ are arbitrary scalars and $f_i \in \mathcal{M}'_i$ if $i > 0$ and zero otherwise.

Proof. It is known (see Page 121 [8]) that, for $k \geq m, k \equiv 0 \pmod 2$,

$$\dim(J_{k,m}) = \sum_{\nu=0}^{m}(\dim(\mathcal{M}'_{k+2\nu}) - \lceil \frac{\nu^2}{4m} \rceil).$$

So, with dimension formula of the space of modular forms M_j, it follows from that

$$\dim(M_{4l-8}) + \dim(\mathcal{M}'_{4l-16}) + \dim(\mathcal{M}'_{4l-24}) = \dim(J_{4l,2}) - 2,$$

and the algebraic independence of $E_{4,2}$ and $E_{4,1}^2$, derived from Theorem 8.2 of [8, p.96] where $A = E_{4,1}, X = E_{4,2}$ and $\Delta Z = \phi_{10,1}^2$. □

Next, we treat forms of even weight.

Theorem 37. *Let $l \geq 4$ be an integer. Every Jacobi form of $J_{2l,1}$ can be written as*

$$f_{2l-4}E_{4,1} + f_{2l-6}E_{6,1},$$

where $f_i \in M_i$ if $i > 0$ and zero otherwise.

Proof. This is essentially Theorem 3.5 of [8]. □

Theorem 38. *Let $l \geq 10$ be an integer. Every Jacobi form of $J_{2l,2}$ can be written as*

$$f_{2l-8}E_{4,1}^2 + f_{2l-10}E_{4,1}E_{6,1} + f_{2l-12}E_{6,1}^2 + \lambda f_{2l-20}\phi_{10,1}^2,$$

where λ is an arbitrary scalar and $f_i \in \mathcal{M}'_i$ if $i > 0$ and zero otherwise.

Proof. This follows (see [8, p. 121])) from

$$\dim(\mathcal{M}'_{2l-8}) + \dim(\mathcal{M}'_{2l-10}) + \dim(\mathcal{M}'_{2l-12}) = \dim(J_{2l,2}) - 1,$$

and the algebraic independence of $E_{4,1}$ and $E_{6,1}$ (Theorem 8.1 of [8, p. 90].) □

4.4 Ozeki Polynomials

4.4.1 *Weight Enumerators*

Let C be a binary code of length n. The Ozeki polynomial $J_{C,v}$ attached to C and an arbitrary binary vector v of length n is essentially the joint weight enumerator [9] of C with v. If $(a_1, b_1)(u)$ (resp. $(a_2, b_2)(u)$) denote the composition (i.e. (number of zeros, number of ones)) of u on the support of v (resp. the support of $\mathbf{1} + v$) then

$$J_{C,v}(w, z, x, y) := \sum_{u \in C} w^{a_1(u)} z^{b_1(u)} x^{a_2(u)} y^{b_2(u)}.$$

(Note that the order of the variables is different from [3, 10]) We see that $a_2 + b_2(u) = w(u)$ and $a_1 + b_1(u) = n - w(u)$ showing that this polynomial is homogeneous in each pair of variables (w, z) and (x, y). The space of homogeneous polynomials of degree t in w, z and total degree n will be denoted by

$$\mathbb{C}[w, z, x, y]_{t,n-t}.$$

The subspace of such polynomials invariant under a group G will be denoted

$$\mathbb{C}[w, z, x, y]^G_{t,n-t}.$$

It is easy to show, using properties of joint weight enumerators that the Ozeki polynomial satisfies a MacWilliams relation. In fact if C is a doubly even self-dual code it is invariant under $G_{II} \oplus G_{II} := \langle M_2, N_2 \rangle$, where

$$M_2 = \begin{pmatrix} M & 0 \\ 0 & M \end{pmatrix},$$

and similarly

$$N_2 = \begin{pmatrix} N & 0 \\ 0 & N \end{pmatrix}.$$

The bi-Molien series for this group can be found in [2] where the main motivation is to construct designs. Thus this group is the same abstract group as G_{II} but with a different representation, the direct sum of the 2-dimensional representation with itself. More generally, if G is a matrix group we denote by $G \oplus G$ the group obtained as

$$G \oplus G := \left\{ \begin{pmatrix} g & 0 \\ 0 & g \end{pmatrix} \; g \in G \right\}.$$

If G acts on the ring R then $G \oplus G$ acts on the tensor product of R with itself. Invariants for $G \oplus G$ are called following Schur **simultaneous invariants** for the group G. The starting point of [3, 10] is that $J_{C,v}$ is a simultaneous invariant for G_{II}. An important tool in invariant theory is *polarization* which maps invariants into simultaneous invariants for the same group.

Lemma 39. *Let G be a finite matrix group $\leq GL(2, \mathbb{C})$. If $f \in \mathbb{C}[x, y]^G$, then $Af \in \mathbb{C}[w, z, x, y]^{G \oplus G}$, where the polarization operator $A := w\partial/\partial x + z\partial/\partial y$.*

We shall give special names to the two most important polarized polynomials. Let $\psi_{8,1} = A(\psi_8)$, $\psi_{8,2} = A^2(\psi_8)$ and $k_{12,1} = A(k_{12})$, as well as $\nu_{24,1} = A(\nu_{24})$.

4.4.2 *Bannai-Ozeki Map*

We recall two of Jacobi theta functions we shall need.

$$\theta_3(\tau, z) := \sum_{n \in \mathbb{Z}} q^{n^2} \zeta^{2n},$$

and

$$\theta_3(\tau, z) := \sum_{n \in \mathbb{Z}} q^{(n+1/2)^2} \zeta^{2n+1},$$

where $q = \exp(\pi\sqrt{-1}\tau)$ and $\zeta = \exp(\pi\sqrt{-1}z)$.
We recall without proof the main result of [3].

Theorem 40. *Let n be a multiple of 4. If $f \in \mathbb{C}[w, z, x, y]^{H_{II} \oplus H_{II}}_{m,n-m}$ then*

$$BO(f) := f(\theta_3(2\tau, 2z), \theta_2(2\tau, 2z), \theta_3(2\tau, 0), \theta_2(2\tau, 0))$$

is a Jacobi form of weight $n/2$ and index m.

We shall need a few special instances of the preceding.

Lemma 41. $BO(\nu_{24,1}) = 32\phi_{12,1}$.

Proof. By the properties of the BO map and inspection of $A(\nu_{24})$ we see that $BO(A(\nu_{24}))$ is a Jacobi cusp form of weight 12 and index unity. By [8, p. 40] we see that it is a multiple of $\phi_{12,1}$. The factor 32 follows on comparing Taylor expansions of both sides using [8, p. 39]. □

The proof of the following three Lemmas is analogous and omitted.

Lemma 42. $BO(\psi_{8,1}) = 128E_{4,1}$ *and* $O(\psi_8) = 16E_4$.

Lemma 43. $BO(\psi_{8,2}) = 2^7 7E_{4,2}$.

Lemma 44. $BO(k_{12,1}) = -768E_{6,1}$ *and* $O(k_{12}) = -64E_6$.
For our coding-theoretic application we shall require the following analogue and strengthening.

Theorem 45. *Let n be a multiple of 8. If $f \in \mathbb{C}[w, z, x, y]^{G_{II} \oplus G_{II}}_{m,n-m}$ then*

$$BO(f) := f(\theta_3(2\tau, 2z), \theta_2(2\tau, 2z), \theta_3(2\tau, 0), \theta_2(2\tau, 0))$$

is a Jacobi form of weight $n/2$ and index m. Furthermore, the map $f \mapsto BO(f)$ of Theorem 40 is injective.

Proof. The modularity follows by Theorem 40 since the map of Theorem 45 is a restriction of the map of Theorem 40. Injectivity follows from [8, Theorem 8.5 p. 99] by observing that, by Lemmas 42 and 44 the image by BO of $\mathbb{C}[w,z,x,y]^{H_{II}\oplus H_{II}}$ contains $\mathbb{C}[E_4, E_6, E_{4,1}, E_{6,1}]$. □

Define $\psi_{20,1} := (-k_{12}\psi_{8,1} + \frac{3}{2}\psi_8 k_{12})$.

Lemma 46. $BO(\psi_{20,1}) = 2^{17}3^2\phi_{10,1}$.

4.4.3 *Gleason Formulas*

We shall prove the following three generalizations of Gleason formula from weight enumerators to Ozeki polynomials.

The following result is the analogue of Theorem 4 for invariants.

Theorem 47. *Every element of* $\mathbb{C}[w,z,x,y]^{G_{II}\oplus G_{II}}_{m,n-m}$ *is a homogeneous polynomial of degree* m *in* $\psi_{8,1}$ *and* $\nu_{24,1}$ *with coefficients in* $\mathbb{C}[x,y]^{G_{II}}[1/\nu_{24}]$.

Proof. For any $f \in \mathbb{C}[w,z,x,y]^{G_{II}\oplus G_{II}}_{m,n-m}$ the quantity $BO(f)$ is a Jacobi form of weight $\frac{n}{2}$ with index m. Since $BO(f)$ can be written uniquely as a polynomial in $E_{4,1}$ and $\phi_{12,1}$ over $\mathcal{M}'[\frac{1}{\Delta}]$(see Theorem 4), the Bannai-Ozeki map BO is injective map from $\mathbb{C}[w,z,x,y]^{G_{II}\oplus G_{II}}_{m,n-m}$ to $\mathcal{M}'[\frac{1}{\Delta}][E_{4,1}, \phi_{12,1}]$. Also, from the fact that $BO(\psi_{8,1}) = 128E_{4,1}$ and that $BO(\nu_{24,1}) = 32\phi_{12,1}$, the theorem follows. □

The following result, the analogue of Theorem 3 for invariants, extends the preceding to formal weight enumerators in the sense of Ozeki [10].

Theorem 48. *Every element of* $\mathbb{C}[w,z,x,y]^{H_{II}\oplus H_{II}}_{m,n-m}$ *is a homogeneous polynomial of degree* m *in* $\psi_{8,1}$ *and* $k_{12,1}$ *with coefficients in* $\mathbb{C}[x,y]^{H_{II}}[1/\nu_{24}]$.

Proof. The same argument as Theorem 47 follows: combine the Bannai-Ozeki map BO of Theorem 40 with Theorem 3. Observe as per Lemmata 2 and 3 that $BO(\psi_{8,1}) = 128E_{4,1}$ and that $BO(k_{12,1}) = -768E_{6,1}$. □

We are now in a position to justify the title of the section.

Theorem 49. *Every element of* $\mathbb{C}[w,z,x,y]_{m,n-m}^{H_{II}\oplus H_{II}}$ *is a homogeneous polynomial of degree* m *in* $\psi_{20,1}/\nu_{24}$ *and* $\nu_{24,1}/\nu_{24}$, *namely*

$$\frac{1}{\nu_{24}^m}\sum_{j=0}^{m} d_j \psi_{20,1}^{m-j}\nu_{24,1}^{j},$$

with $d_j \in \mathbb{C}[x,y]_{n+4m-4j}^{H_{II}}$.

Proof: The proof follows by combining Lemmas 2 and 6 with the properties of the Bannai-Ozeki map in Theorem 30 and Eichler-Zagier's theory of weak Jacobi forms [8, Theorem 3]. □

4.5 Improvements

We now apply Theorems 5 and 7 to invariants.

4.5.1 $m = 1$

Theorem 50. *Let* $l \geq 4$ *be an integer. Every invariant of* $\mathbb{C}[w,z,x,y]_{1,8l-1}^{G_{II}\oplus G_{II}}$ *can be written as*

$$f_{8l-8}\psi_{8,1} + f_{8l-24}\nu_{24,1},$$

where $f_i \in \mathbb{C}[x,y]_i^{G_{II}}$ *if* $i > 0$ *and zero otherwise.*

Proof: Theorem 47 implies that any $g \in \mathbb{C}[w,z,x,y]_{1,8l-1}^{G_{II}\oplus G_{II}}$ can be written as

$$g(w,z,x,y) = \alpha\psi_{8,1} + \beta\nu_{24,1}$$

with $\alpha, \beta \in \mathbb{C}[x,y]^{G_{II}}[1/\nu_{24}]$. By applying Bannai-Ozeki map BO, one has

$$BO(g) = 128 \cdot BO(\alpha)E_{4,1} + 32 \cdot BO(\beta)\phi_{12,1} \in J_{4l,1}.$$

Here, Theorem 35 implies that $128 \cdot BO(\alpha) = 128 \cdot O(\alpha) \in M_{4l-4}, 32BO(\beta) = 128 \cdot O(\beta) \in M_{4l-12}$. Now, the classical Broué-Enguehard map of Chapter 1 implies that there exist unique $f_j \in \mathbb{C}[x,y]_j^{G_{II}}$ such that $f_{8l-8} = \alpha$ and $f_{8l-24} = \beta$. □

In the special casc of Ozeki polynomials of Type II code this result can be checked by combining the polarization lemma of [2, Theorem 4] with the Assmus-Mattson theorem [2, Theorem 3].

The analogue for formal weight enumerators spells out.

Theorem 51. *Let $l \geq 4$ be an integer. Every invariant of $\mathbb{C}[w,z,x,y]_{1,4l-1}^{H_{II}\oplus H_{II}}$ can be written as*

$$f_{4l-8}\psi_{8,1} + f_{4l-12}k_{12,1},$$

where $f_i \in \mathbb{C}[x,y]_i^{H_{II}}$ if $i > 0$ and zero otherwise.

Proof: Theorem 48 implies that any $h \in \mathbb{C}[w,z,x,y]_{1,4l-1}^{H_{II}\oplus H_{II}}$ can be written as

$$h(w,z,x,y) = \alpha\psi_{8,1} + \beta k_{12,1},$$

with $\alpha, \beta \in \mathbb{C}[x,y]^{H_{II}}[1/\nu_{24}]$. By applying Bannai-Ozeki map BO, one has

$$BO(h) = 128 \cdot BO(\alpha)E_{4,1} - 768 \cdot BO(\beta)E_{6,1} \in J_{2l,1}.$$

Moreover, Theorem 37 implies that $128 \cdot BO(\alpha) = 128 \cdot O(\alpha) \in M_{2l-4}, -768 \cdot BO(\beta) = -768 \cdot O(\beta) \in M_{2l-6}$. Now, since $\mathbb{C}[x,y]^{H_{II}}$ is isomorphic onto $\mathbb{C}[E_4, E_6]$ by the Ozeki map, one concludes that there exist unique $f_j \in \mathbb{C}[x,y]_j^{H_{II}}$ such that $f_{4l-8} = \alpha$ and $f_{4l-12} = \beta$. □

4.5.2 $m = 2$

Theorem 52. *Let $l \geq 7$ be an integer. Every invariant of $\mathbb{C}[w,z,x,y]_{2,8l-2}^{G_{II}\oplus G_{II}}$ can be written as*

$$f_{8l-16}\psi_{8,1}^2 + f_{8l-32}\psi_{8,1}\nu_{24,1} + f_{8l-48}\nu_{24,1}^2,$$

where $f_i \in \mathbb{C}[x,y]_i^{G_{II}}$ if $i > 0$ and zero otherwise.

Proof: Theorem 47 implies that, any $g \in \mathbb{C}[w,z,x,y]_{2,8l-2}^{G_{II}\oplus G_{II}}$, can be written as

$$g(w,z,x,y) = a\psi_{8,1}^2 + b\psi_{8,1}\nu_{24,1} + c\nu_{24,1}^2$$

with $a, b, c \in \mathbb{C}[x,y]^{G_{II}}[1/\nu_{24}]$. So, by applying Bannai-Ozeki map BO given in Theorem 45, one has

$$BO(g) = 128^2 \cdot BO(a)E_{4,1}^2 + 128 \cdot 32 \cdot BO(b)E_{4,1}\phi_{12,1} + 32^2 \cdot BO(c)\phi_{12,1}^2 \in J_{4l,2}.$$

Here, the Ozeki map implies that $128^2 \cdot BO(a) = 128^2 \cdot O(a) \in M_{4l-8}, 128 \cdot 32 \cdot BO(b) = 128 \cdot 32 \cdot O(b) \in M_{4l-16}$ and $32^2 \cdot BO(c) = 32^2 \cdot O(c) \in M_{4l-24}$. Therefore, the Broué-Enguehard map shows that there exist unique f_j in $\mathbb{C}[x,y]_j^{G_{II}}$ such that $f_{8l-16} = a$, $f_{8l-32} = b$ and $f_{8l-48} = c$. □

For instance in agreement with [12] we find three degrees of freedom for $n = 40, m = 2$. Analogue for formal weight enumerators is:

Theorem 53. *Let $l \geq 7$ be an integer. Every invariant of $\mathbb{C}[w,z,x,y]_{2,4l-2}^{H_{II}\oplus H_{II}}$ can be written as*

$$f_{4l-16}\psi_{8,1}^2 + f_{4l-20}\psi_{8,1}k_{12,1} + f_{4l-24}k_{12,1}^2$$

where $f_i \in \mathbb{C}[x,y]_i^{H_{II}}$ if $i > 0$ and zero otherwise.

Proof: Theorem 48 implies that, any $h \in \mathbb{C}[w,z,x,y]_{2,4l-2}^{H_{II}\oplus H_{II}}$, can be written as

$$h(w,z,x,y) = a\psi_{8,1}^2 + b\psi_{8,1}k_{12,1} + ck_{12,1}^2,$$

with $a, b, c \in \mathbb{C}[x,y]^{H_{II}}[1/\nu_{24}]$. So, by applying Bannai-Ozeki map BO given in Theorem 40, one has

$$BO(h) = 128^2 \cdot BO(a)E_{4,1}^2 - 128 \cdot 768 \cdot BO(b)E_{4,1}E_{6,1} + 768^2 E_{6,1}^2 \in J_{2l,2}.$$

But, Theorem 38 implies that $128 \cdot BO(a) = 128 \cdot O(a) \in M_{2l-8}, 128 \cdot 768 \cdot BO(b) = 128 \cdot 768 \cdot O(b) \in M_{2l-10}$ and $768^2 \cdot BO(c) = 768^2 \cdot O(c) \in M_{2l-12}$. Now, since $\mathbb{C}[x,y]^{H_{II}}$ is isomorphic onto $\mathbb{C}[E_4, E_6]$ by the Ozeki map one concludes that there exist unique f_j in $\mathbb{C}[x,y]_j^{H_{II}}$ such that $f_{4l-16} = a, f_{4l-20} = b$ and $f_{4l-24} = c$. □

References

[1] T. A. Apostol, *Modular Functions and Dirichlet Series in Number Theory*, Springer-Verlag (1976).

[2] A. Bonnecaze, B. Mourrain, P. Solé, Jacobi Polynomials, Type II codes and designs, Designs, Codes and Cryptography vol 16, (1999) 215–234.

[3] E. Bannai, M. Ozeki, Construction of Jacobi forms from certain combinatorial polynomials, Proceedings of the Japan Ac., Vol. 72, Ser. A, No 1 (1996) 12–15.

[4] Y. Choie and H. Kim, Codes and Jacobi forms of higher genus n, to appear in Journal of Combinatorial Theory-Series A (2001).

[5] Y. Choie, N. Kim, The complete weight enumerator of Type II codes over $\mathbb{Z}_{2m}$ and Jacobi Forms, IEEE Transactions on Information Theory, Vol. IT-47, pp. 396–399 (2001).

[6] Y. Choie, P. Solé, A Gleason Formula for Ozeki Polynomials, Journal of Combinatorial Theory, Series A **98**, (2002), 60–73.

[7] J. H. Conway, N. J. A. Sloane, *Sphere Packings, Lattices and Groups* Springer (1998) 3rd edition.

[8] M. Eichler, D. Zagier, *The theory of Jacobi forms*, Birkhäuser (1985).
[9] F. J. MacWilliams, N. J. A. Sloane, *The theory of Error Correcting Codes*, North Holland second edition (1981).
[10] M. Ozeki, On the notion of Jacobi polynomials for codes, Math. Proc. Cambridge Phil. Soc. (1997).
[11] M. Ozeki, On covering radii and coset weight distributions of extremal binary self-dual codes of length 56. IEEE Trans. Information Theory 46(7), (2000), 2359–2372.
[12] M. Ozeki, On covering radii and coset weight distributions of extremal binary self-dual codes of length 40. Theor. Comput. Sci. 235(2), (2000), 283–308.
[13] M. Ozeki, On the covering radius problem for ternary self-dual codes. Theor. Comput. Sci. 263(1-2), (2001), 311–332.
[14] M. Ozeki, Jacobi polynomials for singly even self-dual codes and the covering radius problems. IEEE Trans. Information Theory 48(2), (2002), 547–557.
[15] D. Zagier, Ramanujan's Mock Theta Functions and their Applications (d'après Zwegers et Bringmann-Ono), Séminaire Bourbaki, No. 986, 2007.
[16] S. P. Zwegers, Mock theta functions, PhD Utrecht University (2002).

Chapter 5

Automorphic Forms over Number Fields

5.1 Introduction

In the previous chapters, we have studied the interplay between codes and modular forms either elliptic (Chapter II), Jacobi (Chapter III) or Siegel (Chapter IV). In this chapter, we tackle Hilbert, Jacobi, and Hilbert-Siegel modular forms. The complete weight enumerators of self-dual codes over fields can be considered as a polynomial invariant under a certain finite group. It is known that one can construct various modular forms from the weight enumerators of the codes by substituting special types of theta-functions into the variables of these polynomials, see for example [1,1,3–5,8]. It turns out that this connection can be made more general. By studying the relationship between finite polynomial rings and lattices over totally real fields we can generalize the construction of integral lattices induced from codes over finite field (the construction A of Chapter I) to codes over rings. This has connections with Jacobi forms over the real quadratic field $K = \mathbb{Q}(\sqrt{5})$ and Hilbert modular forms over K (see [3]).

The codes over $\mathbb{F}_4$ have been studied with respect to their binary image under the so-called Gray map, based on the results of [1] and [6]. The first section follows the paper [1], which explains a connection among Jacobi forms over the real quadratic field $K = \mathbb{Q}(\sqrt{5})$, Hilbert modular forms over K and codes over $\mathbb{F}_4$. The second section explains a more general situation, which is based on the paper in [6]. The lattices constructed in a connection between codes and modular forms usually were defined over either real, complex or quaternionic. There we explain how to extend a construction of integral lattices induced from codes over $\mathbb{F}_4$ to more general rings.

5.2 Codes over the Field of Order 4

The source of this section is from [1]. In this section, we relate self-dual codes over $\mathbb{F}_4$ to real 5-modular lattices and to self-dual codes over $\mathbb{F}_2$ via a Gray map. We construct Jacobi forms over $\mathbb{Q}(\sqrt{5})$ from the complete weight enumerator of self-dual codes over $\mathbb{F}_4$. Furthermore, we relate Hilbert-Siegel forms to joint weight enumerators of self-dual codes over $\mathbb{F}_4$.

5.2.1 *Notations and Definitions*

Let $\mathbb{F}_4 = \mathbb{F}_2(\omega) := \{0, 1, \omega, \overline{\omega} = \omega^2\}$ be the finite field of order 4.

Definition 54.

(1) A code C of length n over $\mathbb{F}_4$ is an $\mathbb{F}_4$-subspace of $\mathbb{F}_4^n$. An additive code is a subgroup of $(\mathbb{F}_4^n, +)$. An element of C is called a codeword.
(2) A code is self-orthogonal if $C \subseteq C^\perp$ and is self-dual if $C = C^\perp$ with respect to the following inner product

$$[u, v] := \sum_{j=1}^{n} u_j v_j, u = (u_j), v = (v_j) \in \mathbb{F}_4^n.$$

Here, $C^\perp = \{v' \mid [v', v] = 0 \text{ for all } v \in C\}$.
(3) The **Lee weight** $w_L(x)$ of x is defined as $n_1(x) + 2n_2(x)$, where $n_0(x)$ is the number of $x_j = 0, n_2(x)$ the number of $x_j = 1$, and $n_1(x) = n - n_0(x) - n_2(x)$ for a vector $x = (x_1, .., x_n) \in \mathbb{F}_4^n$.

It is known that there is a natural Gray map ϕ which is a $\mathbb{F}_2$-linear isometry from ($\mathbb{F}_4^n$, Lee distance) onto ($\mathbb{F}_2^{2n}$, Hamming distance). Here, the Lee distance of two codewords x and y means the Lee weight of $x - y$; we let, for all $x, y \in \mathbb{F}_2^n$,

$$\phi(\omega x + \overline{\omega} y) = (x, y).$$

5.2.1.1 *Weight Enumerators*

The weight enumerators and the complete joint weight enumerators for codes over $\mathbb{F}_4$ are defined as follows:

Definition 55.

(1) For a code C over $\mathbb{F}_4$ define the complete weight enumerator by

$$cwe_C(y_0, y_1, y_\omega, y_{\omega^2}) = \sum_{v \in C} \prod_{a \in \mathbb{F}_4} y_a^{n_a(v)}, \tag{5.1}$$

where $n_a(v) = |\{j \mid v_j = a\}|$. We use the ordering given by $0 < 1 < \omega < \omega^2$ for the elements in $\mathbb{F}_4$.

(2) Let $C_1, C_2, .., C_g$ be codes over $\mathbb{F}_4$. The complete joint weight enumerator for codes $C_1, C_2, .., C_g$ of length n over $\mathbb{F}_4$ is defined as

$$J_{C_1,..,C_g}(X_{\mathbf{a}} | \, \mathbf{a} \in \mathbb{F}_4^g) = \sum_{(v_1, v_2, \cdots, v_g) \in C_1 \times C_2 \times \ldots \times C_g} \prod_{\mathbf{a} \in \mathbb{F}_4^g} X_{\mathbf{a}}^{n_{\mathbf{a}}(v_1, v_2, \cdots, v_g)} \tag{5.2}$$

where $n_{\mathbf{a}}(v_1, \cdots, v_g) = |\{j \, |((v_1)_j, (v_2)_j, \cdots, (v_g)_j) \equiv \mathbf{a}\}|$.

Example 1. Let C_4 be the $[4, 2, 3]$ linear code over $\mathbb{F}_4$ with generator matrix

$$\begin{pmatrix} 1 & 0 & \omega^2 & \omega \\ 0 & 1 & \omega & \omega^2 \end{pmatrix}.$$

The complete weight enumerator ϕ_4 of C_4 is

$$\phi_4(y_0, y_1, y_\omega, y_{\omega^2}) = y_0^4 + y_1^4 + y_\omega^4 + y_{\omega^2}^4 + 12 y_0 y_1 y_\omega y_{\omega^2}.$$

5.2.1.2 *MacWilliam's Relations*

Let $Tr : \mathbb{F}_4 \to \mathbb{F}_2$ be a trace map and M be a 4 by 4 matrix indexed by the elements of $\mathbb{F}_4$, defined by

$$M = (M)_{\nu,\mu} := ((-1)^{Tr(\nu\mu)})_{\nu,\mu} = \begin{pmatrix} 1 & 1 & 1 & 1 \\ 1 & 1 & -1 & -1 \\ 1 & -1 & -1 & 1 \\ 1 & -1 & 1 & -1 \end{pmatrix}. \tag{5.3}$$

Theorem 56. (MacWilliams identities)

(1) Let C be a code over $\mathbb{F}_4$. Then

$$cwe_{C^\perp}(y_\mu | \mu \in \mathbb{F}_4) = \frac{1}{|C|} cwe_C(M \cdot (y_\mu) | \mu \in \mathbb{F}_4). \tag{5.4}$$

(2) *Let $C_1, C_2, .., C_g$ be codes over $\mathbb{F}_4$ and let $\tilde{C}$ denote either C or $C^{\perp}$. Then*

$$J_{\tilde{C}_1,..,\tilde{C}_g}(Y_{\mathbf{a}}|\mathbf{a} \in \mathbb{F}_4^g) = \frac{1}{\prod_{i=1}^{g} |C_i|^{\delta_{\tilde{C}_i}}} J_{C_1,...,C_g}((\otimes_{i=1}^{g} M^{\delta_{\tilde{C}_i}}) \cdot (Y_{\mathbf{a}}|\mathbf{a} \in \mathbb{F}_4^g)), \tag{5.5}$$

where

$$\delta_{\tilde{C}} = \begin{cases} 0 \ \ if\ \tilde{C} = C \\ 1 \ \ if\ \tilde{C} = C^{\perp}. \end{cases}$$

Proof. (1) For each $u \in {\mathbb{F}_4}^n$, consider the map $f_u : {\mathbb{F}_4}^n \to \mathbb{F}_2$ given by $f_u(v) = Tr(v \cdot u)$. It is a homomorphism as additive groups. First note that we have

$$\sum_{u \in C} (-1)^{Tr(v \cdot u)} = \begin{cases} |C| \text{ if } f_u(C) = \{0\} \\ 0 \text{ if } f_u(C) = \mathbb{F}_2. \end{cases}$$

Now, for $cwe_C((y_\mu)) = \sum_{u \in C} \prod_{i=1}^{n} y_{u_i}, u = (u_1, u_2, .., u_i, .., u_n)$, we have

$$\begin{aligned} cwe_C(M \cdot (y - \mu)) &= \sum_{u \in C} \prod_{i=1}^{n} (\sum_{v_i \in \mathbb{F}_4} (-1)^{Tr(v_i u_i)} y_{v_i}) \\ &= \sum_{u \in C} \sum_{v \in \mathbb{F}_4^n} \prod_{i=1}^{n} (-1)^{Tr(v_i u_i)} y_{v_i} \\ &= \sum_{v \in \mathbb{F}_4^n} (\sum_{u \in C} (-1)^{Tr(v \cdot u)}) \prod_{i=1}^{n} y_{v_i} \\ &= \sum_{v \in C^{\perp}} |C| \prod_{i=1}^{n} y_{v_i} = |C| cwe_{C^{\perp}}((y_\mu)). \end{aligned}$$

(2) We can prove this using induction on g with Theorem 98, so we omit the detailed proof.

□

5.2.2 *Type II Codes over* $\mathbb{F}_4$

A self-dual code over $\mathbb{F}_4$ is said to be **Type II** if the Lee weight of every codeword is a multiple of 4 and **Type I** otherwise. In this section we construct some explicit examples of Type II codes over $\mathbb{F}_4$ and we give their complete weight enumerators which will be used to study a structure of invariant ring in the following section. We first recall the following known properties:

Proposition 57. *[16] Let C be a code over $\mathbb{F}_4$. C is Type I (resp. Type II) iff $\phi(C)$ is Type I (resp. Type II).*

Corollary 58. *[16] There exists a Type II code of length n iff $n \equiv 0 \pmod 4$.*

To present the examples let us define some notations for polynomials as follows:

$$m_{(\lambda_1,\lambda_2,\lambda_3,\lambda_4)} = \sum y_0^{\lambda_{i_1}} y_1^{\lambda_{i_2}} y_\omega^{\lambda_{i_3}} y_{\omega^2}^{\lambda_{i_4}}$$

summed over all distinct permutations $(\lambda_{i_1}, \lambda_{i_2}, \lambda_{i_3}, \lambda_{i_4})$ of $(\lambda_1, \lambda_2, \lambda_3, \lambda_4)$,

$$\tilde{m}_{(\lambda_1,\lambda_2,\lambda_3,\lambda_4)} = \sum y_0^{\lambda_{i_1}} y_1^{\lambda_{i_2}} y_\omega^{\lambda_{i_3}} y_{\omega^2}^{\lambda_{i_4}}$$

summed over all distinct even permutations $(\lambda_{i_1}, \lambda_{i_2}, \lambda_{i_3}, \lambda_{i_4})$ of $(\lambda_1, \lambda_2, \lambda_3, \lambda_4)$ and

$$\hat{m}_{(\lambda_1,\lambda_2,\lambda_3,\lambda_4)} = \varepsilon((i_1, i_3, i_2, i_4)) \sum y_0^{\lambda_{i_1}} y_1^{\lambda_{i_2}} y_\omega^{\lambda_{i_3}} y_{\omega^2}^{\lambda_{i_4}}$$

summed over all permutations (i_1, i_3, i_2, i_4) of $(1, 2, 3, 4)$, where ε is the sign of permutation. m_λ is called a *monomial symmetric polynomial* and $\hat{m}_\lambda$ is called a skew symmetric polynomial. We give examples of self-dual codes and Type II codes and derive their complete weight enumerators which will be referred to in the next section.

Example 2. Let C_4 be the $[4, 2, 3]$ linear code over $\mathbb{F}_4$ with generator matrix

$$\begin{pmatrix} 1 & 0 & \omega^2 & \omega \\ 0 & 1 & \omega & \omega^2 \end{pmatrix}.$$

The complete weight enumerator ϕ_4 of C_4 is

$$\begin{aligned} \phi_4 &= cwe_{C_4}(Y) = m_{(4,0,0,0)} + 12m_{(1,1,1,1)} \\ &= y_0^4 + y_1^4 + y_\omega^4 + y_{\omega^2}^4 + 12y_0y_1y_\omega y_{\omega^2}. \end{aligned}$$

The linear code C_4 is a Type II code whose minimum Lee weight is 4.

Example 3. Let C_8 be the $[8,4,4]$ linear code over $\mathbb{F}_4$ with generator matrix

$$\begin{pmatrix} 1\,0\,0\,0\,1\,1\,1\,0 \\ 0\,1\,0\,0\,1\,1\,0\,1 \\ 0\,0\,1\,0\,1\,0\,1\,1 \\ 0\,0\,0\,1\,0\,1\,1\,1 \end{pmatrix}.$$

The complete weight enumerator ϕ_8 of C_8 is

$$\begin{aligned}\phi_8 &= cwe_{C_8}(Y) \\ &= m_{(8,0,0,0)} + 14m_{(4,4,0,0)} + 168m_{(2,2,2,2)} \\ &= y_0^8 + y_1^8 + y_\omega^8 + y_{\omega^2}^8 \\ &\quad +14y_0^4y_1^4 + 14y_0^4y_\omega^4 + 14y_0^4y_{\omega^2}^4 + 14y_1^4y_\omega^4 + 14y_1^4y_{\omega^2}^4 + 14y_\omega^4y_{\omega^2}^4 \\ &\quad +168y_0^2y_1^2y_\omega^2y_{\omega^2}^2.\end{aligned}$$

The linear code C_8 is a Type II code whose minimum Lee weight is 4.

Example 4. Let C_{12} be the $[12,6,6]$ linear code over $\mathbb{F}_4$ with generator matrix

$$\begin{pmatrix} 1\,0\,0\,0\,0\,0 & 0 & \omega & \omega & \omega^2 & \omega^2 & 1 \\ 0\,1\,0\,0\,0\,0 & \omega & 0 & \omega^2 & \omega & 1 & \omega^2 \\ 0\,0\,1\,0\,0\,0 & \omega & \omega^2 & \omega^2 & 0 & \omega & 1 \\ 0\,0\,0\,1\,0\,0 & \omega^2 & \omega & 0 & \omega^2 & 1 & \omega \\ 0\,0\,0\,0\,1\,0 & \omega^2 & 1 & \omega & 1 & 1 & 1 \\ 0\,0\,0\,0\,0\,1 & 1 & \omega^2 & 1 & \omega & 1 & 1 \end{pmatrix}.$$

The complete weight enumerator ϕ_{12} of C_{12} is

$$\begin{aligned}\phi_{12} &= cwe_{C_{12}}(Y) \\ &= m_{(12,0,0,0)} + 330m_{(6,2,2,2)} + 132m_{(5,5,1,1)} + 165m_{(4,4,4,0)} + 1320m_{(3,3,3,3)} \\ &= y_0^{12} + y_1^{12} + y_\omega^{12} + y_{\omega^2}^{12} \\ &\quad +330y_0^6y_1^2y_\omega^2y_{\omega^2}^2 + 330y_0^2y_1^6y_\omega^2y_{\omega^2}^2 + 330y_0^2y_1^2y_\omega^6y_{\omega^2}^2 + 330y_0^2y_1^2y_\omega^2y_{\omega^2}^6 \\ &\quad +132y_0^5y_1^5y_\omega y_{\omega^2} + 132y_0^5y_1y_\omega^5y_{\omega^2} + 132y_0^5y_1y_\omega y_{\omega^2}^5 \\ &\quad +132y_0y_1^5y_\omega^5y_{\omega^2} + 132y_0y_1^5y_\omega y_{\omega^2}^5 + 132y_0y_1y_\omega^5y_{\omega^2}^5 \\ &\quad +165y_0^4y_1^4y_\omega^4 + 165y_0^4y_1^4y_{\omega^2}^4 + 165y_0^4y_\omega^4y_{\omega^2}^4 + 165y_1^4y_\omega^4y_{\omega^2}^4 \\ &\quad +1320y_0^3y_1^3y_\omega^3y_{\omega^2}^3.\end{aligned}$$

The linear code C_{12} is a Type II code whose minimum Lee weight is 8.

See [1] for more examples.

5.2.2.1 *Invariant Rings Arising from Codes*

We study the invariant rings, where the complete weight enumerators of Type II codes over $\mathbb{F}_4$ belong to. Let M^*, g_1, g_2, g_3 and g_4 be matrices indexed by $\mathbb{F}_4$, where

$$M^* := \frac{1}{2}M,$$
$$(M)_{uv} = (-1)^{Tr(uv)},$$

$$(g_1)_{uv} = \begin{cases} 1 & \text{for } v+1=u \\ 0 & \text{otherwise,} \end{cases} \quad (g_2)_{uv} = \begin{cases} 1 \text{ for } \omega v = u \\ 0 \text{ otherwise,} \end{cases}$$

$$(g_3)_{uv} = \begin{cases} i^{\mathrm{wt}_B(u)} & \text{for } v=u \\ 0 & \text{otherwise,} \end{cases} \text{ and } (g_4)_{uv} = \begin{cases} 1 \text{ for } v^2 = u \\ 0 \text{ otherwise.} \end{cases}$$

Here, $\mathrm{wt}_B(u) = Y(Tr(\omega u)) + Y(Tr(\omega^2 u))$ with the map $Y : \mathbb{F}_2 \to \mathbb{Z}$, defined by $Y(1) = 1$ and $Y(0) = 0$. Explicitly, these matrices are

$$M^* = \frac{1}{2}\begin{pmatrix} 1 & 1 & 1 & 1 \\ 1 & 1 & -1 & -1 \\ 1 & -1 & -1 & 1 \\ 1 & -1 & 1 & -1 \end{pmatrix}, g_1 = \begin{pmatrix} 0&1&0&0 \\ 1&0&0&0 \\ 0&0&0&1 \\ 0&0&1&0 \end{pmatrix}, g_2 = \begin{pmatrix} 1&0&0&0 \\ 0&0&0&1 \\ 0&1&0&0 \\ 0&0&1&0 \end{pmatrix},$$

$$g_3 = \begin{pmatrix} 1&0&0&0 \\ 0&-1&0&0 \\ 0&0&i&0 \\ 0&0&0&i \end{pmatrix}, \text{ and } g_4 = \begin{pmatrix} 1&0&0&0 \\ 0&1&0&0 \\ 0&0&0&1 \\ 0&0&1&0 \end{pmatrix}.$$

Proposition 59. *Let G be the group generated by $\{M^*, g_1, g_2, g_3\}$. Then*

$$G = \langle M^*, T_1, T_2, T_3 \rangle,$$

where

$$T_1 := \begin{pmatrix} 1&0&0&0 \\ 0&-1&0&0 \\ 0&0&-i&0 \\ 0&0&0&-i \end{pmatrix}, T_2 := \begin{pmatrix} 1&0&0&0 \\ 0&-i&0&0 \\ 0&0&-i&0 \\ 0&0&0&-1 \end{pmatrix} T_3 := \begin{pmatrix} 1&0&0&0 \\ 0&i&0&0 \\ 0&0&1&0 \\ 0&0&0&-i \end{pmatrix}.$$

Proof. The relation $G \supset \langle M^*, T_1, T_2, T_3 \rangle$ is immediate from the fact that $T_1 = g_3^3$, $T_2 = g_2^2 g_3^3 g_2$ and $T_3 = g_3^2 g_2 g_3 g_2^2$. Moreover, the order of each group is 3840. This completes the proof. □

5.2.2.2 *Self-dual Codes over* $\mathbb{F}_4$

The following results concerning invariant ring for the self-dual codes over $\mathbb{F}_4$ are known.

Theorem 60.

(1) **(Invariant ring for self-dual codes with** $cwe_C = cwe_{\bar{C}}$ **[21])** *Let* G'_{SD} *be the group generated by* $\{M^*, g_1, g_2, g_4\}$*. Then the order of* G'_{SD} *is* 384 *and the Molien series of* G'_{SG} *is*

$$\frac{1}{(1-t^2)(1-t^4)(1-t^6)(1-t^8)}. \tag{5.6}$$

The complete weight enumerator of any self-dual code C *over* $\mathbb{F}_4$ *with* $cwe_C = cwe_{\bar{C}}$ *belongs to* $\mathbb{C}[Y]^{G'_{SD}}$*. Moreover,* $\mathbb{C}[Y]^{G'_{SD}} = \mathbb{C}[\phi_2, \phi_4, \phi_6, \phi_8]$*.*

(2) *(***Invariant ring for self-dual codes (see [21])***) Let* G_{SD} *be the group generated by* $\{M^*, g_1, g_2\}$*. Then the order of* G_{SD} *is* 192 *and the Molien series of* G_{SD} *is*

$$\frac{t^{16}+1}{(1-t^2)(1-t^4)(1-t^6)(1-t^8)}. \tag{5.7}$$

The complete weight enumerator of every self-dual code over $\mathbb{F}_4$ *belongs to* $\mathbb{C}[Y]^{G_{SD}}$*. Moreover,* $\mathbb{C}[Y]^{G_{SD}} = R \bigoplus \psi_{16} R$*, where* $R = \mathbb{C}[\phi_2, \phi_4, \phi_6, \phi_8]$ *and*

$$\psi_{16} = \hat{m}_{(10,4,2,0)} - 3\hat{m}_{(8,6,2,0)} - 8\hat{m}_{(7,5,3,1)}. \tag{5.8}$$

Denote the conjugate code of C over $\mathbb{F}_4$ by $\bar{C} = \{(c_1^2, c_2^2, \ldots, c_n^2) | (c_1, c_2, \ldots, c_n) \in C\}$. Then we have the following result:

Proposition 61. *The minimum length of any self-dual code* C *over* $\mathbb{F}_4$ *with* $cwe_C \neq cwe_{\bar{C}}$ *is* 16*. For any such code* C*,* $cwe_C(Y) - cwe_{\bar{C}}(Y) = k\psi_{16}$ *for some* $k \in \mathbb{Z}$*. Here,* ψ_{16} *is given in (5.8).*

Proof. First, we note $C_{16} \neq \bar{C}_{16}$ and $64\psi_{16} = \bar{\phi}_{16} - \phi_{16}$. Since $cwe_C \neq cwe_{\bar{C}}$, $cwe_C(Y) \in \mathbb{C}[Y]^{G_{SD}} \setminus \mathbb{C}[Y]^{G'_{SG}}$. Let $P_{16} = \{cwe_C(Y) - cwe_{\bar{C}}(Y) | C \neq \bar{C}, \text{ Length } (C) = 16 \}$. (5.6) and (5.7) imply that $\dim_{\mathbb{C}}(\langle P_{16} \rangle) = 1$. Moreover, the coefficients of any element in P_{16} are integers. So, the result follows. □

About the minimal weight of the self-dual code over $\mathbb{F}_4$ we conclude the following:

Corollary 62. *For a self-dual code C of length 16 over $\mathbb{F}_4$ with $cwe_C \neq cwe_{\bar{C}}$, both of the minimum Hamming weight and the minimum Lee weight are at most 6.*

Proof. According to ψ_{16}, the complete weight enumerator of any self-dual $[16, 8]$ code $C \neq \bar{C}$ has the monomial $y_0^{10} y_\omega^4 y_{\omega^2}^2$. It corresponds to codewords of Hamming weight 6 and of Lee weight 6. □

5.2.2.3 *Type II Codes over $\mathbb{F}_4$*

Now we study the structure of invariant ring where the complete weight enumerator of Type II codes over $\mathbb{F}_4$ belong to. We state one of our main results, which were discovered independently in [21].

Theorem 63. *(***Invariant ring for Type II codes with** $cwe_C = cwe_{\bar{C}}$*) Consider the group*

$$G' = \langle M, g_1, g_2, g_3, g_4 \rangle$$

of order 7680. Then

(1) the Molien series of G' is

$$\frac{1}{(t^4-1)(t^8-1)(t^{12}-1)(t^{20}-1)}.$$

Note that G' is the unitary reflection group No.29 in the Sephard-Todd list.

(2) The invariant ring of G' is

$$\mathbb{C}[y_0, y_1, y_\omega, y_{\omega^2}]^{G'} = \mathbb{C}[\phi_4, \phi_8, \phi_{12}, \phi_{20}]$$

as a polynomial ring.

Proof. It is immediate that $\mathbb{C}[y_0, y_1, y_\omega, y_{\omega^2}]^{G'} \supset \mathbb{C}[\phi_4, \phi_8, \phi_{12}, \phi_{20}]$. By Proposition 64, it is shown that the dimension of the homogenous components coincide. Therefore,

$$\mathbb{C}[y_0, y_1, y_\omega, y_{\omega^2}]^{G'} = \mathbb{C}[\phi_4, \phi_8, \phi_{12}, \phi_{20}].$$

□

Proposition 64. *Let $\phi_i, i = 4, 8, 12, 20$, be the complete weight enumerators of the Type II codes given before. Then ϕ_4,ϕ_8,ϕ_{12} and ϕ_{20} are algebraically independent.*

Proof. First, we introduce an order among monomials by the lexicographic order of exponents as

$$y_0^a y_1^b y_\omega^c y_{\omega^2}^d < y_0^{a'} y_1^{b'} y_\omega^{c'} y_{\omega^2}^{d'}$$

if $a < a'$ or $a = a'$ and $b < b'$ or $a = a'$, $b = b'$ and $c < c'$ or $a = a'$, $b = b'$, $c = c'$ and $d < d'$. Let f be a homogenous polynomial. We define $\mathrm{In}(f)$ by the minimal monomial of f with respect to the order. It is easy to see that $\mathrm{In}(fg) = \mathrm{In}(f)\,\mathrm{In}(g)$. We denote

$$\phi'_8 = \phi_8 - \phi_4^2,$$
$$\phi'_{12} = \phi_{12} + \frac{1}{4}\phi_4\phi_8 - \frac{5}{4}\phi_4^3,$$
$$\phi'_{20} = \phi_{20} + \frac{179}{336}\phi_4^5 + \frac{265}{1008}\phi_4^3\phi_8 - \frac{100}{63}\phi_4^2\phi_{12} - \frac{5}{72}\phi_4\phi_8^2 - \frac{5}{36}\phi_8\phi_{12}.$$

Then $\mathrm{In}(\phi_4) = y_{\omega^2}^4$, $\mathrm{In}(\phi'_8) = y_\omega^4 y_{\omega^2}^4$, $\mathrm{In}(\phi'_{12}) = y_1^4 y_\omega^4 y_{\omega^2}^4$ and $\mathrm{In}(\phi'_{20}) = y_0 y_1 y_\omega y_{\omega^2}^{17}$. We assume $g(x_1^4, x_2^8, x_3^{12}, x_4^{20})$ is a non-trivial homogenous polynomial in $\mathbb{C}[x_1, x_2, x_3, x_4]$ such that $g(\phi_4, \phi'_8, \phi'_{12}, \phi'_{20}) = 0$. Let $x_1^a x_2^b x_3^c x_4^d$ and $x_1^{a'} x_2^{b'} x_3^{c'} x_4^{d'}$ be monomials of $g(x_1, x_2, x_3, x_4)$ such that $x_1^a x_2^b x_3^c x_4^d > x_1^{a'} x_2^{b'} x_3^{c'} x_4^{d'}$. Then $\mathrm{In}(\phi_4^a \phi_8'^b \phi_{12}'^c \phi_{20}'^d) < \mathrm{In}(\phi_4^{a'} \phi_8'^{b'} \phi_{12}'^{c'} \phi_{20}'^{d'})$. If we let $x_1^a x_2^b x_3^c x_4^d$ be the maximal monomial of g, the monomial $\mathrm{In}(\phi_4^a \phi_8'^b \phi_{12}'^c \phi_{20}'^d)$ cannot be cancelled in $g(\phi_4, \phi'_8, \phi'_{12}, \phi'_{20})$. This is a contradiction. □

Finally, we state our main theorem regarding the invariant ring for Type II codes.

Theorem 65. *(***Invariant ring for Type II codes***) Consider the group $G = \langle M^*, g_1, g_2, g_3\rangle$ with order 3840. Then*

(1) The Molien series of G is

$$\frac{t^{40} + 1}{(t^4 - 1)(t^8 - 1)(t^{12} - 1)(t^{20} - 1)}.$$

(2) The invariant ring

$$\mathbb{C}[y_0, y_1, y_\omega, y_{\omega^2}]^G = R \oplus \psi_{40} R,$$

where $R := \mathbb{C}[\phi_4, \phi_8, \phi_{12}, \phi_{20}]$ *and*

$$\begin{aligned}\psi_{40} = &\hat{m}_{(28,8,4,0)} - 32\hat{m}_{(25,9,5,1)} - 5\hat{m}_{(24,12,4,0)} + 272\hat{m}_{(22,10,6,2)}\\ &+64\hat{m}_{(21,13,5,1)} + 10\hat{m}_{(20,16,4,0)} - 125\hat{m}_{(20,12,8,0)} - 768\hat{m}_{(19,11,7,3)}\\ &-304\hat{m}_{(18,14,6,2)} + 608\hat{m}_{(17,13,9,1)} + 969\hat{m}_{(16,12,8,4)}.\end{aligned}$$

The following proposition tells about the minimal length of a certain family of Type II codes and about relations between the complete weight enumerators of C and its conjugate $\overline{C}$.

Proposition 66. *The minimum length for Type II codes C over $\mathbb{F}_4$ with $cwe_C \neq cwe_{\bar{C}}$ is* 40. *For any such code C, $cwe_C(Y) - cwe_{\bar{C}}(Y) = k\psi_{40}$ for some $k \in \mathbb{Z}$.*

Proof. Note that

$$40\psi_{40} = \bar{\phi}_{40} - \phi_{40}. \tag{5.9}$$

Since the conclusion can be derived by the similar manner to the proof of Proposition 61, so we omit the detailed proof. □

As a corollary, we can also state about the upper bound of the minimal weight of a certain family of Type II codes over $\mathbb{F}_4$.

Corollary 67. *For any Type II code C of length* 40 *over $\mathbb{F}_4$, both the minimum Hamming weight and the minimum Lee weight are at most* 12.

Proof. From the definition of ψ_{40} given in (5.9), the complete weight enumerator of any Type II $[40, 20]$ code C with $cwe_C \neq cwe_{\bar{C}}$ has the monomial $y_0^{28} y_\omega^8 y_{\omega^2}^4$ that corresponds to codewords of Hamming weight 12 and of Lee weight 12. □

5.2.3 *Lattices*

From now on we let $K = Q(\alpha), \alpha = \frac{1+\sqrt{5}}{2}$. The ring of integers $\mathcal{O}_K$ of K is known to be

$$\mathcal{O}_K = \mathbb{Z}[\alpha] = \mathbb{Z} + \alpha\mathbb{Z}.$$

For each $\beta \in K$, $\beta^{(2)}$ denotes the algebraic conjugate of $\beta = \beta^{(1)}$. Then the trace map $Tr_{K/Q} : K \to Q$ is defined as $Tr_{K/Q}(x) = x^{(1)} + x^{(2)}$. Note that $Tr_{K/Q} : \mathcal{O}_K \to \mathbb{Z}$.

A lattice Λ in K^n can be considered as a free $\mathcal{O}_K$-module. The standard inner product is attached:

$$\langle v \cdot u \rangle = \sum Tr_{K/Q}(v_i u_i) \ \ \forall v = (v_j), u = (u_j) \in \Lambda. \tag{5.10}$$

Define the dual lattice $\Lambda^* = \{u \in K^n \mid u \cdot v \in \mathbb{Z} \text{ for all } v \in \Lambda\}$. A lattice is integral if $\Lambda \subseteq \Lambda^*$. The norm of a vector v is given by $N(v) = v \cdot v$. If Λ is a unimodular lattice and $N(v) \in 2\mathbb{Z}$ for all $v \in \Lambda$, then Λ is said to be a Type II (or an **even**) lattice.

To construct the lattice from a code C over $\mathbb{F}_4$ we recall the following elementary property.

Proposition 68. *Let $\mathbb{F}$ be a real quadratic field with the discriminant $d_{\mathbb{F}} = m$. Then the ideal (2) in its ring of integers $\mathcal{O}_{\mathbb{F}}$ is decomposed as follows;*

$$\left\{ \begin{array}{ll} (2) = \mathcal{P}\mathcal{P}', \mathcal{P} \neq \mathcal{P}', N(\mathcal{P}) = N(\mathcal{P}') = 2, & \textit{if } m \equiv 1 \pmod 8 \\ (2) = \mathcal{P}, N(\mathcal{P}) = 2^2, & \textit{if } m \equiv 5 \pmod 8 \end{array} \right\}.$$

Here, $\mathcal{P}, \mathcal{P}'$ are distinct prime ideals in $\mathcal{O}_K$ with norm $N(\mathcal{P}), N(\mathcal{P}')$, respectively.

Remark 69. According to the above proposition, since $(2) = \mathcal{P}$, we have $N(\mathcal{P}) = |\mathcal{O}_K/(2)| = 4$, where $K = Q(\sqrt{5})$.

Now, consider the following reduction map h modulo the ideal (2):

$$h : \mathcal{O}_K \to \mathbb{F}_4, \tag{5.11}$$

defined by $h(a+b\alpha) = a \pmod 2 + b \pmod 2 \omega$ for all $a+b\alpha \in \mathcal{O}_K$. Note that this map h is a ring homomorphism. Then the following naturally induced homomorphism for a code over $\mathbb{F}_4$ is given by

$$\tilde{h} : \mathcal{O}_K^n \to \mathbb{F}_4^n. \tag{5.12}$$

It is easy to check that $\tilde{h}^{-1}(C)$, the preimage of a code C defined over $\mathbb{F}_4$, is a free $\mathcal{O}_K$-module. So, we define the lattice induced from a code C as follows:

$$\Lambda(C) := \frac{1}{\sqrt{2}} \tilde{h}^{-1}(C) = \frac{1}{\sqrt{2}} \{v \in \mathcal{O}_K^n \mid v \pmod{2\mathcal{O}_K} \in C\}.$$

Theorem 70.

If C is a self-dual code over $\mathbb{F}_4$, then $\Lambda(C)$ is an integral lattice. Moreover, if C is Type II, then $\Lambda(C)$ is an even lattice.

Proof. If $v = (v_j = a_j + b_j\omega)_j$ and $v' = (v'_j = c_j + d_j\omega)_j$ are vectors in C, then $\sum v_j v'_j = 0$. Consider a single coordinate:

$$(a_j + b_j\omega)(a'_j + b'_j\omega) = (a_j a'_j + b_j b'_j) + (b_j a'_j + a_j b'_j + b_j b'_j)\omega = 0.$$

Now $\frac{1}{\sqrt{2}}\tilde{h}^{-1}(v) = \frac{1}{\sqrt{2}}(a_j + b_j\alpha + (2)\mathcal{O}_K)_j$ and $\frac{1}{\sqrt{2}}\tilde{h}^{-1}(v') = \frac{1}{\sqrt{2}}(a'_j + b'_j\alpha + 2\mathcal{O}_K)_j$. We need to show that the vector z in $\tilde{h}^{-1}(v)$ has an integral inner product with any vector z' in $\frac{1}{\sqrt{2}}\tilde{h}^{-1}(v')$, i.e., $\sum Tr_{K/Q}(z_j z'_j) \in \mathbb{Z}$. Consider a single coordinate. Since

$$(a_j + b_j\omega)(a'_j + b'_j\omega) = (a_j a'_j + b_j b'_j) + (b_j a'_j + a_j b'_j + b_j b'_j)\omega = 0$$

if and only if

$$(a_j a'_j + b_j b'_j) \equiv 0 \pmod{2},\ b_j a'_j + a_j b'_j + b_j b'_j \equiv 0 \pmod{2}.$$

Then, we have

$$\begin{aligned}
&\frac{1}{\sqrt{2}}\tilde{h}^{-1}(a_j + b_j\omega)\frac{1}{\sqrt{2}}\tilde{h}^{-1}(a'_j + b'_j\omega)\\
&\quad = \frac{1}{2}Tr_{K/Q}((a_j a'_j + b_j b'_j) + (b_j a'_j + a_j b'_j + b_j b'_j)\alpha + (2)\mathcal{O}_K)\\
&\quad = \frac{1}{2}Tr_{K/Q}(a_j a'_j + b_j b'_j) + \frac{1}{2}Tr_{K/Q}((b_j a'_j + a_j b'_j + b_j b'_j)\alpha) \in \mathbb{Z}.
\end{aligned}$$

This means that $\Lambda(C)$ is an integral lattice. This, together with the above computation, shows that the image of a Type II code is an even lattice. □

5.2.4 *Constructions of Jacobi Forms over K*

In this section we make a variation on the theme of Broué-Enguehard maps by studying the connections between the complete weight enumerators of codes over $\mathbb{F}_4$ and Jacobi forms over K. More precisely, the Jacobi theta series formed from the complete weight enumerators of the codes over $\mathbb{F}_4$ is a Jacobi form over the real quadratic field K. Also, Hilbert-Siegel modular forms of higher genus have been derived from the joint weight enumerators of codes over $\mathbb{F}_4$.

5.2.4.1 *Jacobi Group and Jacobi Forms over the Real Quadratic Field K*

We recall the definition of Jacobi forms over K and theta series. We follow the definitions given in [24].

The Jacobi group of K will be denoted by

$$\Gamma^J(K) := SL_2(\mathcal{O}_K) \propto \mathcal{O}_K^2.$$

This group acts on $\mathcal{H}^2 \times \mathbb{C}^2$, where $\mathcal{H}$ denotes the complex upper half plane. Variables of this space will be listed as, $(\vec{\tau}, \vec{z}) := (\tau_1, \tau_2, z_1, z_2)$. The action of $\Gamma^J(K)$ on the space $\mathcal{H}^2 \times \mathbb{C}^2$ are given by, $\begin{pmatrix} \alpha\ \beta \\ \gamma\ \delta \end{pmatrix} \in SL_2(\mathcal{O}_K)$,

$$\begin{pmatrix} \alpha\ \beta \\ \gamma\ \delta \end{pmatrix} \cdot (\vec{\tau}, \vec{z}) := \left(\frac{\alpha^{(1)}\tau_1 + \beta^{(1)}}{\gamma^{(1)}\tau_1 + \delta^{(1)}}, \frac{\alpha^{(2)}\tau_2 + \beta^{(2)}}{\gamma^{(2)}\tau_2 + \delta^{(2)}}, \frac{z_1}{\gamma^{(1)}\tau_1 + \beta^{(1)}}, \frac{z_2}{\gamma^{(2)}\tau_2 + \beta^{(2)}} \right)$$

and, for all $[\lambda, \mu] \in \mathcal{O}_K^2$,

$$[\lambda, \mu] \cdot (\vec{\tau}, \vec{z}) := \left(\tau_1, \tau_2, z_1 + \lambda^{(1)}\tau_1 + \mu^{(1)}, z_2 + \lambda^{(2)}\tau_2 + \mu^{(2)}\right).$$

Remark 71. It is known (see [18]) that $SL_2(\mathcal{O}_K)$ is generated by the matrices

$$\begin{pmatrix} 0 & -1 \\ 1 & 0 \end{pmatrix}, \begin{pmatrix} 1 & \beta \\ 0 & 1 \end{pmatrix} \quad \forall \beta \in \mathcal{O}_K.$$

We first introduce the following notation for convenience; for $\tau \in \mathcal{H}^2, \gamma, \delta \in \mathcal{O}_K$, write

$$\mathcal{N}(\gamma\tau + \delta) := \prod_{j=1}^{2} (\gamma^{(j)}\tau_j + \delta^{(j)}),$$

$$e^{2\pi i Tr_{K/Q}(m \frac{cz^2}{c\tau + d})} := \prod_{j=1}^{2} e^{2\pi i m^{(j)} \frac{c^{(j)} z_j^2}{c^{(j)}\tau_j + d^{(j)}}},$$

$$e^{-2\pi i Tr_{K/Q}(m(\lambda^2\tau + 2\lambda z))} := \prod_{j=1}^{2} e^{-2\pi i m^{(j)} ({\lambda^{(j)}}^2 \tau_j + 2\lambda^{(j)} z_j)}.$$

Definition 72. Given $k \in \frac{1}{2}\mathbb{Z}$ and $m \in \mathcal{O}_K$, a function $f : \mathcal{H}^2 \times \mathbb{C}^2 \to \mathbb{C}$ is said to be a Jacobi form of weight k and index m for the real quadratic field K if it is an analytic function satisfying

(1)

$$\begin{aligned} (f|_{k,m}M)(\vec{\tau}, \vec{z}) &:= \mathcal{N}(c\tau + d)^{-k} e^{-2\pi i Tr_{K/Q}(m \frac{cz^2}{c\tau + d})} f(M \cdot (\vec{\tau}, \vec{z})) \\ &= f(\vec{\tau}, \vec{z}) \ \text{ for all } \ M = \begin{pmatrix} * & * \\ c & d \end{pmatrix} \in SL_2(\mathcal{O}_K). \end{aligned}$$

(2)

$$(f|_m[\lambda,\mu])(\vec{\tau},\vec{z}) := e^{-2\pi i Tr_{K/Q}(m(\lambda^2\tau+2\lambda z_j))} f(\vec{\tau},[\lambda,\mu]\cdot\vec{z})$$
$$= f(\vec{\tau},\vec{z}) \text{ for all } [\lambda,\mu]\in \mathcal{O}_K^2.$$

And it has the following Fourier expansion:

(3)

$$f(\vec{\tau},\vec{z}) = \sum_{\alpha,\beta\in\delta_K^{-1},\alpha\geq 0} c(\alpha,\beta)e^{2\pi i Tr_{K/Q}(\alpha\tau+\beta z)}.$$

Here δ_K^{-1} is the inverse different of K.

Remark 73.

(1) The $\mathbb{C}$-vector space of Jacobi forms of weight k and index m for the field K is denoted by $\mathcal{J}_{k,m}(\Gamma_1(\mathcal{O}_K))$.
(2) Note that letting $\vec{z}=0$ one obtains a Hilbert modular form $f(\vec{\tau},\vec{0})$ from a Jacobi form.

5.2.4.2 *Theta Series*

The following theta function was first introduced and studied in [24] to show the correspondence between the space of Jacobi forms over the real quadratic field and the space of the vector valued modular forms. Let $\mu\in\mathcal{O}_K$ be the corresponding preimage of $\tilde{\mu}\in\mathbb{F}_4$ under the reduction map h given before, so, $\mu\in S:=\{0,1,\alpha,\alpha^2=\alpha+1\}$ (sometimes we identify the set S as $\{0,1,\omega,\omega+1\}$). Now, for each μ, consider the following theta series:

$$\theta_{1,\mu}(\vec{\tau},\vec{z}) := \sum_{r\in\delta_K^{-1}, r\equiv\mu \pmod{(2)}} e^{2\pi i Tr_{K/Q}(\frac{r^2\tau}{4}+rz)}. \tag{5.13}$$

Then, by the Poisson summation formula, the theta series satisfies the following transformation formulae [24].

Lemma 74.

(1) $(\theta_{1,\mu}|_{\frac{1}{2},1}\begin{pmatrix}1&\beta\\0&1\end{pmatrix})(\vec{\tau},\vec{z}) = e^{2\pi i Tr_{K/Q}(\frac{\mu^2\beta}{4})}\theta_{1,\mu}(\vec{\tau},\vec{z})$ *for all* $\beta\in\mathcal{O}_K$.

(2)

$$\left(\theta_{1,\mu}|_{\frac{1}{2},1}\begin{pmatrix}0 & -1\\ 1 & 0\end{pmatrix}\right)(\vec{\tau},\vec{z}) = \frac{\chi\left(\begin{pmatrix}0 & -1\\ 1 & 0\end{pmatrix}\right)}{2}\sum_{\nu\in\mathcal{O}_K/2\mathcal{O}_K} e^{2\pi i Tr_{K/Q}(\frac{\mu\nu}{4})}\theta_{1,\nu}(\vec{\tau},\vec{z}).$$

Here, $\chi\left(\begin{pmatrix}0 & -1\\ 1 & 0\end{pmatrix}\right)^4 = 1.$

Remark 75. The transformation formula of the theta series given in Lemma 97 implies that, for any $\beta = a + b\alpha \in \mathcal{O}_K, a, b \in \mathbb{Z}$,

$$(\theta_{1,0}(\vec{\tau}+\beta,\vec{z}),\theta_{1,1}(\vec{\tau}+\beta,\vec{z}),\theta_{1,\alpha}(\vec{\tau}+\beta,\vec{z}),\theta_{1,\alpha^2}(\vec{\tau}+\beta,\vec{z}))^t$$
$$= \begin{Bmatrix} T_1(\theta_{1,0}(\vec{\tau},\vec{z}),\theta_{1,1}(\vec{\tau},\vec{z}),\theta_{1,\alpha}(\vec{\tau},\vec{z}),\theta_{1,\alpha^2}(\vec{\tau},\vec{z}))^t, & \text{if } a\equiv 1, b\equiv 0 \pmod 2\\ T_2(\theta_{1,0}(\vec{\tau},\vec{z}),\theta_{1,1}(\vec{\tau},\vec{z}),\theta_{1,\alpha}(\vec{\tau},\vec{z}),\theta_{1,\alpha^2}(\vec{\tau},\vec{z}))^t, & \text{if } a\equiv 1, b\equiv 1 \pmod 2\\ I(\theta_{1,0}(\vec{\tau},\vec{z}),\theta_{1,1}(\vec{\tau},\vec{z}),\theta_{1,\alpha}(\vec{\tau},\vec{z}),\theta_{1,\alpha^2}(\vec{\tau},\vec{z}))^t, & \text{if } a\equiv 0, b\equiv 0 \pmod 2\\ T_3(\theta_{1,0}(\vec{\tau},\vec{z}),\theta_{1,1}(\vec{\tau},\vec{z}),\theta_{1,\alpha}(\vec{\tau},\vec{z}),\theta_{1,\alpha^2}(\vec{\tau},\vec{z}))^t, & \text{if } a\equiv 0, b\equiv 1 \pmod 2 \end{Bmatrix}.$$

Here, T_i are matrices in Proposition 59 and I is an identity matrix.

5.2.4.3 *Complete Weight Enumerators of Codes over* $\mathbb{F}_4$ *and Jacobi Forms over* K

In this subsection we construct a theta series defined over the lattice induced from codes C over $\mathbb{F}_4$. Furthermore, we show that the constructed theta series is a Jacobi form over K if the codes are Type II.

For each Y in the lattice Λ, consider the theta series $\Theta_{\Lambda,Y} : \mathcal{H}^2 \times \mathbb{C}^2 \to \mathbb{C}$ associated with a lattice Λ:

$$\Theta_{\Lambda,Y}(\vec{\tau},\vec{z}) := \sum_{x\in\Lambda} e^{2\pi i Tr_{K/Q}(\frac{x\cdot x}{2}\tau + x\cdot Yz)}. \tag{5.14}$$

The following theorem gives a connection between a theta series defined over the lattices induced from codes and their complete weight enumerators.

Theorem 76. *Let C be a code over $\mathbb{F}_4$. Let $\Lambda(C)$ be a lattice induced from C over $\mathbb{F}_4$, i.e. $\Lambda(C) = \frac{1}{\sqrt{2}}\tilde{h}^{-1}(C)$. The following theta series $\Theta_{\Lambda(C),\sqrt{2}(1,..,1)}(\vec{\tau},\vec{z})$ associated with $\Lambda(C)$ is exactly the same as the complete weight enumerator $cwe_C(y_0, y_1, y_\omega, y_{\omega^2})$ evaluated at $\theta_{1,\mu}(\vec{\tau},\vec{z}), \mu \in S$. In other words,*

$$\Theta_{\Lambda(C),\sqrt{2}(1,..,1)}(\vec{\tau},\vec{z}) = cwe_C(\theta_{1,\mu}(\vec{\tau},\vec{z}) \mid \mu \in S), \tag{5.15}$$

where $\{\theta_{1,\mu}\}$ is given in Lemma 97.

Proof. Note that $\sqrt{\mathbf{2}} = \frac{1}{\sqrt{2}}(2, ..., 2, 2) \in \Lambda(C)$. Let $v = (v_1, .., v_n)$ be any given codeword in C and, for each $\mu \in \mathbb{F}_4$, let $n_\mu(v) = |\{j \mid v_j = \mu\}|$. If we let $\tilde{h}(\tilde{v}) = v$, then the image can be arranged in the following form: $\{\tilde{h}^{-1}(v)\} = \{\tilde{h}^{-1}(0) + \tilde{v} \mid \tilde{v} = (\tilde{v_j}), \tilde{v_j} = a_j + b_j w, 0 \le a_j, b_j < 2\}$ and the number of μ in $(\tilde{v}_1, .., \tilde{v}_n)$ is exactly $n_\mu(v)$. Thus, for each $v \in C$,

$$\sum_{x \in \frac{1}{\sqrt{2}}\tilde{h}^{-1}(v)} e^{2\pi i Tr_{K/Q}(\frac{x\cdot x}{2}\tau + x\cdot\sqrt{\mathbf{2}}z)}$$

$$= \sum_{x\in\tilde{h}^{-1}(0)} e^{2\pi i Tr_{K/Q}(\frac{(x+\tilde{v})\cdot(x+\tilde{v})}{4}\tau + (x+\tilde{v})\cdot\mathbf{1}z)}$$

$$= \sum_{x_1\in 2\mathcal{O}_K+\tilde{v_1}} \cdots \sum_{x_n\in 2\mathcal{O}_K+\tilde{v_n}} e^{2\pi i Tr_{K/Q}(\frac{(x_1^2+x_2^2+\cdots+x_n^2)}{4}\tau + x_1 z+\cdots+x_n z)}$$

$$= \Bigg(\sum_{x_1\in 2\mathcal{O}_K+\tilde{v_1}} e^{2\pi i Tr_{K/Q}(\frac{x_1^2}{4}\tau + x_1 z_1)}\Bigg)\cdots\Bigg(\sum_{x_n\in 2\mathcal{O}_K+\tilde{v_n}} e^{2\pi i Tr_{K/Q}(\frac{x_n^2}{4}\tau + x_n z)}\Bigg)$$

$$= \prod_{\mu\in S} \theta_{1,\mu}(\vec{\tau}, \vec{z})^{n_\mu(v)}.$$

Therefore, we have $\Theta_{\Lambda(C),\sqrt{\mathbf{2}}}(\vec{\tau}, \vec{z}) = cwe_C(\theta_{1,\mu}(\vec{\tau}, \vec{z}) \mid \mu \in S)$. This finishes the proof. ☐

Theorem 77. *Let C be a Type II code of length n over $\mathbb{F}_4$. Then*

$$cwe_C(\theta_{1,\mu}(\vec{\tau}, \vec{z}) \mid \mu \in \mathbb{F}_4)$$

is a Jacobi form of weight $\frac{n}{2}$ and index n over K.

Proof. For convenience, let $g(\vec{\tau}, \vec{z}) := cwe_C(\theta_{1,\mu}(\vec{\tau}, \vec{z}) \mid \mu \in S)$. To check the modularity of $g(\vec{\tau}, \vec{z})$, it is enough to check the transformation formula under the two types of generators (see Remark 94) $\begin{pmatrix} 1 & \beta \\ 0 & 1 \end{pmatrix}$ and $\begin{pmatrix} 0 & -1 \\ 1 & 0 \end{pmatrix}$ of $\Gamma_1(\mathcal{O}_K)$ $\forall \beta \in \mathcal{O}_K$.

First, for any $\beta = a + b\alpha \in \mathcal{O}_K, a, b \in \mathbb{Z}$,

$$\begin{aligned}
&\left(g|_{\frac{n}{2},n}\begin{pmatrix}1 & \beta\\ 0 & 1\end{pmatrix}\right)(\vec{\tau},\vec{z})\\
&\quad = cwe_C(\theta_{1,\mu}(\vec{\tau}+\beta,\vec{z}) \mid \mu \in S)\\
&\quad = cwe_C(e^{2\pi i Tr_{K/Q}(\frac{\mu^2\beta}{4})}\theta_{1,\mu}(\vec{\tau},\vec{z}) \mid \mu \in \mathbb{F}_4)\ \text{(from Lemma 97)}\\
&\quad = \left\{\begin{array}{l} cwe_C(T_1\cdot(\theta_{1,\mu}(\vec{\tau},\vec{z})))\ \text{if } a\equiv 1, b\equiv 0 \pmod 2\\ cwe_C(T_2\cdot(\theta_{1,\mu}(\vec{\tau},\vec{z})))\ \text{if } a\equiv 1, b\equiv 1 \pmod 2\\ cwe_C(I\cdot(\theta_{1,\mu}(\vec{\tau},\vec{z})))\ \text{if } a\equiv 0, b\equiv 0 \pmod 2\\ cwe_C(T_3\cdot(\theta_{1,\mu}(\vec{\tau},\vec{z})))\ \text{if } a\equiv 0, b\equiv 1 \pmod 2\end{array}\right\}\\
&\quad = g(\vec{\tau},\vec{z})\ \text{(from Remark 75).}
\end{aligned}$$

Secondly,

$$\begin{aligned}
&\left(g|_{\frac{n}{2},n}\begin{pmatrix}0 & -1\\ 1 & 0\end{pmatrix}\right)(\vec{\tau},\vec{z})\\
&\quad = \mathcal{N}(\tau)^{-\frac{n}{2}}e^{-2\pi i Tr_{K/Q}(\frac{nz^2}{\tau})}cwe_C\left(\theta_{1,\mu}\left(\frac{-1}{\tau_1},\frac{-1}{\tau_2},\frac{z_1}{\tau_1},\frac{z_2}{\tau_2}\right) \mid \mu \in S\right)\\
&\quad = \mathcal{N}(\tau)^{-\frac{n}{2}}e^{-2\pi i Tr_{K/Q}(\frac{nz^2}{\tau})}cwe_C\Bigg(\chi\left(\begin{pmatrix}0 & -1\\ -1 & 0\end{pmatrix}\right)\\
&\qquad \times\left(\frac{\mathcal{N}(\tau)}{2^2}\right)^{\frac{1}{2}}e^{2\pi i Tr_{K/Q}(\frac{z^2}{\tau})}M\cdot(\theta_{1,\mu}(\vec{\tau},\vec{z})|\mu \in S)\Bigg)\\
&\quad = \frac{1}{2^n}cwe_C(M\cdot(\theta_{1,\mu}(\vec{\tau},\vec{z})|\mu\in S))\\
&\quad = cwe_{C^\perp}((\theta_{1,\mu}|\mu\in S))(\text{since } C \text{ is Type II})\\
&\quad = cwe_C(\theta_{1,\mu}(\vec{\tau},\vec{z})|\mu\in S) = g(\vec{\tau},\vec{z}).
\end{aligned}$$

Next, one needs to check the elliptic property; for any λ, $\mu \in \mathcal{O}_K$,

$$\begin{aligned}
&e^{2\pi i Tr_{K/Q}(n(\lambda^2\tau+\lambda z))}g(\vec{\tau},\vec{z}+\lambda\vec{\tau}+\mu)\\
&\quad = \sum_{x\in\Lambda(C)} e^{2\pi i Tr_{K/Q}(\frac{x\cdot x+2n\lambda^2}{2}\tau+(x+\sqrt{2}\lambda)\cdot\sqrt{\mathbf{2}}z)}\\
&\quad \text{(since } \lambda\in\mathcal{O}_K\text{, replace } x+\sqrt{2}\lambda\to x)\\
&\quad = g(\vec{\tau},\vec{z}).
\end{aligned}$$

Finally, the proper Fourier expansion can be checked easily and we omit the detailed proof. □

5.2.4.4 *Invariant Space and Space of Jacobi forms over K*

In the previous section, we showed how to construct Jacobi forms over K from the complete weight enumerators of the Type II codes C over $\mathbb{F}_4$. This comes from the invariant property of the complete weight enumerator of C over $\mathbb{F}_4$. More generally, we construct an algebra homomorphism from the algebra of the invariant polynomials, where the complete weight enumerators of the Type II codes belong to, to the algebra of Jacobi forms over K.

Let $\mathcal{G}$ be a matrix group and let $\mathbb{C}[X]$ be polynomial ring. Then $\mathcal{G}$ acts on $\mathbb{C}[X]$ in the following way: $g \cdot F(X) := F(g \cdot X)\ for\ all\ g \in \mathcal{G}\ for\ all\ F \in \mathbb{C}[X]$. With this action we can write

$$\mathbb{C}[X]^{\mathcal{G}} := \{F \in \mathbb{C}[X] \,|\, g \cdot F(X) = F(g \cdot X)\ for\ all\ g \in \mathcal{G}\}.$$

Theorem 78. *Let* $\mathbb{C}[y_0, y_1, y_\omega, y_{\omega^2}]^G := \{f(y_0, y_1, y_\omega, y_{\omega^2}) \in \mathbb{C}[y_0, y_1, y_\omega, y_{\omega^2}] \,|\, g \cdot f(y_0, y_1, y_\omega, y_{\omega^2}) = f(y_0, y_1, y_\omega, y_{\omega^2}), \}$. *Then the following map*

$$\Phi : \mathbb{C}[y_0, y_1, y_\omega, y_{\omega^2}]^G \to \bigoplus_{\ell,m} \mathcal{J}_{\ell,m}(\Gamma_1(\mathcal{O}_K)),$$

given by

$$\Phi(F(y_0, y_1, y_\omega, y_{\omega^2})) = F(\theta_{1,0}, \theta_{1,1}, \theta_{1,\alpha}, \theta_{1,\alpha^2})\ \ \forall F \in \mathbb{C}[y_0, y_1, y_\omega, y_{\omega^2}]^G$$

is an algebra homomorphism.

Remark 79. Note that

$$\mathbb{C}[y_0, y_1, y_\omega, y_{\omega^2}]^G \subset \bigoplus_{\ell>0} \mathbb{C}[y_0, y_1, y_\omega, y_{\omega^2}]_{4\ell}$$
$$:= \{F \in \mathbb{C}[y_0, y_1, y_\omega, y_{\omega^2}] \,|\, degree(F) \equiv 0 \pmod 4\}.$$

So, the image Φ in Theorem 107 belongs to the space of Jacobi forms with even integral weight.

Proof. It is enough to check the transformation formula for $g(\vec{\tau}, \vec{z}) := F(\theta_{1,0}(\vec{\tau}, \vec{z}), \theta_{1,1}(\vec{\tau}, \vec{z}), \theta_{1,\alpha}(\vec{\tau}, \vec{z}), \theta_{1,\alpha^2}(\vec{\tau}, \vec{z}))$; with $degree(F) = \ell\ for\ all\ \beta \in \mathcal{O}_K$,

$$g\left(\begin{pmatrix} 1 & \beta \\ 0 & 1 \end{pmatrix} \cdot (\vec{\tau}, \vec{z})\right) = F((\theta_{1,\mu}(\vec{\tau} + \beta, \vec{z}) | \mu \in S)) = F(T_i(\theta_{1,\mu}(\vec{\tau}, \vec{z}) | \mu \in S),$$

for some $T_i \in \{T_1, T_2, T_3\}$ because of transformation formula of $\theta_{1,\mu}(\vec{\tau}, \vec{z})$. Next,

$$\begin{aligned}
& g\left(-\frac{1}{\tau_1}, -\frac{1}{\tau_2}, \frac{z_1}{\tau_1}, \frac{z_2}{\tau_2}\right) \\
& = F\left(\theta_{1,\mu}\left(-\frac{1}{\tau_1}, -\frac{1}{\tau_2}, \frac{z_1}{\tau_1}, \frac{z_2}{\tau_2}\right) | \mu \in S\right) \\
& = F\left(\chi\left(\begin{pmatrix} 0 & -1 \\ 1 & 0 \end{pmatrix}\right) \mathcal{N}\left(\frac{\tau}{2}\right)^{\frac{1}{2}} e^{2\pi i Tr_{K/Q}(\ell \frac{z^2}{\tau})} M \cdot (\theta_{1,\mu}(\vec{\tau}, \vec{z}) | \mu \in S)\right) \\
& = \mathcal{N}(\tau)^{\frac{\ell}{2}} e^{2\pi i Tr_{K/Q}(\ell \frac{z^2}{\tau})} F(M^*(\theta_{1,\mu}(\vec{\tau}, \vec{z}) | \mu \in S)^t) \\
& \qquad (\text{since } \ell = degree(F) \equiv 0 \pmod 4) \\
& = \mathcal{N}(\tau)^{\frac{\ell}{2}} e^{2\pi i Tr_{K/Q}(\ell \frac{z^2}{\tau})} F((\theta_{1,\mu}(\vec{\tau}, \vec{z}) | \mu \in S)^t).
\end{aligned}$$

Since the condition at the cusps and the elliptic property can be also checked in the similar manner as in the proof of Theorem 77, so we omit the detailed proof. □

5.2.4.5 *Explicit Example*

Note that one can construct an elliptic modular form from Jacobi Hilbert forms or from Hilbert forms by specializing variables, namely, $\tau_1 = \tau_2, z_1 = z_2 = 0$. In this subsection we derive some explicit relations among the various theta functions using the map Φ studied in the previous section.

For $K = \mathbb{Q}(\sqrt{5})$, the theta series defined in (5.26) are

$$\theta_{1,0}(\tau, \tau, 0, 0) = \sum_{a,b \in \mathbb{Z}} q^{2a^2+3b^2+2ab},$$

$$\theta_{1,1}(\tau, \tau, 0, 0) = q^{\frac{1}{2}} \sum_{a,b \in \mathbb{Z}} q^{2a^2+3b^2+2ab+2a+b},$$

$$\theta_{1,\alpha}(\tau, \tau, 0, 0) = q^{\frac{3}{4}} \sum_{a,b \in \mathbb{Z}} q^{2a^2+3b^2+2ab+a+3b},$$

$$\theta_{1,\alpha+1}(\tau, \tau, 0, 0) = q^{\frac{7}{4}} \sum_{a,b \in \mathbb{Z}} q^{2a^2+3b^2+2ab+2a+4b}.$$

So,

$$\phi_4(\theta_{1,0}(\tau,\tau,0,0),\theta_{1,1}(\tau,\tau,0,0),\theta_{1,\alpha}(\tau,\tau,0,0),\theta_{1,\alpha+1}(\tau,\tau,0,0))$$

$$=\Big(\sum_{a,b\in\mathbb{Z}} q^{2a^2+3b^2+2ab}\Big)^4+\Big(\sum_{a,b\in\mathbb{Z}} q^{2a^2+3b^2+2ab+2a+b-\frac{1}{2}}\Big)^4$$

$$+\Big(\sum_{a,b\in\mathbb{Z}} q^{2a^2+3b^2+2ab+a+3b+\frac{3}{4}}\Big)^4+\Big(\sum_{a,b\in\mathbb{Z}} q^{2a^2+3b^2+2ab+2a+4b+\frac{3}{4}}\Big)^4$$

$$+12\Big(\sum_{a,b\in\mathbb{Z}} q^{2a^2+3b^2+2ab}\Big)\Big(\sum_{a,b\in\mathbb{Z}} q^{2a^2+3b^2+2ab+2a+b-\frac{1}{2}}\Big)$$

$$\Big(\sum_{a,b\in\mathbb{Z}} q^{2a^2+3b^2+2ab+a+3b+\frac{3}{4}}\Big)\Big(\sum_{a,b\in\mathbb{Z}} q^{2a^2+3b^2+2ab+2a+4b+\frac{3}{4}}\Big)=E_4(\tau).$$

Here, $E_4(\tau)$ is the elliptic Eisenstein series of weight 4 defined by

$$E_4(\tau)=1+240\sum_{n>0}\sigma_3(n)q^n,\sigma_3(n)=\sum_{0<d|n}d^3.$$

5.2.5 *Construction of Hilbert-Siegel Modular Forms of Genus g over K*

In this section, we consider a higher genus Hilbert-Siegel modular form over K and derive a connection with the joint weight enumerators of codes over $\mathbb{F}_4$ and the symmetrized weight enumerators of codes as well.

5.2.5.1 *Hilbert-Siegel Modular Forms of Genus g over K*

Let $\mathcal{H}_g := \{\tau \in M_{g\times g}(\mathbb{C}) | Im(\tau) > 0\}$ be the Siegel upper half plane and let $\Gamma_g(\mathcal{O}_K) := SP_{2g}(\mathcal{O}_K)$ be the symplectic group acting on $\mathcal{H}_g$ by

$$\begin{pmatrix} A & B \\ C & D \end{pmatrix}\cdot\tau := (A\tau+B)(C\tau+D)^{-1}.$$

We define Hilbert-Siegel modular forms of genus g over K.

Definition 80. A holomorphic function $F:\mathcal{H}_g^2\to\mathbb{C}$ is called a Hilbert modular form of weight k and genus g over K if

$$F(M\cdot\tau)=\mathcal{N}(Det(C\tau+D))^{-k}F(\tau)\ \ \forall M=\begin{pmatrix} * & * \\ C & D \end{pmatrix}\in\Gamma_g(\mathcal{O}_K),$$

with a proper holomorphic condition at each cusp in the case of $g=1$.

Consider, for each $\mu \in S^g = S \times S \times \cdots \times S$, the theta series $\theta^{(g)}_{1,\mu} : \mathcal{H}^2_g \to \mathbb{C}$,

$$\theta^{(g)}_{1,\mu}(\tau) = \sum_{r \in (\delta_K^{-1})^g, r \equiv \mu \pmod{(2\mathcal{O}_K)^g}} e^{2\pi i Tr_{K/Q}(\frac{r\tau r^t}{4})}. \tag{5.16}$$

Then the following can be derived.

Lemma 81. *Let $\theta^{(g)}_{1,\mu}(\tau)$ be the function defined in (5.16). Then it satisfies the following transformation formula:*

$$\theta^{(g)}{}_{1,\mu}(-\tau^{-1}) = \frac{\chi\left(\begin{pmatrix} 0 & -1 \\ -1 & 0 \end{pmatrix}\right)^g}{2^g} \mathcal{N}(Det(\tau))^{\frac{1}{2}} \sum_{\nu \in S^g} (-1)^{Tr(\mu \cdot \nu)} \theta_{1,\nu}(\tau).$$

Proof. This can be derived using the Poisson summation formula and we omit the detailed proof. □

5.2.5.2 *Joint Weight Enumerators and Hilbert-Siegel Modular Forms of Genus g over K*

For given lattices $\Lambda_1, \cdots, \Lambda_g$, let us consider the following theta series $\Theta_{\Lambda_1,\cdots,\Lambda_g} : \mathcal{H}^2_g \to \mathbb{C}$ defined as:

$$\Theta_{\Lambda_1,\cdots,\Lambda_g}(\tau, z) = \sum_{x \in \Lambda_1 \times \Lambda_2 \times .. \times \Lambda_g} e^{2\pi i Tr(\sigma(\frac{x\tau x}{4}))}. \tag{5.17}$$

Here σ denotes the trace of the matrix, i.e., $\sigma((x)_{i,j}) = \sum_i x_{ii}$.

The next theorem states a connection between the theta series defined over the lattices induced from codes over $\mathbb{F}_4$ and their joint weight enumerators.

Theorem 82.

Let $C_j, 1 \le j \le g$, be the codes over $\mathbb{F}_4$ and Λ_j be an induced lattice from the codes C_j, i.e., $\Lambda_j = \frac{1}{\sqrt{2}}\tilde{h}^{-1}(C_j)$. Let $J_{C_1,C_2,..,C_g}(X)$ be the complete joint weight enumerator of the codes $C_j, 1 \le j \le g$. Then the following holds:

$$\Theta_{\Lambda_1,\Lambda_2,..,\Lambda_g}(\tau) = J_{C_1,C_2,..,C_g}(\theta^{(g)}_{1,\mu}(\tau) \mid \mu \in S^g).$$

Proof. Let $\tilde{h} : \mathcal{O}_K^n \times ... \times \mathcal{O}_K^n \to \mathbb{F}_4^n \times .. \times \mathbb{F}_4^n$ be the homomorphism induced from the map $\tilde{h}$ in (5.12). For each $v \in C_1 \times C_2 \times ... \times C_g$, let $\tilde{h}^{-1}(v) =$

$\tilde{h}^{-1}(0) + (\tilde{v}_i)$ be a preimage of v, all of whose entries $(\tilde{v}_i)_j = (a_{ij} + b_{ij}\alpha)$ are of the forms such that $0 \le a_{ij}, b_{ij} < 2$.
Then

$$\sum_{x \in \frac{1}{\sqrt{2}}\tilde{h}^{-1}(v)} e^{2\pi i Tr_{K/Q}(\sigma(\frac{x\tau x^t}{2}))}$$

$$= \sum_{x \in \tilde{h}^{-1}(0)} e^{2\pi i Tr_{K/Q}(\sigma(\frac{(x_1+\tilde{v_1},..,x_n+\tilde{v_n})\tau(x_1+\tilde{v_1},..,x_n+\tilde{v_n})^t}{4}))}$$

$$= (\sum_{x_1 \in (2\mathcal{O}_K)^g} e^{2\pi i Tr_{K/Q}(\frac{(x_1+\tilde{v_1})\tau(x_1+\tilde{v_1})^t}{4})})..$$

$$(\sum_{x_n \in (2\mathcal{O}_K)^g} e^{2\pi i Tr_{K/Q}(\frac{(x_n+\tilde{v_n})\tau(x_n+\tilde{v_n})^t}{4})})$$

$$= \prod_{a \in S^g} \theta_{1,\mathbf{a}}(\tau)^{n_{\mathbf{a}}(\tilde{v}_1,..,\tilde{v}_n)},$$

from the fact that the number of $\mathbf{a}$ in $\mathbb{F}_4^g$ which are equal to $\tilde{v}_1,..,\tilde{v}_n$ is exactly $n_{\mathbf{a}}(v_1,..,v_n)$. □

Theorem 83. *Let $C_j, 1 \le j \le g$, be a length n Type II code over $\mathbb{F}_4$. Let $J_{C_1,C_2,..,C_g}(X)$ be the complete joint weight enumerator of the codes $C_j, 1 \le j \le g$. Then*

$$J_{C_1,C_2,...,C_g}(\theta_{1,\mu}^{(g)}(\tau) \mid \mu \in S^g)$$

is a Hilbert-Siegel modular form of weight $\frac{n}{2}$ and genus g over K.

Proof. For simplicity let $H(\tau) := J_{C_1,C_2,...,C_g}(\theta_{1,\mu}^{(g)}(\tau) \mid \mu \in S^g)$. It is enough to check the transformation law of $H(\tau)$ under the three types of generators of $\Gamma_g(\mathcal{O}_K)$ (see Remark 94):

$$\begin{pmatrix} 0 & -I \\ I & 0 \end{pmatrix}, \quad \begin{pmatrix} u & 0 \\ 0 & u^{-1} \end{pmatrix} \; for\ all\ u \in GL(g;\mathcal{O}_K),$$

$$\begin{pmatrix} 1 & \alpha \\ 0 & 1 \end{pmatrix} \; for\ all\ \alpha \in Sym(g;\mathcal{O}_K).$$

Then,

$$
\begin{aligned}
H(-\tau^{-1}) &= J_{C_1,..,C_g}(\theta^{(g)}_{2,\mu}(-\tau^{-1})|\mu \in S^g) \\
&= J_{C_1,..,C_g}\Big(\frac{\chi(\begin{pmatrix} 0 & -I \\ I & 0 \end{pmatrix})^g}{2^g}\mathcal{N}(Det(\tau))^{\frac{1}{2}}(\otimes_g M)(\theta^{(g)}_{1,\mu}(\tau)|\mu \in S^g)\Big) \\
&= \mathcal{N}(Det(\tau))^{\frac{n}{2}}\frac{1}{2^{gn}}J_{C_1,..,C_g}((\otimes_g M)(\theta^{(g)}_{1,\mu}(\tau,)|\mu \in S^g))) \\
&= \mathcal{N}(Det(\tau))^{\frac{n}{2}}\frac{1}{\prod_i^g |C_i|}J_{C_1,..,C_g}(\tau) \\
&= \mathcal{N}(Det(\tau))^{\frac{n}{2}}H(\tau)
\end{aligned}
$$

(since $n \equiv 0 \pmod 4$) and from MacWilliams' identity given in Theorem 1.2). □

5.3 Self-Dual Codes over Polynomial Rings $\mathbb{Z}_{2m}[x]/\langle g(x)\rangle$

This section is based on the paper in [6]. We study codes over polynomial rings and give a connection to Jacobi Hilbert modular forms, in particular, Hilbert modular forms over the totally real field via the complete weight enumerators of codes over polynomial rings.

5.3.1 *Notations and Definitions*

Let p be an odd prime and $K = \mathbf{Q}(\zeta_p + \zeta_p^{-1})$ be the maximal real subfield of a cyclotomic field $\mathbf{Q}(\zeta_p)$, where $\zeta_p = e^{\frac{2\pi i}{p}}$. Then its ring of integers is $\mathcal{O}_K = \mathbb{Z}[\zeta_p + \zeta_p^{-1}]$. For convenience, let $\alpha_p := \zeta_p + \zeta_p^{-1}$, then $\mathcal{O}_K = \mathbb{Z}[\alpha_p]$. Then the elements of $\mathbb{Z}[\alpha_p]$ can be written as follows:

$$\mathcal{O}_K = \mathbb{Z}[\alpha_p] = \{a_0 + a_1\alpha_p + \cdots + a_n\alpha_p^n \mid a_i \in \mathbb{Z}, n \geq 0\}.$$

Let $\mathbb{Z}_m$ denote the residue ring of integers modulo m, and let $\mathbb{Z}_m[x]$ be the polynomial rings. We take the monic irreducible polynomial $g_1(x) \in \mathbb{Z}[x]$ of degree $r = \frac{p-1}{2}$ corresponding to α_p, i.e., $g_1(x) = b_0 + b_1x + \cdots + b_{r-1}x^{r-1} + x^r$ is irreducible and $g_1(\alpha_p) = 0$. Since for any $f(\alpha_p) \in \mathbb{Z}[\alpha_p]$, where $f(x) \in \mathbb{Z}[x]$, there exist unique polynomials $q(x), r(x) \in \mathbb{Z}[x]$ such that

$$f(x) = q(x)g_1(x) + r(x),$$

where $\deg r(x) < \deg g_1(x)$ or $r(x) = 0$. This gives that $f(\alpha_p) = r(\alpha_p)$. Hence we have that

$$\mathcal{O}_K = \mathbb{Z}[\alpha_p] = \{a_0 + a_1\alpha_p + \cdots + a_{r-1}\alpha_p^{r-1} \mid a_i \in \mathbb{Z}\}.$$

Let $g(x)$ be a polynomial in $\mathbb{Z}_{2m}[x]$ such that $g(x) \equiv g_1(x) (\bmod 2m)$. Then $g(x)$ is a monic polynomial, and there is a homomorphism

$$\Psi : \mathcal{O}_K \to \mathbb{Z}_{2m}[x]/\langle g(x)\rangle$$

given by

$$\begin{aligned}\Psi&(a_0 + a_1\alpha_p + a_2\alpha_p^2 + \cdots + a_{\frac{p-3}{2}}\alpha_p^{\frac{p-3}{2}})\\ &= a_0 + a_1x + a_2x^2 + \cdots + a_{\frac{p-3}{2}}x^{\frac{p-3}{2}} (\bmod g(x)).\end{aligned}$$

It is easy to obtain that the kernel of Ψ is generated by $2m$, that is $\mathrm{Ker}(\Psi) = \langle 2m\rangle$. We let $Poly(2m, r)$ denote the ring $\mathbb{Z}_{2m}[x]/\langle g(x)\rangle$.

Example 5. Let $p = 5$ and $m = 3$, then $r = \frac{5-1}{2} = 2$, and $\alpha_5 = e^{\frac{2\pi i}{5}} + e^{\frac{-2\pi i}{5}} = 2\cos\frac{2\pi}{5}$. Let $g_1(x) = x^2 + x - 1$. We have that

$$\begin{aligned}\alpha_5^2 + \alpha_5 - 1 &= (\zeta_5 + \zeta_5^{-1})^2 + (\zeta_5 + \zeta_5^{-1}) - 1\\ &= \zeta_5^2 + \zeta_5^{-2} + 2 + \zeta_5 + \zeta_5^{-1} - 1\\ &= 1 + \zeta_5 + \zeta_5^2 + \zeta_5^3 + \zeta_5^4 = 0,\end{aligned}$$

since $\zeta_5^5 = 1$. This gives that $g(x) = x^2 + x + 5$, and $g(x)$ is irreducible over $\mathbb{Z}_6$. Then $\mathbb{Z}_6[x]/\langle g(x)\rangle = \{a + bx + \langle g(x)\rangle \mid a, b \in \mathbb{Z}_6\}$.

Remark 84. We note that the example above shows that $g_2(x) = x^2+x+1$ and $g_3(x) = x^2 + x + 2$ are both irreducible over $\mathbb{Z}_2$ and $\mathbb{Z}_3$ respectively. But this is not always true. The following is a counter example.

Let $m = 5$ in example above, we get that $g(x) = x^2+x+9$ is irreducible over $\mathbb{Z}_{10}$, but $g_5(x) = x^2 + x + 4 = (x - 2)^2$ is reducible over $\mathbb{Z}_5$.

A code C over the ring $Poly(2m, r)$ of length n is a subset of $Poly(2m, r)^n$. The code is said to be *linear* if it is a submodule. All codes are assumed to be linear unless otherwise stated. To the ring $Poly(2m, r)$ we attach an involution $\overline{z}$ which corresponds to algebraic conjugation in the ring $\mathcal{O}_K/\langle 2m\rangle$. The involution satisfies the usual properties in that it is additive and multiplicative (since the ring is commutative). Additionally, the involution is the identity on $\mathbb{Z}_{2m}$. The ambient space $Poly(2m,r)^n$ is equipped with the following inner product

$$[v,w] = \sum v_i \overline{w_i}.$$

The orthogonal of a code is defined to be

$$C^{\perp} = \{v \mid v \in Poly(2m,r)^n \text{ such that } [v,w]=0 \text{ for all } w \in C\}.$$

The orthogonal of a linear code is linear and satisfies $|C||C^{\perp}| = (2m)^{rn}$.

We say that a code is *self-orthogonal* if $C \subseteq C^{\perp}$ and self-dual if $C = C^{\perp}$. We define the norm of an element $z \in Poly(2m,r)$ by $N(z) = z\overline{z}$ where the computation is done in $Poly(4m,r)$ and each coefficient in the polynomials is read as an element in $\mathbb{Z}_{4m}$ rather than as an element of $\mathbb{Z}_{2m}$. For a vector $v = (v_i)$ we define $N(v) = \sum N(v_i) = \sum v_i\overline{v_i}$. We always read the norm as an element of $Poly(4m,r)$. If a self-dual code C over $Poly(2m,r)$ has $N(v) = 0$ for all $v \in C$ then C is said to be a *Type II code*, otherwise it is said to be *Type I*. Note that the norms of self-orthogonal vectors must either be 0 or $2m$ since their inner product is 0 in $Poly(2m,r)$.

5.3.2 *Codes over the Rings $\mathbb{Z}_{2m}[x]/\langle g(x)\rangle$*

In this section, we first discuss some properties on the ring $\mathbb{Z}_{2m}[x]/\langle g(x)\rangle$ and then show the existence of a basis of codes over the ring $\mathbb{Z}_{2m}[x]/\langle g(x)\rangle$.

5.3.2.1 *Some Properties of the Rings $\mathbb{Z}_{2m}[x]/\langle g(x)\rangle$*

Suppose $2m = p_1^{e_1}\cdots p_s^{e_s}$ with p_i prime and $p_i \neq p_j$ if $i \neq j$. Let

$$\varphi : \mathbb{Z}_{2m} \to \mathbb{Z}_{p_1^{e_1}} \times \cdots \times \mathbb{Z}_{p_s^{e_s}} \tag{5.18}$$
$$a \mapsto (a \pmod{p_1^{e_1}}, \cdots, a \pmod{p_s^{e_s}}) \tag{5.19}$$

be the canonical isomorphism. Let

$$\varphi_{p_i} : \mathbb{Z}_{2m} \to \mathbb{Z}_{p_i^{e_i}}, \tag{5.20}$$
$$a \mapsto a \pmod{p_i^{e_i}}. \tag{5.21}$$

For the function $f(x) = a_0 + a_1x + \cdots + a_tx^t \in \mathbb{Z}_{2m}[x]$, define

$$f_{p_i}(x) = \varphi_{p_i}(a_0) + \varphi_{p_i}(a_1)x + \cdots + \varphi_{p_i}(a_t)x^t.$$

Let $f(x) + \langle g(x)\rangle, f'(x) + \langle g(x)\rangle \in \mathbb{Z}_{2m}[x]/\langle g(x)\rangle$. Suppose $f(x) + \langle g(x)\rangle = f'(x)+\langle g(x)\rangle$ and $\deg f(x), \deg f'(x) < \deg g(x)$, then there exists a polynomial $g'(x)$ such that

$$f(x) - f'(x) = g(x)g'(x).$$

Since $g(x)$ is monic, if $g'(x) \neq 0$ then

$$\deg g(x) > \deg(f(x)-f'(x)) = \deg(g(x)g'(x)) = \deg g(x) + \deg g'(x) > \deg g(x).$$

This is a contradiction. This means that for each element of $\mathbb{Z}_{2m}[x]/\langle g(x)\rangle$, there exists a unique $f(x) + \langle g(x)\rangle \in \mathbb{Z}_{2m}[x]/\langle g(x)\rangle$ such that $\deg f(x) < \deg g(x)$.

Theorem 85. *Assume the notation given above. Then*

$$\mathbb{Z}_{2m}[x]/\langle g(x)\rangle \cong \mathbb{Z}_{p_1^{e_1}}[x]/\langle g_{p_1}(x))\rangle \times \cdots \times \mathbb{Z}_{p_s^{e_s}}[x]/\langle g_{p_s}(x)\rangle,$$

where the isomorphism is given as follows:

$$\varphi(f(x) + \langle g(x)\rangle) = (f_{p_1}(x) + \langle g_{p_1}(x)\rangle, \cdots, f_{p_s}(x) + \langle g_{p_s}(x)\rangle),$$

and $f(x)$ is the unique representative element of $f(x) + \langle g(x)\rangle$ with $\deg f(x) < \deg g(x)$ and $g_{p_i}(x)$ is $g_1(x)$ mod $p_i^{e_i}$.

Proof. It is easy to get that the map above is a homomorphism. Let $f(x)+\langle g(x)\rangle \in \mathrm{Ker}(\varphi)$, where $f(x) = a_0+a_1x+\cdots+a_tx^t$ with $t < \deg g(x)$. Then we get that

$$(f_{p_1}(x) + \langle g_{p_1}(x)\rangle, \cdots, f_{p_s}(x) + \langle g_{p_s}(x)\rangle) = (\langle g_{p_1}(x)\rangle, \cdots, \langle g_{p_s}(x)\rangle).$$

This means that $f_{p_i}(x) + \langle g_{p_i}(x)\rangle = \langle g_{p_i}(x)\rangle$ for all I. This implies that $g_{p_i}(x)|f_{p_i}(x)$. Note that $g_{p_i}(x)$ is a monic polynomial, and $\deg f_{p_i}(x) < \deg g_{p_i}(x)$. This implies that $f_{p_i}(x) = 0$. In fact, suppose there exists a polynomial $h(x) \neq 0$ such that $f_{p_i}(x) = g_{p_i}(x)h(x)$. Without loss of generality, suppose $h(x) = h_0 + h_1x + \cdots + h_lx^l$ with $h_l \neq 0$, then we have that

$$\deg f_{p_i}(x) = \deg g_{p_i}(x)h(x) = \deg g_{p_i}(x) + l \geq \deg g_{p_i}(x) > \deg f_{p_i}(x),$$

since $g_{p_i}(x)$ is a monic polynomial. This is a contradiction. Therefore, for each a_j we have that

$$a_j \equiv 0 \,(\mathrm{mod}\, p_i^{e_i}) \quad \text{for all } i.$$

Since $\gcd(p_1^{e_1}, \cdots, p_s^{e_s}) = 1$, this gives that for each a_j we have that $a_j \equiv 0(\mathrm{mod}\, 2m)$. Hence $a_j = 0$ for all j and we get that $f(x) = 0$. This implies that the homomorphism above is an isomorphism. $\square$

Lemma 86. *Let $\tilde{g}(x)$ be a monic polynomial over $\mathbb{Z}_{p^e}[x]$ with $\tilde{g}(x) = \Pi_{i=1}^{s} p_i^{e_i}(x)$, where $p_i(x)$ and $p_j(x)$ are relatively prime if $i \neq j$. Then*

$$\mathbb{Z}_{p^e}[x]/\langle \tilde{g}(x)\rangle \cong \mathbb{Z}_{p^e}[x]/\langle p_1^{e_1}(x)\rangle \times \cdots \times \mathbb{Z}_{p^e}[x]/\langle p_s^{e_s}(x)\rangle.$$

Proof. Let

$$\varphi_1 : \mathbb{Z}_{p^e}[x]/\langle \tilde{g}(x)\rangle \to \mathbb{Z}_{p^e}[x]/\langle p_1^{e_1}(x)\rangle \times \cdots \times \mathbb{Z}_{p^e}[x]/\langle p_s^{e_s}(x)\rangle, \quad (5.22)$$

$$f(x) + \langle \tilde{g}(x)\rangle \mapsto (f(x) + \langle p_1^{e_1}(x)\rangle, \cdots, f(x) + \langle p_s^{e_s}(x)\rangle). \quad (5.23)$$

If $f(x) + \langle \tilde{g}(x)\rangle = f'(x) + \langle \tilde{g}(x)\rangle$ then $f(x) - f'(x) = \tilde{g}(x)h(x)$ for some $\tilde{h}(x)$ in $\mathbb{Z}_{p^e}[x]$. This means that

$$f(x) - f'(x) = (p_1^{e_1}(x) \cdots p_{i-1}^{e_{i-1}}(x) p_{i+1}^{e_{i+1}}(x) \cdots p_s^{e_s}(x)\tilde{h}(x))p_i^{e_i}(x) \in \langle p_i^{e_i}(x)\rangle.$$

Hence we have that $f(x) + \langle p_i^{e_i}(x)\rangle = f'(x) + \langle p_i^{e_i}(x)\rangle$. This implies that the corresponding φ_1 is a well-defined map. It is easy to see that the map is a homomorphism. We have that

$$\text{Ker}(\varphi_1) = \{f(x) + \langle \tilde{g}(x)\rangle \mid f(x) + \langle p_i^{e_i}(x)\rangle = \langle p_i^{e_i}(x)\rangle \text{ for all } i\}.$$

This gives that $\tilde{g}(x) \big| f(x)$ since $p_i(x)$ and $p_j(x)$ are relatively prime. We have that

$$f(x) + \langle \tilde{g}(x)\rangle = \tilde{g}(x)h(x) + \langle \tilde{g}(x)\rangle = 0 + \langle \tilde{g}(x)\rangle.$$

Therefore the homomorphism is injective. This implies that φ_1 is an isomorphism. □

Lemma 87. *Let α be an arbitrary positive integer. Let $p(x)$ be a monic irreducible polynomial over $\mathbb{Z}_{p^e}[x]$. Then $f(x) + \langle p^\alpha(x)\rangle$ is a zero divisor if and only if $p(x) \big| f(x)$.*

Proof. If $p(x) \big| f(x)$ then there exists a polynomial $h'(x)$ and an integer $\beta \leq \alpha$ such that $f(x) = p^\beta(x)h'(x)$. Then

$$(f(x) + \langle p^\alpha(x)\rangle)(p^{\alpha-\beta}(x) + \langle p^\alpha(x)\rangle) = \langle p^\alpha(x)\rangle.$$

This gives that $f(x) + \langle p^\alpha(x)\rangle$ is a zero divisor.

Now suppose $f(x) + \langle p^\alpha(x)\rangle$ is a zero divisor then there exists a polynomial $q(x)$ such that

$$f(x)q(x) + \langle p^\alpha(x)\rangle = \langle p^\alpha(x)\rangle.$$

This implies that $f(x)q(x) = p^a(x)r(x)$ for some $r(x)$. Hence we have that $p(x) \big| f(x)$ since otherwise $q(x) = p^a(x)l(x)$ and $q(x) + \langle p^\alpha(x)\rangle = \langle p^\alpha(x)\rangle$ is zero in $\mathbb{Z}_{p^e}[x]/\langle p^\alpha(x)\rangle$. □

Lemma 88. *Assume the notation given above. If $p(x)$ is a monic irreducible polynomial over $\mathbb{Z}_{p^e}[x]$, then $\mathbb{Z}_{p^e}[x]/\langle p^\alpha(x)\rangle$ is a chain ring with a maximal ideal $\langle p(x)\rangle$.*

Proof. Let I be an ideal of $\mathbb{Z}_{p^e}[x]/\langle p^\alpha(x)\rangle$. If $I = \{0\}$ then $I = (0)$. Suppose $I \neq \{0\}$. If $I \neq (p(x)+\langle p^\alpha(x)\rangle)^i$ for $i = 1, \cdots, \alpha-1$. Then there exists a $h(x) + \langle p^\alpha(x)\rangle \in I$ such that $p(x) \nmid h(x)$. Since the ring $\mathbb{Z}_{p^e}[x]/\langle p^\alpha(x)\rangle$ is finite, by Lemma 87. $h(x) + \langle p^\alpha(x)\rangle$ is a unit in $\mathbb{Z}_{p^e}[x]/\langle p^\alpha(x)\rangle$. This implies that $I = \mathbb{Z}_{p^e}[x]/\langle p^\alpha(x)\rangle$. So the chain of ideals is

$$0 \subseteq \langle p^{\alpha-1}(x) + \langle p^\alpha(x)\rangle\rangle \subseteq \cdots \subseteq \langle p(x) + \langle p^\alpha(x)\rangle\rangle \subseteq \mathbb{Z}_{p^e}[x]/\langle p^\alpha(x)\rangle.$$

Hence $\mathbb{Z}_{p^e}[x]/\langle p^\alpha(x)\rangle$ is a chain ring. □

Example 6. For example, $\mathbb{Z}_4[x]/\langle (x+1)^2\rangle$ is a chain ring. We know that $\langle x+1+\langle (x+1)^2\rangle\rangle$ is the unique maximal ideal. We have that $\langle x+1+\langle (x+1)^2\rangle\rangle \subseteq \langle x+2+\langle (x+1)^2\rangle\rangle = \mathbb{Z}_4[x]/\langle (x+1)^2\rangle$, since

$$(x+1)(x+2) = x^2+3x+2 = (x^2+2x+1)+(x+1) = (x+1)^2+(x+1).$$

We have that

$$(x+2)(ax+b) = a(x^2+2x+1)+2b-a+bx = a(x+1)^2+2b-a+bx.$$

This gives that $\langle x+2+\langle (x+1)^2\rangle\rangle = \mathbb{Z}_4[x]/\langle (x+1)^2\rangle$.

Corollary 89. *Assume the notation given above. Then the ring $\mathbb{Z}_{2m}[x]/\langle g(x)\rangle$ is a principal ideal ring.*

Proof. By Theorem 85, we have that

$$\mathbb{Z}_{2m}[x]/\langle g(x)\rangle \cong \mathbb{Z}_{p_1^{e_1}}[x]/\langle g_{p_1}(x)\rangle \times \cdots \times \mathbb{Z}_{p_s^{e_s}}[x]/\langle g_{p_s}(x)\rangle. \tag{5.24}$$

Suppose $g_{p_i}(x) = \prod_{j=1}^{s_i} p_{ij}^{e_{ij}}(x)$. By Lemma 86, for each $\mathbb{Z}_{p_i^{e_i}}[x]/\langle g_{p_i}(x)\rangle$, we have that

$$\mathbb{Z}_{p_i^{e_i}}[x]/\langle g_{p_i}(x)\rangle \cong \mathbb{Z}_{p_i^{e_i}}[x]/\langle p_{i1}^{e_{i1}}(x)\rangle \times \cdots \times \mathbb{Z}_{p_i^{e_i}}[x]/\langle p_{is_i}^{e_{is_i}}(x)\rangle. \tag{5.25}$$

Since the product of chain rings is a principal ideal ring, the result follows from Equation (5.24) and Equation (5.25). □

5.3.2.2 *Basis of Codes over Rings* $\mathbb{Z}_{2m}[x]/\langle g(x)\rangle$

It is always important to understand the generating matrix of a code. Unlike codes over fields and chain rings, the generating matrix is not always in a simple form. In this subsection, we show the existence of a basis of a code over the ring $\mathbb{Z}_{2m}[x]/\langle g(x)\rangle$. This basis forms the generator matrix of the code.

Let R be a finite principal ideal ring and let R_i be a chain ring. We begin with some definitions and lemmas. In [14] the following definitions are given with respect to Frobenius and local rings. We specialize the definitions and results to principal ideal rings and chain rings. Note that a principal ideal ring is Frobenius and a chain ring is a local ring.

Definition 90. Let R_i be a chain ring with unique maximal ideal M_i, and let $w_1, \cdots, w_s$ be vectors in R_i^n. Then $w_1, \cdots, w_s$ are modular independent if and only if $\sum \alpha_j w_j = 0$ implies that $\alpha_j \in \mathbf{m}_j$ for all j. The vectors $v_1, \cdots, v_k$ in R^n are called modular independent if $\Phi_i(v_1), \cdots, \Phi_i(v_k)$ are modular independent for some I. Let $v_1, \cdots, v_k$ be vectors in R^n. The vectors $v_1, \cdots, v_k$ are called independent if $\sum \alpha_j v_j = 0$ implies that $\alpha_j v_j = 0$ for all j.

Remark 91. It is possible to have vectors that are independent but not modular independent and to have vectors that are modular independent but not independent. See [14] for examples.

Following the remark above, we have the following definition.

Definition 92. Let C be a code over R. The codewords $c_1, c_2, \cdots, c_k$ are called a *basis* of C if they are independent, modular independent and generate C.

Theorem 93. *Assume the notation given above. Let C be a code over $\mathbb{Z}_{2m}[x]/\langle g(x)\rangle$, then any basis for C contains exactly r codewords, where r is the rank of C.*

Proof. By Corollary 89, we know that the ring $\mathbb{Z}_{2m}[x]/\langle g(x)\rangle$ is a principal ideal ring. Then the result follows from Theorem 4.9 in [14]. □

5.3.3 *Jacobi Forms over the Totally Real Field K*

We recall the definition of Jacobi forms over the totally real field $K = \mathbf{Q}(\zeta_p + \zeta_p^{-1})$ and theta-functions. We follow the definition given in [24].

5.3.3.1 *Jacobi Group*

The Jacobi group of $K = \mathbf{Q}(\zeta_p + \zeta_p^{-1})$ will be denoted by

$$\Gamma^J(K) := SL_2(\mathcal{O}_K) \propto \mathcal{O}_K^2.$$

This group acts on $\mathcal{H}^r \times \mathbb{C}^r$, where $\mathcal{H}$ denotes the complex upper half plane. Variables of this space will be listed as, $(\tau, z) := (\tau_1, .., \tau_r, z_1, .., z_r)$. The action of $\Gamma^J(K)$ on the space $\mathcal{H}^r \times \mathbb{C}^r$ are given by, $\begin{pmatrix} \alpha & \beta \\ \gamma & \delta \end{pmatrix} \in SL_2(\mathcal{O}_K)$,

$$\begin{pmatrix} \alpha & \beta \\ \gamma & \delta \end{pmatrix} \cdot (\tau, z) := \left(\frac{\alpha^{(1)}\tau_1 + \beta^{(1)}}{\gamma^{(1)}\tau_1 + \delta^{(1)}}, .., \frac{\alpha^{(r)}\tau_r + \beta^{(r)}}{\gamma^{(r)}\tau_r + \delta^{(r)}}, \frac{z_1}{\gamma^{(1)}\tau_1 + \beta^{(1)}}, .., \frac{z_r}{\gamma^{(r)}\tau_r + \beta^{(r)}}\right)$$

and, for all $[\lambda, \mu] \in \mathcal{O}_K^2$,

$$[\lambda, \mu] \cdot (\tau, z) := (\tau_1, \tau_2, .., \tau_r, z_1 + \lambda^{(1)}\tau_1 + \mu^{(1)}, .., z_r + \lambda^{(r)}\tau_r + \mu^{(r)}).$$

Remark 94. It is known (see [18]) that $SL_2(\mathcal{O}_K)$ is generated by the matrices

$$\begin{pmatrix} 0 & -1 \\ 1 & 0 \end{pmatrix}, \begin{pmatrix} 1 & b \\ 0 & 1 \end{pmatrix} \quad \forall b \in \mathcal{O}_K.$$

5.3.3.2 *Jacobi forms*

We first introduce the following notations; for $\tau \in \mathcal{H}^r, z \in \mathbb{C}^r, \gamma, \delta, \ell \in \mathcal{O}_K$, denote

$$\mathcal{N}(\gamma\tau + \delta) := \prod_{j=1}^{r} (\gamma^{(j)}\tau_j + \delta^{(j)}),$$

$$e^{2\pi i Tr_{K/Q}(\ell \frac{cz^2}{c\tau + d})} := \prod_{j=1}^{r} e^{2\pi i \ell^{(j)} \frac{c^{(j)} z_j^2}{c^{(j)}\tau_j + d^{(j)}}},$$

$$e^{-2\pi i Tr(\ell(\lambda^2\tau + 2\lambda z))} := \prod_{j=1}^{r} e^{-2\pi i \ell^{(j)} ({\lambda^{(j)}}^2 \tau_j + 2\lambda^{(j)} z_j)}.$$

Definition 95. Given $k \in \mathbb{Z}$ and $\ell \in \mathcal{O}_K$, a function $g : \mathcal{H}^r \times \mathbb{C}^r \to \mathbb{C}$ is said to be a Jacobi forms of weight k and index ℓ for the totally real field K if it is an analytic function satisfying

(1)

$$\begin{aligned}(g|_{k,\ell}M)(\tau, z) &:= \mathcal{N}(c\tau + d)^{-k} e^{-2\pi i Tr(\ell \frac{cz^2}{c\tau+d})} g(M \cdot (\tau, z)) \\ &= g(\tau, z) \ \ for\ all\ M = \begin{pmatrix} * & * \\ c & d \end{pmatrix} \in SL_2(\mathcal{O}_K),\end{aligned}$$

(2)

$$(g|_\ell[\lambda, \mu])(\tau, z) := e^{-2\pi i Tr(\ell(\lambda^2\tau + 2\lambda z))} g(\tau, [\lambda, \mu] \cdot z) = g(\tau, z).$$

It has the following Fourier expansion:

(3)

$$g(\tau, z) = \sum_{n, r \in \delta_K^{-1}, n \geq 0} c(n, r) e^{2\pi i Tr(n\tau + rz)}.$$

Here δ_K^{-1} is the inverse different of K (see a standard textbook for algebraic number theory, for instance [11, p. 203], for a detailed definition of this term).

Remark 96.

(1) The $\mathbb{C}$-vector space of Jacobi forms of weight k and index ℓ for the field K is denoted by $\mathcal{J}_{k,\ell}(\Gamma_1(\mathcal{O}_K))$.
(2) Note that letting $z = 0$ one obtains a Hilbert modular form $g(\tau, 0)$ from a Jacobi form over K.

5.3.3.3 *Theta Series*

The following theta-function was first introduced and studied in [24] to show the correspondence between the space of Jacobi forms over K and that of the vector valued modular forms.

$$\text{For each } \mu \in \mathcal{O}_K,\ \ \theta_{m,\mu}(\tau, z) := \sum_{u \in \delta_K^{-1}, u \equiv \mu \pmod{2m}} e^{2\pi i Tr(\frac{u^2\tau}{4m} + uz)}. \tag{5.26}$$

Then, by the Poisson summation formula, the theta series satisfies the following transformation formula.

Lemma 97.

(1) $\left(\theta_{m,\mu}|_{\frac{1}{2},m}\begin{pmatrix}1 & b\\ 0 & 1\end{pmatrix}\right)(\tau,\vec{z}) = e^{2\pi i Tr(\frac{\mu^2 b}{4m})}\theta_{m,u}(\tau,z)$ *for all* $b \in \mathcal{O}_K$.

(2)

$$\left(\theta_{m,\mu}|_{\frac{1}{2},m}\begin{pmatrix}0 & -1\\ 1 & 0\end{pmatrix}\right)(\tau,z) = \frac{\chi\begin{pmatrix}0 & -1\\ 1 & 0\end{pmatrix}}{(4m)^{\frac{r}{2}}}\sum_{\nu\in\mathcal{O}_K/2m\mathcal{O}_K} e^{2\pi i Tr(\frac{\mu\nu}{4m})}\theta_{m,\nu}(\tau,z),$$

with $\chi^4\left(\begin{pmatrix}0 & -1\\ 1 & 0\end{pmatrix}\right) = 1.$

Proof. The standard tool using the Poisson summation formula gives the result which was stated in [24]. □

5.3.4 *Weight Enumerators and MacWilliams Relations*

We shall define a series of weight enumerators and find the MacWilliams relations for these weight enumerators.

For a code C over $Poly(2m,r)$ define the complete weight enumerator by

$$cwe_C(x_{\alpha_0}, x_{\alpha_1}, \ldots, x_{\alpha_{r-1}}) = \sum_{v\in C}\prod_{a\in Poly(2m,r)} x_a^{n_a(v)}, \tag{5.27}$$

where $n_a(v) = |\{j \mid v_j = a\}|$. The complete weight enumerator is a homogenous polynomial in $(2m)^r$ variables.

On the ring $Poly(2m,r)$ define the relation $a \sim b$ if $a = b\epsilon$ where ϵ is a unit in the ring. Let $P_{2m,r} := Poly(2m,r)/\sim$ denote the equivalence classes of the ring under this relation. The symmetric weight enumerator is given by

$$swe_C(x_{\alpha_0}, x_{\alpha_1}, \ldots) = \sum_{v\in C}\prod_{a\in P_{2m,r}} x_a^{n'_a(v)}, \tag{5.28}$$

where $n'_a(v) = |\{j \mid v_j \sim a\}|$. The symmetric weight enumerator is a homogenous polynomial in $|P_{2m,r}|$ variables.

The Hamming weight enumerator is given by

$$W_C(x,y) = \sum_{v \in C} x^{n-h(v)} y^{h(v)}, \tag{5.29}$$

where $h(v)$ is the number of non-zero elements in the code. The Hamming weight enumerator is a homogenous polynomial in 2 variables.

Note that $W_C(x,y) = cwe(x,y,y,\ldots,y)$ and the symmetric weight enumerator is formed by replacing each occurrence of x_i with $x_{[i]}$, where $[i]$ denotes the equivalence class containing I.

Define the character $\chi_1 : Poly(2m,r) \to \mathbb{C}$ by

$$\chi_1(a_0 + a_1 x + \cdots + a_{r-1}x^{r-1}) = \zeta_{2m}^{\sum a_i} \tag{5.30}$$

and

$$\chi_\alpha(\beta) = \chi_1(\alpha \cdot \beta) \tag{5.31}$$

for any $\alpha, \beta \in Poly(2m,r)$.

Let T be a $(2m)^r$ by $(2m)^r$ matrix indexed lexicographically by the elements of $Poly(2m,r)$, where the α-th row and β-th column of T is given by the values of $\chi_\alpha(\beta)$. Specifically,

$$T_{a_0+a_1x+\cdots+a_rx^r, b_0+b_1x+\cdots+b_{r'}x^{r'}} = \zeta_{2m}^{c_i}, \zeta_{2m} = e^{\frac{2\pi i}{2m}}, \tag{5.32}$$

where $\sum c_i x^i = \sum a_i x^i \overline{\sum b_i x^i} \pmod{g(x)}$.

Essentially the matrix T is a character table of the underlying additive group, with the columns permuted by conjugation, where the characters are canonically associated with multiplication in the ring.

To obtain the MacWilliams relations for the symmetric weight enumerator we define the following matrix. Let S be a $|P_{2m,r}|$ by $|P_{2m,r}|$ matrix indexed by the elements of $P_{2m,r}$ with

$$S_{[a],[b]} = \sum_{c \sim a} T_{c,b}. \tag{5.33}$$

The following notation is used to describe an action of a matrix on a polynomial ring. If $A = (a_{ij})$ is an n by n matrix and $f(x_1,\ldots,x_n)$ a polynomial in $\mathbb{C}[x_1,x_2,\ldots,x_n]$ then

$$A \cdot f(x_1,\ldots,x_n) = f\Big(\sum_{1\le j\le n} a_{1j}x_j, \ldots, \sum_{1\le j\le n} a_{nj}x_j\Big). \tag{5.34}$$

We can now state the MacWilliams relations for the complete and symmetric weight enumerators.

Theorem 98. *Let C be a code over $Poly(2m, r)$ then*

$$cwe_{C^\perp}(X) = \frac{1}{|C|} cwe_C(T \cdot X) \tag{5.35}$$

and

$$swe_{C^\perp}(X) = \frac{1}{|C|} swe_C(S \cdot X) \tag{5.36}$$

Proof. Follows from the results in [13]. □

Then specializing the variables we have the following.

Corollary 99. *Let C be a code over $Poly(2m, r)$ then*

$$W_{C^\perp}(x, y) = \frac{1}{|C|} W_C(x + ((2m)^r - 1)y, x - y). \tag{5.37}$$

Definition 100. For codes C and D over $Poly(2m, r)$ define the complete joint weight enumerator by

$$J_{C,D}(X) = \sum_{v \in C} \sum_{v' \in D} \prod_{(a,b) \in (Poly(2m,r))^2} x_{(a,b)}^{n_{(a,b)}(v,v')} \tag{5.38}$$

where $n_{a,b}(v, v') = |\{j \mid v_j = a, v'_j = b\}|$.

The complete joint weight enumerator is a homogeneous polynomial in $(2m)^{2r}$ variables.

Corollary 101. *Let C and D be codes over $Poly(2m, r)$ then*

$$J_{C^\perp, D^\perp}(X) = \frac{1}{|C|} \frac{1}{|D|} J_{C,D}((T \otimes T) \cdot X), \tag{5.39}$$

$$J_{C^\perp, D}(X) = \frac{1}{|C|} J_{C,D}((T \otimes I) \cdot X), \tag{5.40}$$

$$J_{C, D^\perp}(X) = \frac{1}{|D|} J_{C,D}((I \otimes T) \cdot X). \tag{5.41}$$

Proof. Follows from Theorem 98 and the results in [13]. □

5.3.5 *Shadows*

Let C be a Type I code over $Poly(2m, r)$. A vector v in C is said to be *doubly-even* if $N(v) = 0$ in $Poly(4m, r)$.

Lemma 102. *The sum of two doubly-even vectors in a self-dual code C is doubly-even.*

Proof. Let v and w be two doubly-even vectors in C. Do the following computation in $Poly(4m, r)$:

$$\begin{aligned}(v+w)(\overline{v+w}) &= (v+w)(\overline{v}+\overline{w}) \\ &= v\overline{v} + w\overline{w} + v\overline{w} + \overline{v}w \\ &= v\overline{w} + \overline{v}w\end{aligned}$$

since $v\overline{v}$ and $w\overline{w}$ are both 0 in $Poly(4m, r)$. Now $\overline{\overline{v}w} = v\overline{w}$. Since $v\overline{w}$ and $\overline{v}w$ are both 0 in $Poly(2m, r)$ they are actually equal since $\overline{c} = c$ where c is a constant in $\mathbb{Z}_{2m}$. Hence we have

$$v\overline{w} + \overline{v}w = 2v\overline{w} = 0$$

in $Poly(4m, r)$. □

Let C_0 be the subcode of doubly-even vectors in C. The linear map $v \to N(v)$ has kernel C_0 and an image of size 2, hence C_0 is of index 2 in C. As usual we define the *shadow* to be

$$S = C_0^{\perp} - C = C_1 \cup C_3 \tag{5.42}$$

and

$$C_2 = C - C_0. \tag{5.43}$$

Lemma 103. *Let C be a Type I code over $Poly(2m, r)$. Then*

$$cwe_{C_0}(x_0, \ldots, x_{g(x)-1}) = \frac{1}{2}(cwe_C(x_0, \ldots, x_{g(x)-1}) + cwe_C(y_0, \ldots, y_{g(x)-1})), \tag{5.44}$$

where $y_\alpha = \zeta_{4m}^{N(\alpha)} x_\alpha$.

Proof. If the vector v is doubly-even then it is counted twice, and if it is singly-even, then it is counted once positively, and once negatively. □

Theorem 104. *Let C be a Type I code with shadow S, then*

$$cwe_S(x_0,\ldots,x_{g(x)-1}) = \frac{1}{|C|}(T\cdot cwe_C(y_0,\ldots,y_{g(x)-1})), \tag{5.45}$$

where T is the matrix that gives the MacWilliams relations.

Proof. Simply apply the MacWilliams relations to both sides of equation (5.44). That is

$$\begin{aligned}
&cwe_S(x_0,\ldots,x_{g(x)-1})\\
&= cwe_{C_0^\perp}(x_0,\ldots,x_{g(x)-1}) - cwe_C(x_0,\ldots,x_{g(x)-1})\\
&= \frac{1}{|C_0|}(\frac{1}{2}(cwe_C(T\cdot(x_0,\ldots,x_{g(x)-1}))\\
&\quad + cwe_C(T\cdot(y_0,\ldots,y_{g(x)-1}))) - cwe_C((x_0,\ldots,x_{g(x)-1}))\\
&= \frac{1}{|C|}cwe_C(T\cdot(x_0,\ldots,x_{g(x)-1})) - cwe_C(x_0,\ldots,x_{g(x)-1})\\
&\quad + \frac{1}{|C|}cwe_C(T\cdot(y_0,\ldots,y_{g(x)-1}))\\
&= \frac{1}{|C|}cwe_C(T\cdot(y_0,\ldots,y_{g(x)-1}))
\end{aligned}$$

□

There exists vectors $\mathbf{s}$ and $\mathbf{t}$ with

$$C_2 = C_0 + \mathbf{t}, \quad C_1 = C_0 + \mathbf{s}, \quad C_3 = C_0 + \mathbf{s} + \mathbf{t}.$$

Let $\alpha = [\mathbf{s},\mathbf{s}]$ and $\beta = [\mathbf{s},\mathbf{t}]$ then it is clear that the orthogonality relations are given in Table 5.1.

Table 5.1 Orthogonality Relations.

	C_0	C_1	C_2	C_3
C_0	0	0	0	0
C_1	0	α	β	$\alpha+\beta$
C_2	0	β	0	β
C_3	0	$\alpha+\beta$	β	$\alpha+2\beta$

The glue group of $C_0^\perp/C_0$ can be either the cyclic group of order 4 or the Klein-4 group. We see that in either case $s+s = 2c \in C$ and hence $[2s,t] = 0$ and so $2[s,t] = 0$. This implies that $[s,t] = 0$ or m. But $[t,s] \neq 0$, since otherwise s would be in C. Therefore we have that $\beta = m$. We notice that $N(s) \equiv \alpha \pmod{2m}$. If the glue group is the Klein-4 group then $2s \in C_0$ and $N(2s) \equiv 0 \pmod{4m}$. Then $4N(s) \equiv 0 \pmod{4m}$. This implies that α is either 0 or m. If the glue group is cyclic then $2s \in C_2$ and $N(2s) \equiv 2m \pmod{4m}$. Then $4N(s) \equiv 2m \pmod{4m}$ and we have $2\alpha \equiv m \pmod{2m}$ and so $\alpha = \frac{m}{2}$. This case can happen only when m is even.

5.3.5.1 *Modular Lattices*

Let "Tr" denote the trace function $Tr : K \to \mathbf{Q}$. Note that $Tr(\mathcal{O}_K) \subset \mathbb{Z}$.

We attach to K^n the inner product

$$\langle a, b\rangle = \sum Tr(a_i \cdot b_i), \tag{5.1}$$

The dual lattice is defined as

$$L^* = \{v \mid v \in K^n, \quad \langle v, w\rangle \in \mathcal{O}_K \text{ for all } w \in L\}. \tag{5.2}$$

For a lattice L in K^n we say that L is integral if $L \subseteq L^*$ and unimodular if $L = L^*$. Additionally, if $Tr(\langle v, v\rangle) \in 2\mathbb{Z}$ for all $v \in L$ then the lattice is said to be even.

We denote the inverse image $\tilde{u}$ of $u \in Poly(2m, r)$ under the reduction map modulo an ideal $(2m)$, $\Psi : \mathcal{O}_K \to Poly(2m, r)$.

For a code C over $Poly(2m, r)$ of length n define

$$\Lambda(C) = \{\frac{1}{\sqrt{(2m)^r}}\tilde{u} \mid u \in C\}. \tag{5.3}$$

Theorem 105. *If C is a self-dual code over $Poly(2m, r)$ then $\Lambda(C)$ is a unimodular lattice. Moreover, if C is Type II then $\Lambda(C)$ is even.*

Proof. Let v and w be vectors in C, then

$$\begin{aligned}\langle \frac{1}{\sqrt{(2m)^r}}\tilde{v}, \frac{1}{\sqrt{(2m)^r}}\tilde{w}\rangle &= \frac{1}{(2m)^r} Tr(\sum \tilde{v}_i\tilde{w}_i) \\ &= \frac{1}{(2m)^r}\sum Tr(\tilde{v}_i\tilde{w}_i).\end{aligned}$$

Note that $Tr(\tilde{v}_i\tilde{w}_i) \equiv v_i\overline{w_i} \pmod{(2m)}$ and we have that the lattice is integral. If the code is Type II, then reading $v_i\overline{w_i} \pmod{4m}$ we see that $Tr(\tilde{v}_i\tilde{w}_i) \equiv 0 \pmod{4m}$ and so $\frac{1}{2m}Tr(\tilde{v}_i\tilde{w}_i) \in 2\mathbb{Z}$, giving that the lattice is even.

The standard proof shows that the code is unimodular, i.e., we have

$$2m\mathcal{O}_K^n \subseteq \sqrt{2m}\Lambda(C) \subseteq \mathcal{O}_K^n$$

and $V(2m\mathcal{O}_K^n) = (2m)^n$ and $|\sqrt{2m}\Lambda(C)/2m\mathcal{O}_K^n| = (2m)^{\frac{n}{2}}$. Which gives that $V(\sqrt{2m}\Lambda(C)) = (2m)^{\frac{n}{2}}$ and then $V(\Lambda(C)) = 1$. □

Let L be a lattice that is not even and let $L_0 = \{v \mid v \in L,\ \ Tr(v), v \in 2\mathbb{Z}\}$. Then L_0 is of index 2 in L and

$$L_0^* = L_0 \cup L_1 \cup L_2 \cup L_3 \tag{5.4}$$

with $L = L_0 \cup L_2$. The shadow is defined by $\Sigma = L_1 \cup L_3$. The next theorem follows naturally from the definition.

Theorem 106. *Let C be a Type I code over $Poly(2m, r)$ with $\Lambda(C) = L$. Then $\Lambda(C_0) = L_0$, $\Lambda(C_2) = L_2$ and $\Lambda(S) = \Sigma$.*

The theta series for a lattice is defined by

$$\Theta_L(q) = \sum_{v \in L} q^{\frac{<v,v>}{2}}. \tag{5.5}$$

As usual, the variable $q = e^{2\pi i z}$.

The standard proof gives that

$$\Theta_{L^*}(z) = (det\ L)^{\frac{1}{2}}(\frac{i}{z})^{\frac{n}{2}}\Theta_L(\frac{-1}{z}). \tag{5.6}$$

It is clear that

$$\Theta_{L_0} = \frac{1}{2}(\Theta_L(z) + \Theta_L(z+1)). \tag{5.7}$$

The standard computation gives that

$$\Theta_\sigma(z) = (\frac{i}{z})^{\frac{n}{2}}\Theta_L(1 - \frac{1}{z}). \tag{5.8}$$

5.3.5.2 *Main Theorems*

Theorem 107. *Let $Inv(G_{II}(cwe))$ be the invariant ring of the group defined before.*

Then the following map

$$\Phi : Inv(G_{II}(cwe)) \to \bigoplus_{\ell \in \mathbb{Z}} \mathcal{J}_{4\ell,(8m\ell)}(\Gamma_1(\mathcal{O}_K)),$$

given by

$$\Phi(H(x_a \mid a \in Poly(2m, r))) = H(\theta_{m,\mu} \mid \mu \in \mathcal{O}_K/(2m))$$

for all $H \in Inv(G_{II}(cwe))$, is an algebra homomorphism.

Before we prove the main theorem we need the following lemma:

Lemma 108. *Let $G_{m,r}$ be a group*

$$G_{m,r} := \langle h_m, A_\gamma | \gamma \in \mathcal{O}_K \rangle,$$

where each of A_γ is a matrix indexed by $Poly(2m,r)$ such that

$$(A_\gamma)_{uv} = \delta_{u,v} \cdot \zeta_{2m}^{\frac{Tr(\gamma \bar{u}^2)}{2}}, (h_m)_{uv} = \zeta_{4m}^{Tr(uv)}.$$

Then the group $G_{m,r}$ and the group $G_{II}(cwe)$ are the same.

Proof. of Theorem 107 It is enough to check the transformation formula for
$g(\tau,z) := H(\theta_{m,\mu}(\tau,z) \,|\, \mu \in Poly(m,r))$ with $degree(H) = \ell$ for all $b \in \mathcal{O}_K$,

$$\begin{aligned} g\left(\begin{pmatrix} 1 & b \\ 0 & 1 \end{pmatrix} \cdot (\tau,z)\right) &= H((\theta_{m,\mu}(\tau+b,z)|\mu \in Poly(m,r))) \\ &= H((\zeta_{2m}^{\frac{Tr(\mu^2 b)}{2}} \cdot \theta_{m,\mu}(\tau,z)|\mu \in Poly(m,r))) \\ &= H(A_b(\theta_{m,\mu}(\tau,z)|\mu \in Poly(m,r))) \\ &= H(\theta_{m,\mu}(\tau,z)|\mu \in Poly(m,r)). \end{aligned}$$

Last equality follows from the fact that $H(x_a) \in Inv(G_{II}(cwe))$ and $A_b = (A_b)_{\mu\mu} = (\zeta_{2m}^{\frac{Tr(\mu^2 b)}{2}}) \in Inv(G_{m,r})$ *for all* $b \in \mathcal{O}_K$ from Lemma 97.
Next,

$$\begin{aligned} & g(-\frac{1}{\tau}, \frac{z}{\tau}) = H(\theta_{m,\mu}(-\frac{1}{\tau}, \frac{z}{\tau})|\mu \in Poly(m,r)) \\ &= H\left(\chi\begin{pmatrix} 0 & -1 \\ 1 & 0 \end{pmatrix} \mathcal{N}\left(\frac{\tau}{2}\right)^{\frac{1}{2}} e^{2\pi i Tr(m\frac{z^2}{\tau})} 2^{\frac{r}{2}} h_m \cdot (\theta_{m,\mu}(\tau,z)|\mu \in Poly(m,r))\right) \\ &= \mathcal{N}(\tau)^{\frac{\ell}{2}} e^{2\pi i Tr(m\ell\frac{z^2}{\tau})} H(h_m \cdot (\theta_{m,\mu}(\tau,z)|\mu \in Poly(m,r))) \\ &\quad (\text{since } \ell = deg(F) \equiv 0 \pmod 4) \\ &= \mathcal{N}(\tau)^{\frac{\ell}{2}} e^{2\pi i Tr(m\ell\frac{z^2}{\tau})} H((\theta_{m,\mu}(\tau,z)|\mu \in Poly(m,r))). \end{aligned}$$

Here, $M \cdot (\theta_{m,\mu}|\mu \in Poly(m,r))$ denotes matrix multiplication. Next, to check the elliptic property, first note that, for any $(\lambda_1, \lambda_2) \in \mathcal{O}_K^2$, and for each $\mu \in Poly(m,r)$,

$$\theta_{m,\mu}(\tau, z+\lambda_1\tau+\lambda_2) = \sum_{r\in\delta_K^{-1}, r\equiv\mu \pmod{2m}} e^{2\pi i Tr(\frac{r^2\tau}{4m}+r(z+\lambda_1\tau+\lambda_2))}$$

$$= e^{-2\pi i Tr(m(\lambda_1^2\tau+2\lambda_1 z))} \sum_{r\in\delta_K^{-1}, r\equiv\mu \pmod{2m}} e^{2\pi i Tr(\frac{(r+2\lambda_1)^2}{4m}\tau+(r+2\lambda_1)z)}$$

$$= e^{-2\pi i Tr(m(\lambda_1^2\tau+2\lambda_1 z))}\theta_{m,\mu}(\tau, z).$$

So, the elliptic property of $g(\tau, z)$ is now immediate. The condition at the cusps can also be checked from that of each theta series $\theta_{m,\mu}(\tau, z)$. We omit the detailed proof. □

References

[1] E. Bannai, S.T. Dougherty, M. Harada, and M. Oura, Type II Codes, Even Unimodular Lattices, and Invariant Rings, *IEEE Trans. Information Theory*, **45**, (1999), 1194–1205.

[2] K. Betsumiya and Y. Choie, Codes over $\mathbb{F}_4$, Jacobi forms and Hilbert-Siegel modular forms over $Q(\sqrt{5})$ European J. Combin. **26**, (2005), 629–650.

[3] K. Betsumiya and Y. Choie, Jacobi Forms over Totally Real Fields and Type II Codes over Galois Rings $GR(2^m, f)$, Euro. Journal of Combinatorics, **25**, (2004), 475–486.

[4] Y. Choie and S. T. Dougherty, Codes over Rings, Complex Lattices and Hermitian Modular Forms, Euro. Journal of Combinatorics, **26**, (2005), 145–165.

[5] Y. Choie and S. T. Dougherty, Codes over Σ_{2m} and Jacobi Forms over the Quaternions, Appl. Algebra Engr. Com. Comput., **15**, (2004), 129–147.

[6] Y. Choie, S. T. Dougherty and H. Liu, Jacobi forms and Hilbert-Siegel modular forms over totally real fields and self-dual codes over polynomial rings $Z_{2m}[x]/(g(x))$, Ars Combin. **107**, (2012), 141–160.

[7] Y. Choie and E. Lee, Jacobi forms over the totally real number fields and Codes over F_p, Illinois Journal of Math, **46** (2), (2002), 627–643.

[8] Y. Choie and N. Kim, The Complete Weight Enumerator of Type II Code over $\mathbb{Z}_4$ and Jacobi Forms, *IEEE Trans. Information Theory*, **4**, (2001), 396–399.

[9] Y. Choie and E. Lee, Jacobi forms over the totally real number fields and Codes over F_p, Illinois Journal of Math, **46**, (2), (2002), 627–643.

[10] Y. Choie and P. Sole, Self-dual codes over Z_4 and half-integral weight modular forms, Proceeding of A.M.S, Vol. **130**, (11), (2002) 3125–313.

[11] H. Cohen, *A Course in Computational Algebraic Number Theory*, Springer, 1995.

[12] J. H. Conway and N. J. A. Sloane, *Sphere Packing, Lattices and Groups*, Springer, (1999).

[13] S. T. Dougherty, MacWilliams Relations for Codes over Groups and Rings, preprint.

[14] S. T. Dougherty, H. Liu, Independence of Vectors in Codes over Rings, Designs,Codes and Cryptography, **51**, (2009), 55–68.

[15] W. Ebeling, *Lattices and Codes, A course partially based on Lectures by F. Hirtzebruch*, Advanced Lectures in Mathematics, Vieweg, (1994).

[16] P. Gaborit, V. Pless, P. Solé and O. Atkin, Type II codes over $\mathbb{F}_4$, Finite Fields and Their Applications, **8**, (2002) 171–183.

[17] A. M. Gleason, Weight polynomials of self-dual codes and the MacWilliams identities, in Actes, Congrés International de Mathématiques (Nice, 1970), Gauthiers-Villars, Paris, Vol. 3, 221–215, 1971.

[18] B. Liehl, On the Group $S\ell_2$ over Orders or Arithmetic Type, J. Reine Angew. Math., **323**, (1981), 153–171.

[19] F. J. MacWilliams and N. J. A. Sloane, *The Theory of Error-Correcting Codes*, North-Holland, Amsterdam, 1977.

[20] F. J. MacWilliams, A. M. Odlyzko and N. J. A. Sloane, Self-Dual Codes over $GF(4)$, J. Comb. Th.(Ser. A), **25**, (1978), 288–318.

[21] G. Nebe, H.-G. Quebbemann, E. M. Rains, and N. J. A. Sloane, Complete Weight Enumerators of Generalized Doubly-Even Self-Dual Codes, Finite Fields and Their Applications, **10**, (2004), 540–550.

[22] E. M. Rains and N. J. A. Sloane, Self-Dual Codes, *The Handbook of Coding Theory*, North-Holland, New York, (1998).

[23] H. L. Resnikoff, On the graded ring of Hilbert modular forms associated with $\mathbb{Q}(\sqrt{5})$, Math. Ann. **208**, (1974), 161–170.

[24] H. Skogman, Jacobi Forms over Totally Real Number Fields, Results Math., **39**, (2001), 169–182.

Chapter 6

Construction of Self-dual Codes of Higher Lengths over $\mathbb{Z}_{2^m}$ and Determination of Jacobi Forms

6.1 Introduction

Shadows and generalized shadows of self-dual codes over finite rings have nice connections with the theory of unimodular lattices and modular forms (such as elliptic modular forms, Siegel modular forms), and are also useful in the construction of self-dual codes of higher lengths. This motivated many researchers to study shadows and generalized shadows of self-dual codes over various finite rings.

The shadow of a binary self-dual code was first defined by Conway and Sloane [7] to obtain an upper bound on minimum distances of binary self-dual codes. Using the concept of the shadow of a binary Type I code of length n, a binary self-dual code C_1 of length $n+2$ when $n \equiv 2$ or $6 \pmod 8$ and a binary self-dual code C_2 of length $n+4$ when $n \equiv 0$ or $4 \pmod 8$ are constructed by Brualdi and Pless [2]. Apart from this, weight enumerators of the codes C_1 and C_2 are determined. It is also observed that the code C_1 is Type I when $n \equiv 2 \pmod 8$ and is Type II when $n \equiv 6 \pmod 8$, whereas the code C_2 is Type I when $n \equiv 0 \pmod 8$ and is Type II when $n \equiv 4 \pmod 8$. In the same work, the shadow of a binary Type II code $\mathcal{C}$ of length n is also studied by considering a subcode $\mathcal{C}_0$ of $\mathcal{C}$ with codimension 1. A binary self-dual code of length $n+4$ provided $\mathbf{1} \in \mathcal{C}_0$ and a binary self-dual code of length $n+2$ provided $\mathbf{1} \notin \mathcal{C}_0$ are constructed, and their weight enumerators are also determined. Later, the generalized shadow of a binary self-dual code $\mathcal{C}$ of length n with respect to a vector $s \in \mathbb{Z}_2^n \setminus \mathcal{C}$ is defined and studied by Tsai [14] to construct a binary self-dual code of length $n+2$ when $\langle s,s\rangle \equiv 1 \pmod 2$ and a binary self-dual code of length $n+4$ when $\langle s,s\rangle \equiv 0 \pmod 2$.

In another related direction, self-dual codes over the ring $\mathbb{Z}_{2m}$ and shadows of Type I codes over $\mathbb{Z}_{2m}$ are studied in [1] to obtain Siegel modular forms from complete and symmetrized weight enumerators in genus g of Type II codes over $\mathbb{Z}_{2m}$. Later, Jacobi forms are determined from complete weight enumerators of Type II codes over $\mathbb{Z}_{2m}$ in [3–6]

This chapter follows the paper [13], which generalizes the construction method employed by Brualdi and Pless [2] and Tsai [14]. More precisely, Type I and Type II codes (of higher lengths) over $\mathbb{Z}_{2^m}$ are obtained from shadows of Type I codes over $\mathbb{Z}_{2^m}$. Moreover, for each positive integer n, self-dual codes (of higher lengths) over $\mathbb{Z}_{2^m}$ are constructed from the generalized shadow of a self-dual code $\mathcal{C}$ of length n over $\mathbb{Z}_{2^m}$ with respect to a vector $s \in \mathbb{Z}_{2^m}^n \setminus \mathcal{C}$ satisfying either $\langle s,s\rangle \equiv 0 \pmod{2^m}$ or $\langle s,s\rangle \equiv 2^{m-1} \pmod{2^m}$. Furthermore, complete weight enumerators of the codes constructed above are determined. As an application of these results, some Jacobi forms on the modular group $\Gamma = SL_2(\mathbb{Z})$ are also determined.

In Section 6.2, the shadow of a Type I lattice and the shadow of a Type I code over $\mathbb{Z}_{2k}$ are discussed. Apart from this, generalized shadow of a self-dual code $\mathcal{C}$ over $\mathbb{Z}_{2k}$ is discussed with respect to a vector $s \in \mathbb{Z}_{2k}^n \setminus \mathcal{C}$ satisfying either $\langle s,s\rangle \equiv 0 \pmod{2k}$ or $\langle s,s\rangle \equiv k \pmod{2k}$. In Section 6.3, a self-orthogonal code (of higher length) over $\mathbb{Z}_{2^m}$ is constructed from a self-dual code over $\mathbb{Z}_{2^m}$ (Proposition 122). From there, Type I and Type II codes (of higher lengths) over $\mathbb{Z}_{2^m}$ are constructed from shadows of Type I codes over $\mathbb{Z}_{2^m}$, and their complete weight enumerators are also obtained (Theorems 124 and 129). Some Jacobi forms on the modular group Γ are also determined from complete weight enumerators of Type II codes constructed in Theorems 124-129 (Theorems 128 and 131). In Section 6.4, self-dual codes (of higher lengths) over $\mathbb{Z}_{2^m}$ are constructed using the generalized shadow of a self-dual code $\mathcal{C}$ of length n over $\mathbb{Z}_{2^m}$ with respect to a vector $s \in \mathbb{Z}_{2^m}^n \setminus \mathcal{C}$ for each positive integer n, provided $\langle s,s\rangle \equiv 0$ or $2^{m-1} \pmod{2^m}$ (Theorem 132). Moreover, complete weight enumerators of self-dual codes constructed in the respective cases are also determined. These results are illustrated by constructing Type II codes of length 8 over $\mathbb{Z}_8$ and over $\mathbb{Z}_{16}$ from Type I codes of lengths 4 and 7 over

$\mathbb{Z}_8$ and over $\mathbb{Z}_{16}$ respectively, and by constructing a self-dual code of length 8 over $\mathbb{Z}_8$ from a self-dual code of length 6 over $\mathbb{Z}_8$ using the generalized shadow of the code with respect to $s = (2, 2, 2, 0, 0, 0) \in \mathbb{Z}_8^6$.

6.2 Some Preliminaries

For positive integers k and n, let $\mathbb{Z}_{2k} = \{0, 1, 2, \cdots, 2k-1\}$ denote the ring of integers modulo $2k$, and let $\mathbb{Z}_{2k}^n$ denote the $\mathbb{Z}_{2k}$-module consisting of all n-tuples over the ring $\mathbb{Z}_{2k}$. Now recall that a linear code $\mathcal{C}$ of length n over $\mathbb{Z}_{2k}$ is an additive subgroup of $\mathbb{Z}_{2k}^n$. The dual code of $\mathcal{C}$ is defined as the set

$$\mathcal{C}^{\perp} = \{v \in \mathbb{Z}_{2k}^n : \langle u, v\rangle = 0 \text{ for all } u \in \mathcal{C}\},$$

where

$$\langle u, v\rangle = \sum_{j=1}^{n} u_j v_j$$

for all $u = (u_1, u_2, \cdots, u_n), v = (v_1, v_2, \cdots, v_n) \in \mathbb{Z}_{2k}^n$. Observe that the dual code $\mathcal{C}^{\perp}$ is also a linear code of length n over $\mathbb{Z}_{2k}$. Recall that the linear code $\mathcal{C}$ is called a self-orthogonal code if it satisfies $\mathcal{C} \subseteq \mathcal{C}^{\perp}$, whereas the code $\mathcal{C}$ is called a self-dual code if it satisfies $\mathcal{C} = \mathcal{C}^{\perp}$. Then the following holds.

Lemma 109. [8] *If $\mathcal{C}$ is a self-dual code of length n over $\mathbb{Z}_{2k}$, then $(2k)^n$ is a perfect square. In particular, for an integer $m \geq 1$, if $\mathcal{C}$ is a self-dual code of length n over $\mathbb{Z}_{2^m}$, then mn is an even integer.*

Proof. It is well-known that $|\mathcal{C}||\mathcal{C}^{\perp}| = (2k)^n$. As $\mathcal{C} = \mathcal{C}^{\perp}$, we get $|\mathcal{C}|^2 = (2k)^n$, which implies that the integer $(2k)^n$ is a perfect square. □

Next to define Type I and Type II codes over $\mathbb{Z}_{2k}$, let us first recall the definition of the Euclidean weight in $\mathbb{Z}_{2k}$.

The Euclidean weight of an element $a \in \mathbb{Z}_{2k}$, denoted by $wt_E(a)$, is defined as

$$wt_E(a) = \min\{a^2, (2k-a)^2\}.$$

Note that $wt_E(a) \equiv a^2 \pmod{4k}$ for each $a \in \mathbb{Z}_{2k}$. Furthermore, the Euclidean weight of a vector $v = (v_1, v_2, \cdots, v_n) \in \mathbb{Z}_{2k}^n$ is defined as

$$wt_E(v) = \sum_{j=1}^{n} wt_E(v_j).$$

It is clear that

$$wt_E(v) \equiv \sum_{j=1}^{n} v_j^2 \equiv \langle v, v\rangle \pmod{4k}.$$

From this, one can easily observe that

$$wt_E(u+v) \equiv wt_E(u) + wt_E(v) + 2\langle u, v\rangle \pmod{4k} \tag{6.1}$$

for each $u, v \in \mathbb{Z}_{2k}^n$. Then a self-dual code $\mathcal{C}$ is said to be a Type II code if the Euclidean weight of each of its codewords is divisible by $4k$, otherwise the code $\mathcal{C}$ is called a Type I code.

To describe various properties (e.g. error-detection capability, error-correction capability, etc.) of a linear code, certain specific polynomials are associated with the code, which are called its weight enumerators. One such polynomial is called the complete weight enumerator of a linear code over $\mathbb{Z}_{2k}$, which is defined as follows.

The complete weight enumerator of a linear code $\mathcal{C}$ of length n over $\mathbb{Z}_{2k}$ is defined

$$cwe_{\mathcal{C}}(X_\mu : \mu \in \mathbb{Z}_{2k}) = \sum_{c\in\mathcal{C}} X_0^{n_0(c)} X_1^{n_1(c)} \cdots X_{2k-1}^{n_{2k-1}(c)},$$

where for each $c \in \mathcal{C}$, the number $n_\mu(c)$ $(0 \le \mu \le 2k-1)$ equals the total number of components of c that are equal to μ.

In the following theorem, MacWilliams identity for the complete weight enumerator of a linear code over $\mathbb{Z}_{2k}$ is derived.

Theorem 110. [1, Theorem 5.1] *Let $\mathcal{C}$ be a linear code of length n over $\mathbb{Z}_{2k}$. Then the complete weight enumerator of the dual code $\mathcal{C}^\perp$ of $\mathcal{C}$ is given by*

$$cwe_{\mathcal{C}^\perp}(X_\mu : \mu \in \mathbb{Z}_{2k}) = \frac{1}{|\mathcal{C}|} cwe_{\mathcal{C}}(M_{2k} \cdot (X_\mu : \mu \in \mathbb{Z}_{2k})),$$

where M_{2k} is a $2k \times 2k$ matrix whose (h, j)th entry is given by $M_{2k}(h, j) = e^{\frac{2\pi i}{2k}hj}$ for $0 \le h, j \le 2k-1$.

Proof. To prove the result, we first note that

$$\begin{aligned} cwe_{\mathcal{C}}(X_\mu : \mu \in \mathbb{Z}_{2k}) &= \sum_{c\in\mathcal{C}} X_0^{n_0(c)} X_1^{n_1(c)} \cdots X_{2k-1}^{n_{2k-1}(c)} \\ &= \sum_{(c_1,c_2,\cdots,c_n)\in\mathcal{C}} X_{c_1} X_{c_2} \cdots X_{c_n}. \end{aligned}$$

Now let us define

$$f(v) = X_{v_1} X_{v_2} \cdots X_{v_n}$$

for each $v = (v_1, v_2, \cdots, v_n) \in \mathbb{Z}_{2k}^n$. Let χ be the canonical character on the additive group of the ring $\mathbb{Z}_{2k}$, which is defined as

$$\chi(j) = \xi^j \quad \text{for} \quad 0 \le j \le 2k-1,$$

where $\xi = e^{\frac{2\pi i}{2k}}$. Further, for $u = (u_1, u_2, \cdots, u_n) \in \mathbb{Z}_{2k}^n$, let us define

$$\begin{aligned}\tilde{f}(u) &= \sum_{v \in \mathbb{Z}_{2k}^n} \chi(\langle u, v\rangle) f(v) \\ &= \prod_{\ell=1}^{n} \Big(\sum_{v_\ell \in \mathbb{Z}_{2k}} \chi(u_\ell v_\ell) X_{v_\ell} \Big).\end{aligned}$$

When $u_\ell = 0$, we note that $\sum_{v_\ell \in \mathbb{Z}_{2k}} \chi(u_\ell v_\ell) X_{v_\ell} = X_0 + X_1 + \cdots + X_{2k-1}$. On the other hand, when $1 \le u_\ell \le 2k-1$, we see that

$$\begin{aligned}\sum_{v_\ell \in \mathbb{Z}_{2k}} \chi(j v_\ell) X_{v_\ell} &= X_0 + \chi(u_\ell) X_1 + \cdots + \chi(u_\ell(2k-1)) X_{2k-1} \\ &= X_0 + \xi^{u_\ell} X_1 + \xi^{2u_\ell} X_2 + \cdots + \xi^{u_\ell(2k-1)} X_{2k-1}.\end{aligned}$$

From this, we obtain

$$\tilde{f}(u) = \prod_{j=0}^{2k-1} (X_0 + \xi^j X_1 + \cdots + \xi^{j(2k-1)} X_{2k-1})^{n_j(u)}.$$

Next by Lemma 1.4 of Klemm [12], we have

$$\sum_{v \in \mathcal{C}^\perp} f(v) = \frac{1}{|\mathcal{C}|} \sum_{u \in \mathcal{C}} \tilde{f}(u).$$

From this, we obtain

$$\begin{aligned}&cwe_{\mathcal{C}^\perp}(X_0, X_1, \cdots, X_{2k-1}) \\ &\quad = \sum_{v \in \mathcal{C}^\perp} f(v) \\ &\quad = \frac{1}{|\mathcal{C}|} \sum_{u \in \mathcal{C}} \tilde{f}(u) \\ &\quad = \frac{1}{|\mathcal{C}|} \sum_{u \in \mathcal{C}} \prod_{j=0}^{2k-1} (X_0 + \xi^j X_1 + \cdots + \xi^{j(2k-1)} X_{2k-1})^{n_j(u)} \\ &\quad = \frac{1}{|\mathcal{C}|} cwe_{\mathcal{C}}(M_{2k} \cdot (X_\mu : \mu \in \mathbb{Z}_{2k})).\end{aligned}$$

□

Choie and Kim [4] related complete weight enumerators of Type II codes over $\mathbb{Z}_{2^m}$ with Jacobi forms, which are as defined below:

Definition 111. [11] Let $\mathcal{H}$ be the complex upper-half plane. A Jacobi form of weight k and index u ($k, u \in \mathbb{N}$) on the modular group $\Gamma = SL_2(\mathbb{Z})$ is a holomorphic function $\phi : \mathcal{H} \times \mathbb{C} \to \mathbb{C}$ satisfying the following:

(i) $(c\tau+d)^{-k}e^{-2\pi iu(\frac{cz^2}{c\tau+d})}\phi\left(\frac{a\tau+b}{c\tau+d}, \frac{z}{c\tau+d}\right) = \phi(\tau, z)$ for all $\begin{pmatrix} a & b \\ c & d \end{pmatrix} \in \Gamma$,

(ii) $e^{2\pi iu(\lambda^2\tau+2\lambda z)}\phi(\tau, z+\lambda\tau+\epsilon) = \phi(\tau, z)$ for all $(\lambda, \epsilon) \in \mathbb{Z}^2$, and

(iii) $\phi(\tau, z)$ has a Fourier series expansion of the form

$$\sum_{v=0}^{\infty} \sum_{\substack{r\in\mathbb{Z} \\ r^2 \le 4uv}} c(v, r) q^v \xi^r, \text{ where } q = e^{2\pi i\tau} \text{ and } \xi = e^{2\pi iz}.$$

6.2.1 *Shadow of a Type I Lattice*

For a positive integer n, let $\mathbb{R}^n$ be an n-dimensional inner product space with respect to the Euclidean inner product, defined as

$$\langle u, v\rangle = \sum_{j=1}^{n} u_j v_j$$

for all $u = (u_1, u_2, \cdots, u_n), v = (v_1, v_2, \cdots, v_n) \in \mathbb{R}^n$. The norm of a vector $u = (u_1, u_2, \cdots, u_n) \in \mathbb{R}^n$ is defined as

$$\langle u, u\rangle = \sum_{j=1}^{n} u_j^2.$$

Recall that an n-dimensional lattice L in $\mathbb{R}^n$ is defined as a free $\mathbb{Z}$-module generated by n linearly independent vectors $v_1, v_2, \cdots, v_n$ in $\mathbb{R}^n$. An $n \times n$ matrix G whose rows are the vectors $v_1, v_2, \cdots, v_n$ is called the generator matrix of L. The fundamental volume (determinant) of the lattice L, denoted by $V(L)$, is defined as the determinant of the generator matrix G. For a sublattice L' of L, it is well-known that

$$V(L') = V(L)\ \det(L/L'). \tag{6.2}$$

The dual lattice L^* of the lattice L is defined as $L^* = \{u \in \mathbb{R}^n : \langle u, v\rangle \in \mathbb{Z} \text{ for all } v \in L\}$. A lattice L is said to be integral if it satisfies $L \subseteq L^*$, while L is called a unimodular lattice if $L = L^*$. Equivalently, the lattice L

is unimodular if and only if $V(L) = 1$. Note that when L is a unimodular lattice, we have $\langle u, v\rangle \in \mathbb{Z}$ for all $u, v \in L$. Further, a unimodular lattice L is said to be Type II (even) if the norm of each of its vectors is even, otherwise L is called a Type I lattice.

Lemma 112. *[9] Let L be a Type I lattice, and let $L_0 = \{u \in L : \langle u, u\rangle \text{ is even}\}$. Then the following hold.*

(i) The set L_0 is a sublattice of L of index 2.
(ii) The set L_0 is a sublattice of L_0^* of index 4.

Proof. (i) One can easily observe that L_0 is a sublattice of L. Further, as L is a Type I lattice, there exists $x \in L$ such that $\langle x, x\rangle$ is odd. Now for each $y \in L$, the norm of the vector $y - x \in L$ is even, and hence $y - x \in L_0$. From this, it follows that there are exactly two distinct cosets of L_0 in L, namely L_0 and $x + L_0$.

(ii) We first observe that $L_0 \subseteq L \subseteq L_0^*$. As L is unimodular, we have $V(L) = 1$. By (6.2), we see that $V(L_0) = V(L)\det(L/L_0) = 2$, which implies that $V(L_0^*) = \frac{1}{2}$. Further, by applying (6.2) again, we see that $V(L) = V(L_0^*)\det(L_0^*/L)$. Now as $V(L) = 1$, we obtain $\det(L_0^*/L) = 2$, which implies that L is a sublattice of L_0^* of index 2. From this and by part (i), part (ii) follows immediately. □

Let L be a Type I lattice, and let L_0 be the sublattice of L consisting of even norm vectors. Let L_2 be the unique non-zero coset of L_0 in L. Let us write $L_0^* = L_0 \cup L_2 \cup L_1 \cup L_3$. The shadow of the lattice L is defined as $S_L = L_1 \cup L_3$. The glue group of L_0 is defined as L_0^*/L_0. By Lemma 112(ii), the glue group L_0^*/L_0 of L_0 is a group of order 4, from which it follows that the glue group of L_0 is either the Klein 4-group or a cyclic group of order 4.

Lemma 113. *[9] Let L be a Type I lattice, and let S_L be the shadow of L.*

(i) For each $x \in S_L$, we have

$$\langle x, x\rangle \equiv \frac{n}{4} \ (mod\ 8).$$

(ii) The glue group of L_0 is isomorphic to the Klein 4-group if n is even and is isomorphic to the cyclic group of order 4 if n is odd.

Proof. (i) It follows immediately from Theorem 1 of Dougherty et al. [9].

(ii) Let $x \in S_L$. As S_L is a coset of L, we see that $2x \in L$. By part (i), we see that $\langle 2x, 2x\rangle \equiv n \pmod 8$. From this, it follows that $2x \in L_2$ if n is odd, while $2x \in L_0$ if n is even. □

6.2.2 *Shadow of a Type I Code*

Let $\mathcal{C}$ be a Type I code of length n over $\mathbb{Z}_{2k}$. Let us define

$$\mathcal{C}_0 = \{c \in \mathcal{C} : wt_E(c) \equiv 0 \pmod{4k}\}.$$

Then we observe the following:

Lemma 114. *[9] The following hold.*

(i) The code $\mathcal{C}_0$ is a linear subcode of index 2 in $\mathcal{C}$.
(ii) The code $\mathcal{C}_0$ is a subcode of index 4 in $\mathcal{C}_0^{\perp}$.

Proof. (i) First of all, we shall prove that $\mathcal{C}_0$ is a linear subcode of $\mathcal{C}$. For this, let $u, v \in \mathcal{C}_0$. Here we have $wt_E(u) \equiv wt_E(v) \equiv 0 \pmod{4k}$. As the code $\mathcal{C}$ is Type I and $u, v \in \mathcal{C}$, we see that $\langle u, v\rangle \equiv 0 \pmod{2k}$. From this and by (6.1), we obtain

$$wt_E(u+v) \equiv wt_E(u) + wt_E(v) + 2\langle u, v\rangle \equiv 0 \pmod{4k}.$$

This implies that $u+v \in \mathcal{C}_0$. This shows that the code $\mathcal{C}_0$ is a linear subcode of $\mathcal{C}$.

Further, since the code $\mathcal{C}$ is Type I, there exists $t \in \mathcal{C} \setminus \mathcal{C}_0$. We observe that $wt_E(t) \not\equiv 0 \pmod{4k}$ and $wt_E(t) \equiv \langle t, t\rangle \equiv 0 \pmod{2k}$. From this, it follows that $wt_E(t) \equiv 2k \pmod{4k}$. Further, one can easily show that $\mathcal{C}_0$ and $t+\mathcal{C}_0$ are two distinct cosets of $\mathcal{C}_0$ in $\mathcal{C}$ and that $\mathcal{C} = \mathcal{C}_0 \cup (t+\mathcal{C}_0)$. This shows that the index of $\mathcal{C}_0$ in $\mathcal{C}$ is 2.

(ii) Here we first observe that $\mathcal{C} \subseteq \mathcal{C}_0^{\perp}$. Now as $\mathcal{C}$ is self-dual, we have $|\mathcal{C}|^2 = (2k)^n$. Using this and by part (i), we get $\frac{|\mathcal{C}_0^{\perp}|}{|\mathcal{C}|} = \frac{(2k)^n}{|\mathcal{C}_0||\mathcal{C}|} = \frac{|\mathcal{C}|}{|\mathcal{C}_0|} = 2$. This implies that the index of $\mathcal{C}$ in $\mathcal{C}_0^{\perp}$ is 2, which further implies, by using part (i) again, that the index of $\mathcal{C}_0$ in $\mathcal{C}_0^{\perp}$ is 4. □

By Lemma 114, we see that $\mathcal{C}_0$ is a subcode of index 2 in $\mathcal{C}$, and $\mathcal{C}_0$ is a subcode of index 4 in $\mathcal{C}_0^{\perp}$. From this, it follows that $\mathcal{C} = \mathcal{C}_0 \cup \mathcal{C}_2$ and

$\mathcal{C}_0^\perp = \mathcal{C}_0 \cup \mathcal{C}_2 \cup \mathcal{C}_1 \cup \mathcal{C}_3$, where $\mathcal{C}_2 = t + \mathcal{C}_0$, $\mathcal{C}_1 = s + \mathcal{C}_0$ and $\mathcal{C}_3 = s + t + \mathcal{C}_0$ with $t \in \mathcal{C} \setminus \mathcal{C}_0$ and $s \in \mathcal{C}_0^\perp \setminus \mathcal{C}$. Now the shadow of $\mathcal{C}$ is defined as the set

$$\mathcal{S} = \mathcal{C}_0^\perp \setminus \mathcal{C} = \mathcal{C}_1 \cup \mathcal{C}_3.$$

Further, for each $u \in \mathcal{C}_2$, it is easy to observe that

$$wt_E(u) \equiv \langle u, u \rangle \equiv 2k \pmod{4k}. \tag{6.3}$$

The following theorem provides a method to construct Type II lattices from Type II codes over $\mathbb{Z}_{2^m}$. For this, let $\rho : \mathbb{Z}_{2k} \to \mathbb{Z}$ be the map, defined as

$$\rho(a) = \begin{cases} a & \text{if } 0 \le a \le k; \\ a - 2k & \text{if } k+1 \le a \le 2k-1. \end{cases}$$

Theorem 115. *[1] Let $\mathcal{C}$ be a self-dual code of length n over $\mathbb{Z}_{2k}$.*

(i) (Construction A_{2k}) The lattice

$$\Lambda(\mathcal{C}) = \frac{1}{\sqrt{2k}} \{\rho(\mathcal{C}) + 2k\mathbb{Z}^n\}$$

is an n-dimensional unimodular lattice with the minimum norm as $\min\{2k, \frac{d_E}{2k}\}$, where d_E is the minimum Euclidean distance of $\mathcal{C}$. Moreover, if the code $\mathcal{C}$ is Type II, then the lattice $\Lambda(\mathcal{C})$ is a Type II lattice.

(ii) Let $\mathcal{C}$ be a Type I code of length n over $\mathbb{Z}_{2k}$, and let $\Lambda(\mathcal{C})$ be the Type I lattice. Then we have $\Lambda(\mathcal{C})_0 = \Lambda(\mathcal{C}_0)$ and $\Lambda(\mathcal{C})_2 = \Lambda(\mathcal{C}_2)$. Further, up to relabelling, we have $\Lambda(\mathcal{C})_j = \Lambda(\mathcal{C}_j)$ for $j \in \{1, 3\}$.

Proof. (i) Let $c, d \in \mathcal{C}$. As $\mathcal{C}$ is self-dual, we have $\langle c, d\rangle = 0$ in $\mathbb{Z}_{2k}$, which implies that $\langle \rho(c), \rho(d)\rangle \equiv 0 \pmod{2k}$. We note that $\frac{1}{\sqrt{2k}}(\rho(c) + 2ku), \frac{1}{\sqrt{2k}}(\rho(d) + 2kv) \in \Lambda(\mathcal{C})$ for all $u, v \in \mathbb{Z}^n$. Furthermore, we see that

$$\begin{aligned} &\langle \frac{1}{\sqrt{2k}}(\rho(c) + 2ku), \frac{1}{\sqrt{2k}}(\rho(d) + 2kv)\rangle \\ &\quad = \frac{1}{2k}\left(\langle \rho(c), \rho(d)\rangle + 2k\langle \rho(c), v\rangle + 2k\langle u, \rho(d)\rangle + 4k^2\langle u, v\rangle\right), \end{aligned}$$

which is an integer. From this, it follows that $\Lambda(\mathcal{C})$ is an integral lattice.

Now to show that $\Lambda(\mathcal{C})$ is a unimodular lattice, we note that $2k\mathbb{Z}^n \subseteq \sqrt{2k}\Lambda(\mathcal{C}) \subseteq \mathbb{Z}^n$. We also observe that $V(2k\mathbb{Z}^n) = (2k)^n$, $\det\left(\sqrt{2k}\Lambda(\mathcal{C})/2k\mathbb{Z}^n\right) = (2k)^{\frac{n}{2}}$ and $V(\sqrt{2k}\Lambda(\mathcal{C})) = (2k)^{\frac{n}{2}} V(\Lambda(\mathcal{C}))$. Now by (6.2), we get $V(2k\mathbb{Z}^n) = V(\sqrt{2k}\Lambda(\mathcal{C}))\det\left(\sqrt{2k}\Lambda(\mathcal{C})/2k\mathbb{Z}^n\right)$. From this, it follows that $V(\Lambda(\mathcal{C})) = 1$. This shows that $\Lambda(\mathcal{C})$ is a unimodular lattice.

Now let $\mathcal{C}$ be a Type II code, which implies that $\langle c, c\rangle = wt_E(c) \equiv 0 \pmod{4k}$ for all $c \in \mathcal{C}$. Now for all $c \in \mathcal{C}$ and $u \in \mathbb{Z}^n$, we have

$$\begin{aligned}&\langle\frac{1}{\sqrt{2k}}(\rho(c)+2ku), \frac{1}{\sqrt{2k}}(\rho(c)+2ku)\rangle\\ &\quad= \frac{1}{2k}\left(\langle\rho(c),\rho(c)\rangle + 4k\langle\rho(c),u\rangle + 4k^2\langle u,u\rangle\right),\end{aligned}$$

which is an even integer. From this, it follows that the lattice $\Lambda(\mathcal{C})$ is Type II when $\mathcal{C}$ is a Type II code.

Further, for each $c \in \mathcal{C}$ and $u \in \mathbb{Z}^n$, we see that $a = \frac{1}{\sqrt{2k}}(\rho(c) + 2ku) \in \Lambda(\mathcal{C})$. The norm of a is given by

$$\langle a, a\rangle = \frac{1}{2k}\left(\langle\rho(c),\rho(c)\rangle + 4k\langle\rho(c),u\rangle + 4k^2\langle u,u\rangle\right).$$

On taking $c = 0$, we get $\langle a, a\rangle = 2k\langle u, u\rangle \geq 2k$. Furthermore, when $u = (1, 0, \cdots, 0)$, we have $\langle a, a\rangle = 2k$.

When $c \neq 0$, one can easily observe that

$$\langle a, a\rangle = \frac{1}{2k}\left(\langle\rho(c),\rho(c)\rangle + 4k\langle\rho(c),u\rangle + 4k^2\langle u,u\rangle\right) \geq \frac{1}{2k}\langle\rho(c),\rho(c)\rangle.$$

As d_E is the minimum Euclidean distance of the code $\mathcal{C}$, there exists $c' \in \mathcal{C}$ such that $wt_E(c') = d_E$. We also note that $\langle\rho(c'),\rho(c')\rangle = wt_E(c') = d_E$. Hence there exists $a' = \frac{1}{2k}\rho(c') \in \Lambda(\mathcal{C})$ whose norm is equal to $d_E/2k$. This proves (i).

(ii) As the Euclidean weight of each element in $\mathcal{C}_0$ is congruent to 0 modulo $4k$, working as in part (i), we see that the norm of each element in the corresponding lattice $\Lambda(\mathcal{C}_0)$ is an even integer. This implies that $\Lambda(\mathcal{C}_0) \subseteq \Lambda(\mathcal{C})_0$. Further, one can show that $\det(\Lambda(\mathcal{C})/\Lambda(\mathcal{C}_0)) = \det(\Lambda(\mathcal{C})/\Lambda(\mathcal{C})_0)$, we get $\Lambda(\mathcal{C}_0) = \Lambda(\mathcal{C})_0$.

On the other hand, since the Euclidean weight of each element in $\mathcal{C}_2$ is congruent to $2k$ modulo $4k$, one can show that the norm of each element in the lattice $\Lambda(\mathcal{C}_2)$ is odd, which implies that $\Lambda(\mathcal{C}_2) \subseteq \Lambda(\mathcal{C})_2$. Now it is easy to see that $\det(\Lambda(\mathcal{C})/\Lambda(\mathcal{C}_2)) = \det(\Lambda(\mathcal{C})/\Lambda(\mathcal{C})_2)$, which implies that $\Lambda(\mathcal{C}_2) = \Lambda(\mathcal{C})_2$.

Furthermore, on relabelling $\mathcal{C}_1$ and $\mathcal{C}_3$ (if required), we can assume that $\Lambda(\mathcal{C}_3) = \Lambda(\mathcal{C})_3$ and $\Lambda(\mathcal{C}_4) = \Lambda(\mathcal{C})_4$. This completes the proof of (ii). □

The following lemma provides the Euclidean weight modulo $4k$ of any vector in the shadow of a Type I code.

Lemma 116. [9] *Let C be a Type I code of length n over $\mathbb{Z}_{2k}$ and $S = C_1 \cup C_3$ be its shadow code. Then for every $a \in S$, we have $wt_E(a) \equiv \frac{nk}{2} \pmod{4k}$.*

Proof. Let $a \in S$. The image of a under the map Λ (as defined in Theorem 115) is given by $\frac{1}{\sqrt{2k}}\{\rho(a)+2k\mathbb{Z}^n\}$, which contains the vector $\frac{1}{\sqrt{2k}}\rho(a)$. Note that the vector $\frac{1}{\sqrt{2k}}\rho(a)$ lies in $S_{\Lambda(C)}$ and the norm of $\frac{1}{\sqrt{2k}}\rho(a)$ is given by $\frac{1}{2k}\langle\rho(a),\rho(a)\rangle$. By Lemma 113(i), we get $\frac{1}{2k}\langle\rho(a),\rho(a)\rangle \equiv \frac{n}{4} \pmod 2$, which implies that $wt_E(a) \equiv \langle a, a\rangle \equiv \langle\rho(a),\rho(a)\rangle \equiv \frac{nk}{2} \pmod{4k}$. □

By Lemma 114(ii), the group $C_0^{\perp}/C_0$ has order 4 and is called the glue group of C_0. Hence the glue group $C_0^{\perp}/C_0$ is either the Klein 4-group or a cyclic group of order 4.

Lemma 117. [9] *Let C be a Type I code of length n over $\mathbb{Z}_{2k}$.*

(i) If n is an even integer, then the glue group $C_0^{\perp}/C_0$ is the Klein 4-group.
(ii) If n is an odd integer, then the glue group $C_0^{\perp}/C_0$ is a cyclic group of order 4.

Proof. By Theorem 115(ii), we see that the glue group $\Lambda(C_0)^*/\Lambda(C_0)$ of $\Lambda(C_0)$ is isomorphic to the glue group $C_0^{\perp}/C_0$ of C_0. Now by Lemma 113(ii), we get the desired result. □

The following theorem states orthogonality relations between the cosets of C_0 in $C_0^{\perp}$.

Theorem 118. *Let C be a Type I code of length n over $\mathbb{Z}_{2k}$ with C_h $(0 \leq h \leq 3)$ as defined above.*

(i) If $n \equiv 2 \pmod 4$, then we have the following:

$\langle\cdot,\cdot\rangle$	C_0	C_1	C_2	C_3
C_0	$\perp$	$\perp$	$\perp$	$\perp$
C_1	$\perp$	$\not\perp$	$\not\perp$	$\perp$
C_2	$\perp$	$\not\perp$	$\perp$	$\not\perp$
C_3	$\perp$	$\perp$	$\not\perp$	$\not\perp$

where for $0 \le h, j \le 3$, the symbol $\perp$ in the (h,j)th position means $\langle x, y\rangle \equiv 0 \pmod{2k}$ for each $x \in \mathcal{C}_h$ and $y \in \mathcal{C}_j$, whereas the symbol $\not\perp$ means $\langle x, y\rangle \equiv k \pmod{2k}$ for each $x \in \mathcal{C}_h$ and $y \in \mathcal{C}_j$.

(ii) If $n \equiv 0 \pmod 4$, then we have the following:

$\langle \cdot, \cdot \rangle$	$\mathcal{C}_0$	$\mathcal{C}_1$	$\mathcal{C}_2$	$\mathcal{C}_3$
$\mathcal{C}_0$	$\perp$	$\perp$	$\perp$	$\perp$
$\mathcal{C}_1$	$\perp$	$\perp$	$\not\perp$	$\not\perp$
$\mathcal{C}_2$	$\perp$	$\not\perp$	$\perp$	$\not\perp$
$\mathcal{C}_3$	$\perp$	$\not\perp$	$\not\perp$	$\perp$

where for $0 \le h, j \le 3$, the symbol $\perp$ in the (h,j)th position means $\langle x, y\rangle \equiv 0 \pmod{2k}$ for each $x \in \mathcal{C}_h$ and $y \in \mathcal{C}_j$, whereas the symbol $\not\perp$ means $\langle x, y\rangle \equiv k \pmod{2k}$ for each $x \in \mathcal{C}_h$ and $y \in \mathcal{C}_j$.

(iii) If $n \equiv a \pmod 4$ with $a = 1$ or 3, then we have $k \equiv 0 \pmod 2$. Further, we have the following:

$\langle \cdot, \cdot \rangle$	$\mathcal{C}_0$	$\mathcal{C}_1$	$\mathcal{C}_2$	$\mathcal{C}_3$
$\mathcal{C}_0$	0	0	0	0
$\mathcal{C}_1$	0	$k - \frac{ka}{2}$	k	$-\frac{ka}{2}$
$\mathcal{C}_2$	0	k	0	k
$\mathcal{C}_3$	0	$-\frac{ka}{2}$	k	$k - \frac{ka}{2}$

where for $0 \le h, j \le 3$, the (h,j)th entry in the table represents the value of $\langle x, y\rangle$ ($x \in \mathcal{C}_h$ and $y \in \mathcal{C}_j$) modulo $2k$.

Proof. By applying Lemmas 116 and 117 and using equations (6.1) and (6.3), parts (i) and (ii) follow immediately.

To prove (iii), we first note that n is odd. As $\mathcal{C}$ is a Type I code, by Lemma 109, the integer $(2k)^n$ is a perfect square, which holds if and only if $2k$ is a perfect square. So in this case, k must be an even integer.

Now by Lemma 117, we see that the glue group $\mathcal{C}_0^{\perp}/\mathcal{C}_0$ is a cyclic group of order 4, which implies that $\mathcal{C}_h + \mathcal{C}_j = \mathcal{C}_{h+j}$ for $0 \le h, j \le 3$, where the subscript $h + j$ is to be considered modulo 4. Now we shall distinguish the following five cases: (a) $x \in \mathcal{C}_0$ and $y \in \mathcal{C}_h$, where $h \in \{0, 1, 2, 3\}$; (b) $x, y \in \mathcal{C}_2$; (c) $x \in \mathcal{C}_2$ and $y \in S = \mathcal{C}_1 \cup \mathcal{C}_3$; (d) $x, y \in \mathcal{C}_h$, where $h \in \{1, 3\}$; and (e) $x \in \mathcal{C}_1$ and $y \in \mathcal{C}_3$.

(a) Let $x \in \mathcal{C}_0$ and $y \in \mathcal{C}_h$, where $h \in \{0, 1, 2, 3\}$. Since $\mathcal{C}_0^{\perp} = \mathcal{C}_0 \cup \mathcal{C}_1 \cup \mathcal{C}_2 \cup \mathcal{C}_3$, we have $\langle x, y \rangle \equiv 0 \pmod{2k}$.

(b) Let $x, y \in \mathcal{C}_2$. Here we have $x + y \in \mathcal{C}_0$. Further, by (6.1), we have

$$wt_E(x+y) \equiv wt_E(x) + wt_E(y) + 2\langle x, y \rangle \pmod{4k}. \tag{6.4}$$

As $wt_E(x) \equiv wt_E(y) \equiv 2k \pmod{4k}$ and $wt_E(x+y) \equiv 0 \pmod{4k}$, we obtain $0 \equiv 2k + 2k + 2\langle x, y \rangle \pmod{4k}$, which implies that $\langle x, y \rangle \equiv 0 \pmod{2k}$.

(c) Let $x \in \mathcal{C}_2$ and $y \in S$. Here $x + y \in S$, and further by Lemma 116, we get $wt_E(y) \equiv wt_E(x+y) \equiv \frac{nk}{2} \pmod{4k}$. As $wt_E(x) \equiv 2k \pmod{4k}$, the congruence (6.4) gives $\frac{nk}{2} \equiv 2k + \frac{nk}{2} + 2\langle x, y \rangle \pmod{4k}$, which implies that $\langle x, y \rangle \equiv k \pmod{2k}$.

(d) Let $x, y \in \mathcal{C}_h$, where $h \in \{1, 3\}$. Here we have $x + y \in \mathcal{C}_2$, which implies that $wt_E(x+y) \equiv 2k \pmod{4k}$. As $x, y \in S$, by Lemma 116, we have $wt_E(x) \equiv wt_E(y) \equiv \frac{nk}{2} \pmod{4k}$. Thus the congruence (6.4) gives $2k \equiv \frac{nk}{2} + \frac{nk}{2} + 2\langle x, y \rangle \pmod{4k}$, which implies that $\langle x, y \rangle \equiv k - \frac{nk}{2} \equiv k - \frac{ak}{2} \pmod{2k}$, as $n \equiv a \pmod 4$.

(e) Let $x \in \mathcal{C}_1$ and $y \in \mathcal{C}_3$. Here we have $x + y \in \mathcal{C}_0$, which implies that $wt_E(x+y) \equiv 0 \pmod{4k}$. Further, by Lemma 116, we have $wt_E(x) \equiv wt_E(y) \equiv \frac{nk}{2} \pmod{4k}$. Now since $n \equiv a \pmod 4$, the congruence (6.4) implies $\langle x, y \rangle \equiv -\frac{ka}{2} \pmod{2k}$.

This completes the proof of the theorem. □

As the shadow of a Type II code is an empty set, the shadow is defined only for Type I codes. Dougherty et al. [10] further generalized the notion of shadow for any self-dual linear code over $\mathbb{Z}_4$, which is as discussed below.

6.2.3 *Generalized Shadow of a Self-dual Code*

Let $\mathcal{C}$ be a self-dual code of length n over $\mathbb{Z}_{2k}$. For a vector $s \in \mathbb{Z}_{2k}^n \setminus \mathcal{C}$, let us define a map $\psi_s : \mathcal{C} \to \mathbb{Z}_{2k}$ as $\psi_s(u) = \langle u, s \rangle$ for each $u \in \mathcal{C}$. Note that ψ_s is a group homomorphism from the additive group $\mathcal{C}$ into the additive group $\mathbb{Z}_{2k}$ and the set $\mathcal{C}_0 = ker(\psi_s)$ is a proper subcode of $\mathcal{C}$. Hence by the first isomorphism theorem, we have $\mathcal{C}/\mathcal{C}_0 \simeq Im(\psi_s)$. This gives $[\mathcal{C} : \mathcal{C}_0] = |Im(\psi_s)| = r$, where $r > 1$ is a divisor of $2k$. Further, choose a vector $s \in \mathbb{Z}_{2k}^n$ such that $Im(\psi_s) = \{0, k\}$. With this choice of s, we have $\langle u, s\rangle \equiv 0$ or $k \pmod{2k}$ for each $u \in \mathcal{C}$. Further, one can show that $[\mathcal{C} : \mathcal{C}_0] = 2$. Let $\mathcal{C}_2 = \mathcal{C} \setminus \mathcal{C}_0$, so that $\mathcal{C} = \mathcal{C}_0 \cup \mathcal{C}_2$. Also note that $[\mathcal{C}_0^\perp : \mathcal{C}_0] = 4$, and let us write $\mathcal{C}_0^\perp = \mathcal{C}_0 \cup \mathcal{C}_2 \cup \mathcal{C}_1 \cup \mathcal{C}_3$. Then the generalized shadow of $\mathcal{C}$ with respect to the vector s is defined as $\mathcal{S}_g(s) = \mathcal{C}_1 \cup \mathcal{C}_3$. The group $\mathcal{C}_0^\perp/\mathcal{C}_0$ is a group of order 4 and is called the glue group of $\mathcal{C}_0$. The following lemma states some important properties of the generalized shadow of $\mathcal{C}$ with respect to the vector s.

Lemma 119. *[10] Let $\mathcal{C}$ be a self-dual code of length n over $\mathbb{Z}_{2k}$. Let $s \in \mathbb{Z}_{2k}^n \setminus \mathcal{C}$ be such that $Im(\psi_s) = \{0, k\}$. With respect to the vector s, let $\mathcal{C} = \mathcal{C}_0 \cup \mathcal{C}_2$ and $\mathcal{C}_0^\perp = \mathcal{C}_0 \cup \mathcal{C}_2 \cup \mathcal{C}_1 \cup \mathcal{C}_3$. Then the following hold.*

(i) $\mathcal{C}_1 = s + \mathcal{C}_0$, $\mathcal{C}_2 = t + \mathcal{C}_0$, $\mathcal{C}_3 = s + t + \mathcal{C}_0$ for some $t \in \mathcal{C} \setminus \mathcal{C}_0$.
(ii) When $\langle s, s\rangle \equiv 0$ or $k \pmod{2k}$, the glue group $\mathcal{C}_0^\perp/\mathcal{C}_0$ of $\mathcal{C}_0$ is the Klein 4-group.

Proof. (i) Proof is trivial.

(ii) Suppose, on the contrary, that $\mathcal{C}_0^\perp/\mathcal{C}_0$ is a cyclic group of order 4. Without any loss of generality, the following two cases arise: (a) $\mathcal{C}_2 = 2x + \mathcal{C}_0$, $\mathcal{C}_1 = x + \mathcal{C}_0$ and $\mathcal{C}_3 = 3x + \mathcal{C}_0$ and (b) $\mathcal{C}_2 = x + \mathcal{C}_0$, $\mathcal{C}_1 = 2x + \mathcal{C}_0$ and $\mathcal{C}_3 = 3x + \mathcal{C}_0$.

(a) Let $c_0 \in \mathcal{C}_0$. Here as $s \in \mathcal{C}_0^\perp$, we note that $\langle s, c_0\rangle \equiv 0 \pmod{2k}$. Further, as $c_0 + 2x \in \mathcal{C}_2$, we have $\langle s, c_0 + 2x\rangle \equiv k \pmod{2k}$, which implies that $\langle s, c_0\rangle + 2\langle s, x\rangle \equiv k \pmod{2k}$, which gives

$$2\langle s, x\rangle \equiv k \pmod{2k}. \tag{6.5}$$

On the other hand, as $\mathcal{C}_1 = x + \mathcal{C}_0 = s + \mathcal{C}_0$, we see that $2\langle s, x\rangle \equiv 2\langle s, s\rangle \equiv 0 \pmod{2k}$. This, by (6.5), implies that $k \equiv 0 \pmod{2k}$, which is a contradiction.

(b) As $\mathcal{C}$ is a self-dual code and $x \in \mathcal{C}$, we see that $\langle x, x\rangle \equiv 0 \pmod{2k}$. Now for all $c_0, c_0' \in \mathcal{C}_0$ and $0 \le h, j \le 3$, we note that $\langle hx + c_0, jx + c_0'\rangle \equiv 0 \pmod{2k}$. This implies that $\mathcal{C}_h \subseteq \mathcal{C}^\perp$ for $0 \le h \le 3$. This further implies that $\mathcal{C}_0^\perp \subseteq \mathcal{C}^\perp = \mathcal{C}$, which is a contradiction.

□

The following lemma states orthogonality relations between the cosets of $\mathcal{C}_0$ in $\mathcal{C}_0^\perp$.

Lemma 120. *[10] Let C be a self-dual code of length n over $\mathbb{Z}_{2k}$. Let $s \in \mathbb{Z}_{2k}^n \backslash C$ be such that $Im(\psi_s) = \{0, k\}$. Let $C = C_0 \cup C_2$ and $C_0^\perp = C_0 \cup C_2 \cup C_1 \cup C_3$ with respect to the vector s. Then we have the following:*

(i) If s is a vector in $\mathbb{Z}_{2k}^n \setminus C$ satisfying $\langle s, s \rangle \equiv 0 \pmod{2k}$, then the orthogonality relations between the cosets of C_0 in $C_0^\perp$ are given by

$\langle \cdot, \cdot \rangle$	C_0	C_1	C_2	C_3
C_0	$\perp$	$\perp$	$\perp$	$\perp$
C_1	$\perp$	$\perp$	$\not\perp$	$\not\perp$
C_2	$\perp$	$\not\perp$	$\perp$	$\not\perp$
C_3	$\perp$	$\not\perp$	$\not\perp$	$\perp$

where the symbol $\perp$ means $\langle x, y \rangle \equiv 0 \pmod{2k}$ and the symbol $\not\perp$ means $\langle x, y \rangle \equiv k \pmod{2k}$.

(ii) If s is a vector in $\mathbb{Z}_{2k}^n \setminus C$ satisfying $\langle s, s \rangle \equiv k \pmod{2k}$, then the orthogonality relations between the cosets of C_0 in $C_0^\perp$ are given by

$\langle \cdot, \cdot \rangle$	C_0	C_1	C_2	C_3
C_0	$\perp$	$\perp$	$\perp$	$\perp$
C_1	$\perp$	$\not\perp$	$\not\perp$	$\perp$
C_2	$\perp$	$\not\perp$	$\perp$	$\not\perp$
C_3	$\perp$	$\perp$	$\not\perp$	$\not\perp$

where the symbol $\perp$ means $\langle x, y \rangle \equiv 0 \pmod{2k}$ and the symbol $\not\perp$ means $\langle x, y \rangle \equiv k \pmod{2k}$.

Proof. Let $t \in C \setminus C_0$. By Lemma 119, we have $C_1 = s + C_0$, $C_2 = t + C_0$ and $C_3 = s + t + C_0$. Note that $t \in C$ and $\langle t, t \rangle \equiv 0 \pmod{2k}$. We further assert that $\langle s, t \rangle \equiv k \pmod{2k}$.

For, if this is not so, then by our choice of s, we must have $\langle s, t \rangle \equiv 0 \pmod{2k}$. Now as C is a self-dual code, we observe that that $s \in C^\perp = C$, which is a contradiction.

From this, the desired result follows immediately. □

From now on, throughout this chapter, we shall focus our attention on linear codes over $\mathbb{Z}_{2^m}$. In the following sections, the construction method employed by Brualdi and Pless [2] and Tsai [14] is generalized to construct self-dual codes (of higher lengths) over $\mathbb{Z}_{2^m}$ using shadows of Type I codes over $\mathbb{Z}_{2^m}$ and generalized shadows of self-dual codes of length n over $\mathbb{Z}_{2^m}$ with respect to a vector $s \in \mathbb{Z}_{2^m}^n$, which is not a codeword and satisfies $\langle s, s\rangle \equiv 0$ or 2^{m-1} (mod 2^m).

6.3 Construction of Type I and Type II Codes from Shadows of Type I Codes

First of all, let us fix some notations. For a subset A of $\mathbb{Z}_{2^m}^n$, let $\langle A\rangle$ denote the $\mathbb{Z}_{2^m}$-submodule of $\mathbb{Z}_{2^m}^n$ generated by the set A. Let $o(a)$ denote the additive order of an element $a \in \mathbb{Z}_{2^m}^k$, where $k \geq 1$ is an integer. Further, let the function $\eta : \mathbb{Z}_{2^m} \times \mathbb{Z}_{2^m} \to \{0, 1, 2, 3\}$ be defined as

$$\eta(h, j) = \begin{cases} [h]_2 + 2\,[j]_2 & \text{if } C_0^\perp/C_0 \text{ is the Klein 4-group;} \\ [h + 2j]_4 & \text{if } C_0^\perp/C_0 \text{ is a cyclic group of order 4,} \end{cases} \tag{6.6}$$

where $[a]_r$ is the remainder obtained when the integer a is divided by the integer $r \geq 1$. Then we make the following observation.

Lemma 121. *For $1 \leq h, j \leq 2^m - 1$, we have $hC_1 + jC_2 = C_{\eta(h,j)}$.*

Proof. Its proof is a straightforward exercise. □

In the following proposition, a self-orthogonal code (of higher length) over $\mathbb{Z}_{2^m}$ is constructed from a self-dual code of length n over $\mathbb{Z}_{2^m}$.

Proposition 122. *Let C be a self-dual code of length n over $\mathbb{Z}_{2^m}$, and let C_0 be a proper subcode of index 2 in C. Let $C = C_0 \cup C_2$ and $C_0^\perp = C_0 \cup C_2 \cup C_1 \cup C_3$, where $C_2 = t + C_0$, $C_1 = s + C_0$ and $C_3 = s + t + C_0$ for some $t \in C \setminus C_0$ and $s \in C_0^\perp \setminus C$. Suppose that for some integer $k \geq 1$, there exist vectors $v_1, v_2 \in \mathbb{Z}_{2^m}^k$ satisfying the following three properties:*

(P$_1$) *If there exist $\alpha, \beta \in \mathbb{Z}_{2^m}$ ($0 \leq \alpha < o(v_1)$ and $0 \leq \beta < o(v_2)$) such that $\alpha v_1 + \beta v_2 = 0$, then $\alpha = \beta = 0$.*

(P$_2$) *$\langle v_1, v_1\rangle \equiv -\langle s, s\rangle \pmod{2^m}$, $\langle v_1, v_2\rangle \equiv -\langle t, s\rangle \pmod{2^m}$ and $\langle v_2, v_2\rangle \equiv -\langle t, t\rangle \pmod{2^m}$.*

(P$_3$) $$o(v_1) \equiv \begin{cases} 0 \pmod 2 & \text{if } C_0^\perp/C_0 \text{ is the Klein 4-group;} \\ 0 \pmod 4 & \text{if } C_0^\perp/C_0 \text{ is a cyclic group of order 4} \end{cases} \quad \text{and}$$

$o(v_2) \equiv 0 \pmod 2$ provided $m \geq 2$.

Then the set

$$\mathcal{C}^{\maltese} = \{(hv_1 + jv_2, c_{hj}) : 1 \leq h \leq o(v_1),\ 1 \leq j \leq o(v_2), c_{hj} \in \mathcal{C}_{\eta(h,j)}\}$$

is a self-orthogonal code of length $n + k$ over $\mathbb{Z}_{2^m}$ with cardinality $|\mathcal{C}^{\maltese}| = o(v_1)o(v_2)2^{\frac{mn}{2}-1}$.
(Here the positive integer k is to be chosen suitably depending upon the existence of vectors v_1 and v_2 in $\mathbb{Z}_{2^m}^k$.)

Proof. To prove the result, we first observe that

$$\mathcal{C}^{\maltese} = \bigcup_{h=1}^{o(v_1)} \bigcup_{j=1}^{o(v_2)} (hv_1 + jv_2, \mathcal{C}_{\eta(h,j)}),$$

where $(hv_1 + jv_2, \mathcal{C}_{\eta(h,j)}) = \{(hv_1 + jv_2, c_{hj}) : c_{hj} \in \mathcal{C}_{\eta(h,j)}\}$ for each h and j.

Now we assert that $\mathcal{C}^{\maltese}$ is a $\mathbb{Z}_{2^m}$-submodule of $\mathbb{Z}_{2^m}^{n+k}$ generated by the set $(v_1, \mathcal{C}_1) \cup (v_2, \mathcal{C}_2)$.

To prove this assertion, we first note that

$$o(\mathcal{C}_1) = \begin{cases} 2 & \text{if } \mathcal{C}_0^{\perp}/\mathcal{C}_0 \text{ is the Klein 4-group;} \\ 4 & \text{if } \mathcal{C}_0^{\perp}/\mathcal{C}_0 \text{ is a cyclic group of order 4} \end{cases}$$

and $o(\mathcal{C}_2) = 2$, where $o(\mathcal{C}_h)$ $(h = 1, 2)$ denotes the order of $\mathcal{C}_h \in \mathcal{C}_0^{\perp}/\mathcal{C}_0$. Now as the vectors v_1 and v_2 satisfy the property $(\mathbf{P}_3)$, we see that $o(\mathcal{C}_1)$ divides $o(v_1)$ and $o(\mathcal{C}_2)$ divides $o(v_2)$, which gives $o(v_1)\mathcal{C}_1 = \mathcal{C}_0$ and $o(v_2)\mathcal{C}_2 = \mathcal{C}_0$. We also note that $\mathcal{C}_1 + \mathcal{C}_0 = \mathcal{C}_1$ and $\mathcal{C}_2 + \mathcal{C}_0 = \mathcal{C}_2$. From this, it follows that for every $h \equiv h_1 \pmod{o(v_1)}$ and $j \equiv j_1 \pmod{o(v_2)}$ with $1 \leq h_1 \leq o(v_1)$ and $1 \leq j_1 \leq o(v_2)$, we have $h(v_1, \mathcal{C}_1) = h_1(v_1, \mathcal{C}_1)$ and $j(v_2, \mathcal{C}_2) = j_1(v_2, \mathcal{C}_2)$. Therefore the $\mathbb{Z}_{2^m}$-submodule $\langle (v_1, \mathcal{C}_1) \cup (v_2, \mathcal{C}_2) \rangle$ equals $\{h(v_1, c_1) + j(v_2, c_2) : 1 \leq h \leq o(v_1),\ 1 \leq j \leq o(v_2),\ c_1 \in \mathcal{C}_1, c_2 \in \mathcal{C}_2\}$. Moreover, by Lemma 121, for every $c_1 \in \mathcal{C}_1$ and $c_2 \in \mathcal{C}_2$, we note that $hc_1 + jc_2 \in \mathcal{C}_{\eta(h,j)}$ for each h and j, which proves the assertion.

Now to show that the code $\mathcal{C}^{\maltese}$ is self-orthogonal, it suffices to show that

$$\langle (v_1, c_1), (v_1, c_1) \rangle = \langle (v_1, c_1), (v_2, c_2) \rangle = \langle (v_2, c_2), (v_2, c_2) \rangle = 0 \tag{6.7}$$

for all $c_1 \in \mathcal{C}_1$ and $c_2 \in \mathcal{C}_2$.

To prove this, let $c_1 \in \mathcal{C}_1$ and $c_2 \in \mathcal{C}_2$ be fixed arbitrarily. Since $\mathcal{C}_1 = s + \mathcal{C}_0$ and $\mathcal{C}_2 = t + \mathcal{C}_0$, we write $c_1 = s + c_0$ and $c_2 = t + c_0'$ for some $c_0,\ c_0' \in \mathcal{C}_0$. Further, $\mathcal{C}_0^{\perp} = \mathcal{C}_0 \cup \mathcal{C}_1 \cup \mathcal{C}_2 \cup \mathcal{C}_3$, $s \in \mathcal{C}_1$ and $t \in \mathcal{C}_2$ imply that

$$\langle c_0, c_0 \rangle \equiv \langle c_0', c_0' \rangle \equiv \langle c_0, c_0' \rangle \equiv \langle c_0, t \rangle \equiv \langle c_0', t \rangle \equiv \langle c_0, s \rangle \equiv \langle c_0', s \rangle \equiv 0 \pmod{2^m}.$$

This gives $\langle c_1, c_1\rangle \equiv \langle s, s\rangle \pmod{2^m}$, $\langle c_1, c_2\rangle \equiv \langle s, t\rangle \pmod{2^m}$ and $\langle c_2, c_2\rangle \equiv \langle t, t\rangle \pmod{2^m}$. Now as the vectors v_1 and v_2 satisfy the property $(\mathbf{P}_2)$, we have

$$\langle v_1, v_1\rangle \equiv -\langle s, s\rangle \pmod{2^m},\ \langle v_1, v_2\rangle \equiv -\langle t, s\rangle \pmod{2^m}$$

and

$$\langle v_2, v_2\rangle \equiv -\langle t, t\rangle \pmod{2^m}.$$

From this, we obtain $\langle v_1, v_1\rangle \equiv -\langle c_1, c_1\rangle \pmod{2^m}$, $\langle v_1, v_2\rangle \equiv -\langle c_2, c_1\rangle \pmod{2^m}$ and $\langle v_2, v_2\rangle \equiv -\langle c_2, c_2\rangle \pmod{2^m}$, which proves (6.7).

Next to prove that

$$|\mathcal{C}^{\maltese}| = o(v_1)o(v_2)|\mathcal{C}_0| = o(v_1)o(v_2)2^{\frac{mn}{2}-1},$$

we need to show that the sets $(hv_1 + jv_2, \mathcal{C}_{\eta(h,j)})$ $(1 \le h \le o(v_1),\ 1 \le j \le o(v_2))$ are distinct (and hence disjoint) in $\mathbb{Z}_{2^m}^{n+k}$, as $|\mathcal{C}_{\eta(h,j)}| = |\mathcal{C}_0| = \frac{1}{2}|\mathcal{C}| = 2^{\frac{mn}{2}-1}$ for each h and j. To prove this, we shall distinguish the following three cases: **(a)** $h = o(v_1)$ and $1 \le j \le o(v_2)$, **(b)** $1 \le h < o(v_1)$ and $j = o(v_2)$, **(c)** $1 \le h < o(v_1)$ and $1 \le j < o(v_2)$.

Case (a): Let $h = o(v_1)$ and $1 \le j \le o(v_2)$. Here we have

$$(hv_1 + jv_2, h\mathcal{C}_1 + j\mathcal{C}_2) = (jv_2, h\mathcal{C}_1 + j\mathcal{C}_2).$$

Working as earlier, we see that $o(\mathcal{C}_1)$ divides $o(v_1)$, which gives $o(v_1)\mathcal{C}_1 = \mathcal{C}_0$. This implies that $(hv_1 + jv_2, h\mathcal{C}_1 + j\mathcal{C}_2) = (jv_2, j\mathcal{C}_2)$, as $\mathcal{C}_0 + j\mathcal{C}_2 = j\mathcal{C}_2$ for each j. Further, it is easy to see that the sets $(jv_2, j\mathcal{C}_2)$ $(1 \le j \le o(v_2))$ are disjoint in $\mathbb{Z}_{2^m}^{n+k}$.

Case (b): Let $1 \le h < o(v_1)$ and $j = o(v_2)$. Here we have

$$(hv_1 + jv_2, h\mathcal{C}_1 + j\mathcal{C}_2) = (hv_1, h\mathcal{C}_1 + j\mathcal{C}_2).$$

As the vector v_2 satisfies the property $(\mathbf{P_3})$, $o(\mathcal{C}_2) = 2$ divides $o(v_2)$. This gives $o(v_2)\mathcal{C}_2 = \mathcal{C}_0$, which implies that $(hv_1 + jv_2, h\mathcal{C}_1 + j\mathcal{C}_2) = (hv_1, h\mathcal{C}_1)$, as $h\mathcal{C}_1 + \mathcal{C}_0 = h\mathcal{C}_1$ for each h. Further, it is easy to see that the sets $(hv_1, h\mathcal{C}_1)$ $(1 \le h < o(v_1))$ are disjoint in $\mathbb{Z}_{2^m}^{n+k}$.

Case (c): Let $1 \le h < o(v_1)$ and $1 \le j < o(v_2)$.

Here we need to show that the sets $(hv_1 + jv_2, \mathcal{C}_{\eta(h,j)})$ $(1 \le h < o(v_1)$ and $1 \le j < o(v_2))$ are distinct (and hence disjoint) in $\mathbb{Z}_{2^m}^{n+k}$.

For this, suppose (if possible) that

$$(hv_1 + jv_2, \mathcal{C}_{\eta(h,j)}) = (h'v_1 + j'v_2, \mathcal{C}_{\eta(h'j')})$$

holds for some $1 \leq h' < o(v_1)$ and $1 \leq j' < o(v_2)$. This gives

$$((h-h')v_1 + (j-j')v_2, \mathcal{C}_{\eta(h,j)} - \mathcal{C}_{\eta(h',j')}) = (\mathbf{0}, \mathcal{C}_0),$$

which implies that $(h-h')v_1 + (j-j')v_2 = 0$. This further implies that $r_1v_1 + r_2v_2 = 0$, where $[h-h']_{o(v_1)} = r_1$ and $[j-j']_{o(v_2)} = r_2$. As the vectors v_1 and v_2 satisfy the property $(\mathbf{P}_1)$, we have $r_1 = 0$ and $r_2 = 0$. Further, since $1 \leq h, h' < o(v_1)$ and $1 \leq j, j' < o(v_2)$, we must have $r_1 = |h-h'|$ and $r_2 = |j-j'|$, which gives $h = h'$ and $j = j'$. From this, it follows that the sets $(hv_1 + jv_2, \mathcal{C}_{\eta(h,j)})$, $1 \leq h < o(v_1)$, $1 \leq j < o(v_2)$, are mutually disjoint in $\mathbb{Z}_{2^m}^{n+k}$.

This completes the proof of the proposition. □

Remark 123. When $m = 1$, the binary code $\mathcal{C}^{\maltese}$ (if it exists) is given by $\mathcal{C}^{\maltese} = (\mathbf{0}, \mathcal{C}_0) \cup (v_1, \mathcal{C}_1) \cup (v_2, \mathcal{C}_2) \cup (v_1 + v_2, \mathcal{C}_3)$, as $\mathcal{C}_1 + \mathcal{C}_2 = \mathcal{C}_3$. In general, the code $\mathcal{C}^{\maltese}$ (if it exists) is a $\mathbb{Z}_{2^m}$-submodule of $\mathbb{Z}_{2^m}^{n+k}$ generated by $(v_1, \mathcal{C}_1) \cup (v_2, \mathcal{C}_2)$.

From now on, we shall follow the same notations as in Proposition 122, and we shall consider each subscript μ of the variable X_μ in the complete weight enumerator modulo 2^m.

In the following theorem, Type I and Type II codes (of higher lengths) over $\mathbb{Z}_{2^m}$ are constructed from Type I codes of even length over $\mathbb{Z}_{2^m}$ for all $m \geq 1$, and their complete weight enumerators are also determined.

Theorem 124. *Let $\mathcal{C}$ be a Type I code of even length n over $\mathbb{Z}_{2^m}$ ($m \geq 1$ is an integer), and let $\mathcal{C}_0 = \{c \in \mathcal{C} : wt_E(c) \equiv 0 \pmod{2^{m+1}}\}$. The following hold.*

(i) *Let $n \equiv 2 \pmod 4$. Let us choose $v_1, v_2 \in \mathbb{Z}_{2^m}^2$ as*

$$v_1 = \begin{cases} (2^{\frac{k}{2}-1}, 2^{\frac{k}{2}-1}) & \text{if } m \text{ is even;} \\ (2^{\frac{k-1}{2}}, 0) & \text{if } m \text{ is odd} \end{cases} \quad \text{and} \quad v_2 = \begin{cases} (2^{\frac{k}{2}}, 0) & \text{if } m \text{ is even;} \\ (2^{\frac{k-1}{2}}, 2^{\frac{k-1}{2}}) & \text{if } m \text{ is odd.} \end{cases}$$

Then the code $\mathcal{C}^{\maltese}$ is a self-dual code of length $n+2$ over $\mathbb{Z}_{2^m}$. Furthermore, the code $\mathcal{C}^{\maltese}$ is a Type II code when $n \equiv 6 \pmod 8$ and $\mathcal{C}^{\maltese}$ is a Type I code when $n \equiv 2 \pmod 8$. The complete weight enumerator

$cwe_{\mathcal{C}^{\maltese}}(X_\mu : \mu \in \mathbb{Z}_{2^m})$ of $\mathcal{C}^{\maltese}$ is given by

$$\begin{cases} \sum_{h=1}^{2^{\frac{k}{2}+1}} \sum_{j=1}^{2^{\frac{k}{2}}} X_{(h+2j)2^{\frac{k}{2}-1}} X_{h2^{\frac{k}{2}-1}} cwe_{\mathcal{C}_{\eta(h,j)}}(X_\mu : \mu \in \mathbb{Z}_{2^m}) & \text{if } m \text{ is even;} \\ \sum_{h=1}^{2^{\frac{k+1}{2}}} \sum_{j=1}^{2^{\frac{k+1}{2}}} X_{(h+j)2^{\frac{k-1}{2}}} X_{j2^{\frac{k-1}{2}}} cwe_{\mathcal{C}_{\eta(h,j)}}(X_\mu : \mu \in \mathbb{Z}_{2^m}) & \text{if } m \text{ is odd.} \end{cases}$$

(ii) *Let $n \equiv 0 \pmod 4$. Let $v_1, v_2 \in \mathbb{Z}_{2^m}^4$ and $w_1, w_2 \in \mathbb{Z}_{2^m}^{n+4}$ be chosen as*

$$v_1 = \begin{cases} (2^{\frac{k}{2}-1}, 2^{\frac{k}{2}-1}, 2^{\frac{k}{2}-1}, 2^{\frac{k}{2}-1}) & \text{if } m \text{ is even;} \\ (2^{\frac{k-1}{2}}, 0, 2^{\frac{k-1}{2}}, 0) & \text{if } m \text{ is odd,} \end{cases}$$

$$v_2 = \begin{cases} (2^{\frac{k}{2}}, 0, 0, 0) & \text{if } m \text{ is even;} \\ (2^{\frac{k-1}{2}}, 2^{\frac{k-1}{2}}, 0, 0) & \text{if } m \text{ is odd,} \end{cases}$$

$$w_1 = \begin{cases} (2^{\frac{k}{2}}, 2^{\frac{k}{2}}, 0, 0, 0, \cdots, 0) & \text{if } m \text{ is even;} \\ (2^{\frac{k-1}{2}}, 2^{\frac{k-1}{2}}, 2^{\frac{k-1}{2}}, 2^{\frac{k-1}{2}}, 0, \cdots, 0) & \text{if } m \text{ is odd,} \end{cases}$$

$$w_2 = \begin{cases} (0, 2^{\frac{k}{2}}, 2^{\frac{k}{2}}, 0, 0, \cdots, 0) & \text{if } m \text{ is even;} \\ (2^{\frac{k+1}{2}}, 0, 0, 0, 0, \cdots, 0) & \text{if } m \text{ is odd.} \end{cases}$$

Then the code $\mathcal{C}^{\clubsuit} = \langle \mathcal{C}^{\maltese} \cup \{w_1, w_2\} \rangle$ is a self-dual code of length $n+4$ over $\mathbb{Z}_{2^m}$. Furthermore, $\mathcal{C}^{\clubsuit}$ is a Type II code when $n \equiv 4 \pmod 8$ and $\mathcal{C}^{\clubsuit}$ is a Type I code when $n \equiv 0 \pmod 8$. The complete weight enumerator $cwe_{\mathcal{C}^{\clubsuit}}(X_\mu : \mu \in \mathbb{Z}_{2^m})$ of $\mathcal{C}^{\clubsuit}$ is given by

$$\begin{cases} \sum_1 X_{(h+2j+2k_1)2^{\frac{k}{2}-1}} X_{(h+2k_1+2k_2)2^{\frac{k}{2}-1}} \\ X_{(h+2k_2)2^{\frac{k}{2}-1}} X_{h2^{\frac{k}{2}-1}} cwe_{\mathcal{C}_{\eta(h,j)}}(X_\mu : \mu \in \mathbb{Z}_{2^m}) & \text{if } m \text{ is even;} \\ \sum_2 X_{(h+j+k_1+2k_2)2^{\frac{k-1}{2}}} X_{(j+k_1)2^{\frac{k-1}{2}}} X_{(h+k_1)2^{\frac{k-1}{2}}} \\ X_{k_1 2^{\frac{k-1}{2}}} cwe_{\mathcal{C}_{\eta(h,j)}}(X_\mu : \mu \in \mathbb{Z}_{2^m}) & \text{if } m \text{ is odd,} \end{cases}$$

where the summation $\sum_1$ runs over all integral 4-tuples (h, j, k_1, k_2) satisfying $1 \le j, k_1, k_2 \le 2^{\frac{k}{2}}$ and $1 \le h \le 2^{\frac{k}{2}+1}$, whereas the summation $\sum_2$ runs over all integral 4-tuples (h, j, k_1, k_2) satisfying $1 \le h, j, k_1 \le 2^{\frac{k+1}{2}}$ and $1 \le k_2 \le 2^{\frac{k-1}{2}}$.

Proof. As n is even, by Lemma 117(i), we see that the glue group $\mathcal{C}_0^\perp/\mathcal{C}_0$ of $\mathcal{C}_0$ is the Klein 4-group. Hence by (6.6), we have $\eta(h,j) = [h]_2 + 2\,[j]_2$ for each h and j. Now we will apply Proposition 122 to construct a self-dual code in each case.

(i) Here we have $n \equiv 2 \pmod 4$. First of all, we will construct a self-orthogonal code $\mathcal{C}^{\maltese}$ by applying Proposition 122. For this, we need to choose a suitable positive integer k and vectors v_1, v_2 in $\mathbb{Z}_{2^m}^k$ satisfying the properties $(\mathbf{P_1}), (\mathbf{P_2})$ and $(\mathbf{P_3})$. Now as $s \in \mathcal{C}_1$ and $t \in \mathcal{C}_2$, by Theorem 118(i), we have $\langle s, s\rangle \equiv 2^{m-1} \pmod{2^m}$, $\langle s, t\rangle \equiv 2^{m-1} \pmod{2^m}$ and $\langle t, t\rangle \equiv 0 \pmod{2^m}$. Therefore $(\mathbf{P_2})$ becomes

$$\langle v_1, v_1\rangle \equiv 2^{m-1} \pmod{2^m}, \quad \langle v_1, v_2\rangle \equiv 2^{m-1} \pmod{2^m}, \\ \langle v_2, v_2\rangle \equiv 0 \pmod{2^m}. \tag{6.8}$$

It is clear that the vectors

$$v_1 = \begin{cases} (2^{\frac{k}{2}-1}, 2^{\frac{k}{2}-1}) & \text{if } m \text{ is even;} \\ (2^{\frac{k-1}{2}}, 0) & \text{if } m \text{ is odd} \end{cases} \quad \text{and} \quad v_2 = \begin{cases} (2^{\frac{k}{2}}, 0) & \text{if } m \text{ is even;} \\ (2^{\frac{k-1}{2}}, 2^{\frac{k-1}{2}}) & \text{if } m \text{ is odd.} \end{cases}$$

satisfy the property $(\mathbf{P_1})$ and all the congruences in (6.8). Further, since

$$o(v_1) = \begin{cases} 2^{\frac{m}{2}+1} & \text{if } m \text{ is even;} \\ 2^{\frac{m+1}{2}} & \text{if } m \text{ is odd,} \end{cases} \quad o(v_2) = \begin{cases} 2^{\frac{m}{2}} & \text{if } m \text{ is even;} \\ 2^{\frac{m+1}{2}} & \text{if } m \text{ is odd} \end{cases}$$

and $\mathcal{C}_0^{\perp}/\mathcal{C}_0$ is the Klein 4-group by Lemma 117(i), we see that vectors v_1 and v_2 also satisfy the property $(\mathbf{P_3})$. Therefore by Proposition 122, the code $\mathcal{C}^{\maltese}$ is a self-orthogonal code of length $n+2$ over $\mathbb{Z}_{2^m}$. As $o(v_1)o(v_2) = 2^{m+1}$, we have $|\mathcal{C}^{\maltese}| = 2^{m+1}2^{\frac{mn}{2}-1} = (2^m)^{\frac{n+2}{2}}$ using Proposition 122 again. From this, it follows that $\mathcal{C}^{\maltese}$ is a self-dual code of length $n+2$ over $\mathbb{Z}_{2^m}$.

Next we observe that $wt_E(v_1) = 2^{m-1}$ and $wt_E(v_2) = 2^m$. Also for every $c_2 \in \mathcal{C}_2$, we note that $wt_E(c_2) \not\equiv 0 \pmod{2^{m+1}}$ and $wt_E(c_2) \equiv \langle c_2, c_2\rangle \equiv 0 \pmod{2^m}$, by applying Theorem 118(i). This implies that $wt_E(c_2) \equiv 2^m \pmod{2^{m+1}}$, which gives $wt_E(v_2, c_2) = wt_E(v_2) + wt_E(c_2) \equiv 2^m + 2^m \equiv 0 \pmod{2^{m+1}}$ for every $c_2 \in \mathcal{C}_2$. On the other hand, by Lemma 116, we have $wt_E(c_1) \equiv 2^{m-2}n \pmod{2^{m+1}}$ for every $c_1 \in \mathcal{C}_1$. This implies that for every $c_1 \in \mathcal{C}_1$, we have $wt_E(v_1, c_1) = wt_E(v_1) + wt_E(c_1) \equiv 2^{m-1} + 2^{m-2}n \pmod{2^{m+1}}$, which gives

$$wt_E(v_1, c_1) \equiv \begin{cases} 0 \pmod{2^{m+1}} & \text{if } n \equiv 6 \pmod 8; \\ 2^m \pmod{2^{m+1}} & \text{if } n \equiv 2 \pmod 8). \end{cases}$$

From this, it follows that the code $\mathcal{C}^{\maltese}$ is a Type II code when $n \equiv 6 \pmod 8$ and $\mathcal{C}^{\maltese}$ is a Type I code when $n \equiv 2 \pmod 8$.

In order to compute the complete weight enumerator of $\mathcal{C}^{\maltese}$, by Proposition 122, we note that any element $c^{\maltese} \in \mathcal{C}^{\maltese}$ is of the form $c^{\maltese} =$

$(hv_1 + jv_2, c_{hj})$, where $c_{hj} \in \mathcal{C}_{\eta(ih,j)}$, $1 \leq h \leq o(v_1)$ and $1 \leq j \leq o(v_2)$. This gives

$$c^{\maltese} = \begin{cases} ((h+2j)2^{\frac{k}{2}-1}, h2^{\frac{k}{2}-1}, c_{hj}) & \text{if } m \text{ is even;} \\ ((h+j)2^{\frac{k-1}{2}}, j2^{\frac{k-1}{2}}, c_{hj}) & \text{if } m \text{ is odd,} \end{cases}$$

where $c_{hj} \in \mathcal{C}_{\eta(h,j)}$, $1 \leq h \leq o(v_1)$ and $1 \leq j \leq o(v_2)$. From this, the desired result follows immediately.

(ii) Here we have $n \equiv 0 \pmod 4$. In this case also, we will first construct a self-orthogonal code $\mathcal{C}^{\maltese}$ by choosing a suitable positive integer k and vectors v_1, v_2 satisfying the properties $(\mathbf{P_1})$, $(\mathbf{P_2})$ and $(\mathbf{P_3})$. Now as $s \in \mathcal{C}_1$ and $t \in \mathcal{C}_2$, by Theorem 118(ii), we see that $\langle s, s\rangle \equiv 0 \pmod{2^m}$, $\langle s, t\rangle \equiv 2^{m-1} \pmod{2^m}$ and $\langle t, t\rangle \equiv 0 \pmod{2^m}$. Therefore the property $(\mathbf{P_2})$ becomes

$$\langle v_1, v_1\rangle \equiv 0 \pmod{2^m}, \quad \langle v_1, v_2\rangle \equiv 2^{m-1} \pmod{2^m}, \quad \langle v_2, v_2\rangle \equiv 0 \pmod{2^m}. \tag{6.9}$$

Here we observe that the vectors

$$v_1 = \begin{cases} (2^{\frac{k}{2}-1}, 2^{\frac{k}{2}-1}, 2^{\frac{k}{2}-1}, 2^{\frac{k}{2}-1}) & \text{if } m \text{ is even;} \\ (2^{\frac{k-1}{2}}, 0, 2^{\frac{k-1}{2}}, 0) & \text{if } m \text{ is odd,} \end{cases}$$

$$v_2 = \begin{cases} (2^{\frac{k}{2}}, 0, 0, 0) & \text{if } m \text{ is even;} \\ (2^{\frac{k-1}{2}}, 2^{\frac{k-1}{2}}, 0, 0) & \text{if } m \text{ is odd} \end{cases}$$

satisfy the properties $(\mathbf{P_1})$ and $(\mathbf{P_3})$, and all the congruences in (6.9). Therefore in view of Proposition 122, we see that the code $\mathcal{C}^{\maltese}$ is a self-orthogonal code of length $n+4$ over $\mathbb{Z}_{2^m}$. Also since $o(v_1)o(v_2) = 2^{m+1}$, we have $|\mathcal{C}^{\maltese}| = 2^{m+1}2^{\frac{mn}{2}-1} = (2^m)^{\frac{n+2}{2}}$, by applying Proposition 122 again. Now we need to show that the code $\mathcal{C}^{\clubsuit} = \left\langle \mathcal{C}^{\maltese} \cup \{w_1, w_2\}\right\rangle$ is a self-dual code of length $n+4$ over $\mathbb{Z}_{2^m}$. To prove this, we first observe that $w_p \cdot w_q = w_p \cdot c^{\maltese} = 0$ for $1 \leq p, q \leq 2$ and for each $c^{\maltese} \in \mathcal{C}^{\maltese}$. This implies that the code $\mathcal{C}^{\clubsuit}$ is a self-orthogonal code of length $n+4$ over $\mathbb{Z}_{2^m}$. Also it is easy to observe that $|\mathcal{C}^{\clubsuit}| = o(w_1)o(w_2)|\mathcal{C}^{\maltese}|$, where $o(w_p)$ is the additive order of w_p for each p. Now as $o(w_1)o(w_2) = 2^m$, we obtain $|\mathcal{C}^{\clubsuit}| = 2^m|\mathcal{C}^{\maltese}| = (2^m)^{\frac{n+4}{2}}$. This implies that $\mathcal{C}^{\clubsuit}$ is a self-dual code of length $n+4$ over $\mathbb{Z}_{2^m}$. Further, working in a similar way as in part (i) and using the fact that $wt_E(w_1) \equiv wt_E(w_2) \equiv 0 \pmod{2^{m+1}}$, we see that the code $\mathcal{C}^{\clubsuit}$ is a Type II code when $n \equiv 4 \pmod 8$ and $\mathcal{C}^{\clubsuit}$ is a Type I code when $n \equiv 0 \pmod 8$.

Next to compute the complete weight enumerator of $\mathcal{C}^{\clubsuit}$, by Proposition 122, we note that any element $c^{\clubsuit} \in \mathcal{C}^{\clubsuit}$ is of the form $c^{\clubsuit} = (hv_1+jv_2, c_{hj})+k_1w_1 + k_2w_2$, where $c_{hj} \in \mathcal{C}_{\eta(h,j)}$, $1 \le h \le o(v_1)$, $1 \le j \le o(v_2)$ and $1 \le k_p \le o(w_p)$ for $1 \le p \le 2$. Now working in a similar way as in part (i), we obtain the desired result. □

In the following corollary, Theorems 1 and 2 of Brualdi and Pless [2] are derived as a consequence of Theorem 124.

Corollary 125. *Let $\mathcal{C}$ be a Type I code of even length n over $\mathbb{Z}_2$, and let $\mathcal{C}_0 = \{c \in \mathcal{C} : wt_E(c) \equiv 0 \pmod 4\}$.*

(i) *Let $n \equiv 2 \pmod 4$. Let $v_1 = (1,0)$ and $v_2 = (1,1)$. Then the code $\mathcal{C}^{\maltese}$ is a self-dual code of length $n+2$ over $\mathbb{Z}_2$. Furthermore, $\mathcal{C}^{\maltese}$ is a Type II code when $n \equiv 6 \pmod 8$ and $\mathcal{C}^{\maltese}$ is a Type I code when $n \equiv 2 \pmod 8$. The complete weight enumerator $cwe_{\mathcal{C}^{\maltese}}(X_0, X_1)$ of $\mathcal{C}^{\maltese}$ is given by*

$$X_0^2 cwe_{\mathcal{C}_0}(X_0,X_1) + X_1^2 cwe_{\mathcal{C}_2}(X_0,X_1) + X_0X_1\{cwe_{\mathcal{C}_1}(X_0,X_1) + cwe_{\mathcal{C}_3}(X_0,X_1)\}.$$

(ii) *Let $n \equiv 0 \pmod 4$. Let $v_1, v_2 \in \mathbb{Z}_2^4$ and $w_1 \in \mathbb{Z}_2^{n+4}$ be chosen as $v_1 = (1,0,1,0)$, $v_2 = (1,1,0,0)$ and $w_1 = (1,1,1,1,0,\cdots,0)$. Then the code $\mathcal{C}^{\clubsuit} = \langle \mathcal{C}^{\maltese} \cup \{w_1\}\rangle$ is a self-dual code of length $n+4$ over $\mathbb{Z}_2$. Furthermore, $\mathcal{C}^{\clubsuit}$ is a Type II code when $n \equiv 4 \pmod 8$ and $\mathcal{C}^{\clubsuit}$ is a Type I code when $n \equiv 0 \pmod 8$. The complete weight enumerator $cwe_{\mathcal{C}^{\clubsuit}}(X_0, X_1)$ of $\mathcal{C}^{\clubsuit}$ is given by*

$$(X_0^4 + X_1^4)cwe_{\mathcal{C}_0}(X_0,X_1) + 2X_0^2X_1^2\{cwe_{\mathcal{C}_2}(X_0,X_1) + cwe_{\mathcal{C}_1}(X_0,X_1) + cwe_{\mathcal{C}_3}(X_0,X_1)\}.$$

In the following corollary, Type I and Type II codes (of higher lengths) over $\mathbb{Z}_4$ are constructed from Type I codes of even length n over $\mathbb{Z}_4$, and their complete weight enumerators are also determined.

Corollary 126. *Let $\mathcal{C}$ be a Type I code of even length n over $\mathbb{Z}_4$, and let $\mathcal{C}_0 = \{c \in \mathcal{C} : wt_E(c) \equiv 0 \pmod 8\}$.*

(i) *If $n \equiv 2 \pmod 4$, let $v_1 = (1,1)$ and $v_2 = (2,0)$. Then the code $\mathcal{C}^{\maltese}$ is a self-dual code of length $n+2$ over $\mathbb{Z}_4$. Furthermore, $\mathcal{C}^{\maltese}$ is a Type II code when $n \equiv 6 \pmod 8$ and $\mathcal{C}^{\maltese}$ is a Type I code when $n \equiv 2 \pmod 8$.*

The complete weight enumerator $cwe_{C^{\maltese}}(X_0, X_1, X_2, X_3)$ *of* $C^{\maltese}$ *is given by*

$$(X_0^2 + X_2^2)cwe_{C_0}(X_0, X_1, X_2, X_3) + 2X_0X_2cwe_{C_2}(X_0, X_1, X_2, X_3)$$
$$+(X_1^2 + X_3^2)cwe_{C_1}(X_0, X_1, X_2, X_3) + 2X_1X_3cwe_{C_3}(X_0, X_1, X_2, X_3).$$

(ii) *If* $n \equiv 0 \pmod 4$, *let* $v_1, v_2 \in \mathbb{Z}_4^4$ *and* $w_1, w_2 \in \mathbb{Z}_4^{n+4}$ *be chosen as* $v_1 = (1,1,1,1)$, $v_2 = (2,0,0,0)$, $w_1 = (2,2,0,0,0,\cdots,0)$ *and* $w_2 = (0,2,2,0,0,\cdots,0)$. *Then the code* $C^{\clubsuit} = \langle C^{\maltese} \cup \{w_1, w_2\}\rangle$ *is a self-dual code of length* $n+4$ *over* $\mathbb{Z}_4$. *Furthermore,* $C^{\clubsuit}$ *is a Type II code when* $n \equiv 4 \pmod 8$ *and* $C^{\clubsuit}$ *is a Type I code when* $n \equiv 0 \pmod 8$. *The complete weight enumerator* $cwe_{C^{\clubsuit}}(X_0, X_1, X_2, X_3)$ *of* $C^{\clubsuit}$ *is given by*

$$(X_0^4+X_2^4+6X_0^2X_2^2)cwe_{C_0}(X_0, X_1, X_2, X_3)+4(X_0X_2^3+X_0^3X_2)cwe_{C_2}(X_0, X_1, X_2, X_3)$$
$$+(X_1^4+X_3^4+6X_1^2X_3^2)cwe_{C_1}(X_0, X_1, X_2, X_3)+4(X_1X_3^3+X_1^3X_3)cwe_{C_3}(X_0, X_1, X_2, X_3).$$

Next by applying MacWilliams identity for complete weight enumerator of a linear code over $\mathbb{Z}_{2^m}$, we will determine Jacobi forms from complete weight enumerators of Type II codes over $\mathbb{Z}_{2^m}$ that are constructed in Theorem 124. For this purpose, let us first recall the following theta series defined by Choie and Kim [4].

Definition 127. [4] For each $\mu \in \mathbb{Z}_{2^m}$, the theta series $\theta_{2^{m-1},\mu}$ is a function from $\mathcal{H} \times \mathbb{C}$ into $\mathbb{C}$, defined as

$$\theta_{2^{m-1},\mu}(\tau, z) = \sum_{\substack{r \in \mathbb{Z} \\ r \equiv \mu \pmod{2^m}}} q^{\frac{r^2}{2^{m+1}}} \xi^r, \quad \text{where } q = e^{2\pi i \tau} \text{ and } \xi = e^{2\pi i z}.$$

In the following theorem, some Jacobi forms on the modular group $\Gamma(1)$ are obtained.

Theorem 128. *Let* C *be a Type I code of even length* n *over* $\mathbb{Z}_{2^m}$, *and let* $C_0 = \{c \in C : wt_E(c) \equiv 0 \pmod{2^{m+1}}\}$. *Then the following hold.*

(i) *When* $n \equiv 6 \pmod 8$, *let* $C^{\maltese}$ *be the Type II code of length* $n+2$ *over* $\mathbb{Z}_{2^m}$ *as constructed in Theorem 124 (i). Let* $cwe_{C^{\maltese}}(X_\mu : \mu \in \mathbb{Z}_{2^m})$ *be as obtained in Theorem 124 (i). Then* $cwe_{C^{\maltese}}(\theta_{2^{m-1},\mu}(\tau, z) : \mu \in \mathbb{Z}_{2^m})$ *is a Jacobi form of weight* $\frac{n+2}{2}$ *and index* $(n+2)2^{m-1}$ *on* Γ.

(ii) *When* $n \equiv 4 \pmod 8$, *let* $C^{\clubsuit}$ *be the Type II code of length* $n+4$ *over* $\mathbb{Z}_{2^m}$ *as constructed in Theorem 124 (ii). Let* $cwe_{C^{\clubsuit}}(X_\mu : \mu \in \mathbb{Z}_{2^m})$ *be as obtained in Theorem 124 (ii). Then* $cwe_{C^{\clubsuit}}(\theta_{2^{m-1},\mu}(\tau, z) : \mu \in \mathbb{Z}_{2^m})$ *is a Jacobi form of weight* $\frac{n+4}{2}$ *and index* $(n+4)2^{m-1}$ *on* Γ.

Proof. (i) Since $C^{\maltese}$ is a Type II code of length $n+2$, by applying MacWilliams identity, one can easily see that the complete weight enumerator of the code $C^{\maltese}$ belongs to the space of polynomials that are invariant under the group generated by $\frac{1}{2^{m/2}}M_{2^m}$ and N_{2^m}, where M_{2^m} is the transformation matrix in the MacWilliams identity and N_{2^m} is a $2^m \times 2^m$ matrix such that $N_{2^m}(j,j) = e^{\frac{\pi i}{2^m}j^2}$, $N_{2^m}(j,k) = 0$ for $0 \leq j,k \leq 2^m - 1, j \neq k$. Now by Theorem 4.6 of Choie and Kim [4], we see that $cwe_{C^{\maltese}}(\theta_{2^{m-1},\mu}(\tau,z) : \mu \in \mathbb{Z}_{2^m})$ is a Jacobi form of weight $\frac{n+2}{2}$ and index $(n+2)2^{m-1}$ on Γ.
(ii) Working in a similar manner as in part (i), the desired result follows. □

In the following theorem, Type I and Type II codes (of higher lengths) over $\mathbb{Z}_{2^m}$ are constructed from Type I codes of odd length n over $\mathbb{Z}_{2^m}$. Here by Lemma 109, m must be an even integer.

Theorem 129. *Let C be a Type I code of odd length n over $\mathbb{Z}_{2^m}$ (m is an even integer), and let $C_0 = \{c \in C : wt_E(c) \equiv 0 \pmod{2^{m+1}}\}$. The following hold.*

(i) *Let $n \equiv 3 \pmod 4$. Let $v_1 = (2^{\frac{k}{2}-1})$ and $v_2 = (2^{\frac{k}{2}})$. Then the code $C^{\maltese}$ is a self-dual code of length $n+1$ over $\mathbb{Z}_{2^m}$. Furthermore, the code $C^{\maltese}$ is a Type II code when $n \equiv 7 \pmod 8$ and $C^{\maltese}$ is a Type I code when $n \equiv 3 \pmod 8$. The complete weight enumerator $cwe_{C^{\maltese}}(X_\mu : \mu \in \mathbb{Z}_{2^m})$ of $C^{\maltese}$ is given by*

$$cwe_{C^{\maltese}}(X_\mu : \mu \in \mathbb{Z}_{2^m}) = \sum_{h=1}^{2^{\frac{k}{2}+1}} X_{h2^{\frac{k}{2}-1}} cwe_{C_{\eta(h,0)}}(X_\mu : \mu \in \mathbb{Z}_{2^m}).$$

(ii) *Let $n \equiv 1 \pmod 4$. Let $v_1, v_2 \in \mathbb{Z}_{2^m}^3$ and $w_1 \in \mathbb{Z}_{2^m}^{n+3}$ be chosen as*

$$v_1 = (2^{\frac{k}{2}-1}, 2^{\frac{k}{2}-1}, 2^{\frac{k}{2}-1}),\ v_2 = (2^{\frac{k}{2}}, 0, 0),\ w_1 = (2^{\frac{k}{2}}, 2^{\frac{k}{2}}, 0, 0, \cdots, 0).$$

Then the code $C^{\clubsuit} = \langle C^{\maltese} \cup \{w_1\} \rangle$ is a self-dual code of length $n+3$ over $\mathbb{Z}_{2^m}$. Furthermore, $C^{\clubsuit}$ is a Type II code when $n \equiv 5 \pmod 8$ and $C^{\clubsuit}$ is a Type I code when $n \equiv 1 \pmod 8$. The complete weight enumerator $cwe_{C^{\clubsuit}}(X_\mu : \mu \in \mathbb{Z}_{2^m})$ of $C^{\clubsuit}$ is given by

$$\sum X_{(i+2j+2k_1)2^{\frac{k}{2}-1}} X_{(h+2k_1)2^{\frac{k}{2}-1}} X_{h2^{\frac{k}{2}-1}} cwe_{C_{\eta(h,j)}}(X_\mu : \mu \in \mathbb{Z}_{2^m}),$$

where the summation $\sum$ runs over all integral 3-tuples (h, j, k_1) satisfying $1 \leq j, k_1 \leq 2^{\frac{k}{2}}$ and $1 \leq h \leq 2^{\frac{k}{2}+1}$.

Proof. As n is odd, by Lemma 117(ii), the glue group $\mathcal{C}_0^\perp/\mathcal{C}_0$ of $\mathcal{C}_0$ is a cyclic group of order 4. So by (6.6), we have $\eta(h,j) = [h+2j]_4$ for each h and j.

(i) Here we first observe that $(v_2, \mathcal{C}_2) = 2(v_1, \mathcal{C}_1)$, so the code $\mathcal{C}^{\maltese}$ is a $\mathbb{Z}_{2^m}$-submodule of $\mathbb{Z}_{2^m}^{n+1}$ generated by $(v_1, \mathcal{C}_1) = (2^{\frac{k}{2}-1}, \mathcal{C}_1)$. Next we observe that $\mathcal{C}^{\maltese}$ is a self-orthogonal code of length $n+1$ over $\mathbb{Z}_{2^m}$ and has cardinality $|\mathcal{C}^{\maltese}| = 2^{\frac{k}{2}+1} \times |\mathcal{C}_0| = 2^{\frac{k}{2}}|\mathcal{C}| = 2^{\frac{k}{2}} 2^{\frac{mn}{2}} = (2^m)^{\frac{n+1}{2}}$, which implies that the code $\mathcal{C}^{\maltese}$ is self-dual. Further, one can easily show that the code $\mathcal{C}^{\maltese}$ is Type II if and only if $n \equiv 7 \pmod 8$.

(ii) Working in a similar manner as in Theorem 124 and by applying Proposition 122, part (ii) follows immediately. □

In the following corollary, Theorem 3.15 of Dougherty et al. [10] is deduced from the above theorem.

Corollary 130. *Let $\mathcal{C}$ be a Type I code of odd length n over $\mathbb{Z}_4$, and let $\mathcal{C}_0 = \{c \in \mathcal{C} : wt_E(c) \equiv 0 \pmod 8\}$. The following hold.*

(i) *Let $n \equiv 3 \pmod 4$. Let $v_1 = (1)$ and $v_2 = (2)$. The code $\mathcal{C}^{\maltese} = \langle (1, \mathcal{C}_1) \rangle$ is a self-dual code of length $n+1$ over $\mathbb{Z}_4$. Furthermore, $\mathcal{C}^{\maltese}$ is a Type II code when $n \equiv 7 \pmod 8$ and $\mathcal{C}^{\maltese}$ is a Type I code when $n \equiv 3 \pmod 8$. The complete weight enumerator $cwe_{\mathcal{C}^{\maltese}}(X_0, X_1, X_2, X_3)$ of $\mathcal{C}^{\maltese}$ is given by*

$$X_0 cwe_{\mathcal{C}_0}(X_0,X_1,X_2,X_3) + X_1 cwe_{\mathcal{C}_1}(X_0,X_1,X_2,X_3) \\ + X_2 cwe_{\mathcal{C}_2}(X_0,X_1,X_2,X_3) + X_3 cwe_{\mathcal{C}_3}(X_0,X_1,X_2,X_3).$$

(ii) *Let $n \equiv 1 \pmod 4$. Let $v_1, v_2 \in \mathbb{Z}_4^3$ and $w_1 \in \mathbb{Z}_4^{n+3}$ be chosen as $v_1 = (1,1,1)$, $v_2 = (2,0,0)$ and $w_1 = (2,2,0,0,\cdots,0)$. Then the code $\mathcal{C}^{\clubsuit} = \langle \mathcal{C}^{\maltese} \cup \{w_1\} \rangle$ is a self-dual code of length $n+3$ over $\mathbb{Z}_4$. Furthermore, $\mathcal{C}^{\clubsuit}$ is a Type II code when $n \equiv 5 \pmod 8$ and $\mathcal{C}^{\clubsuit}$ is a Type I code when $n \equiv 1 \pmod 8$. The complete weight enumerator $cwe_{\mathcal{C}^{\clubsuit}}(X_0, X_1, X_2, X_3)$ of $\mathcal{C}^{\clubsuit}$ is given by*

$$(3X_0X_2^2 + X_0^3) cwe_{\mathcal{C}_0}(X_0,X_1,X_2,X_3) + (3X_0^2X_2 + X_2^3) cwe_{\mathcal{C}_2}(X_0,X_1,X_2,X_3) \\ +(3X_1X_3^2 + X_1^3) cwe_{\mathcal{C}_1}(X_0,X_1,X_2,X_3) + (3X_1^2X_3 + X_3^3) cwe_{\mathcal{C}_3}(X_0,X_1,X_2,X_3).$$

In the following theorem, Jacobi forms are obtained from complete weight enumerators of Type II codes over $\mathbb{Z}_{2^m}$, which are constructed in Theorem 129.

Theorem 131. *Let $\mathcal{C}$ be a Type I code of odd length n over $\mathbb{Z}_{2^m}$, and let $\mathcal{C}_0 = \{c \in \mathcal{C} : wt_E(c) \equiv 0 \pmod{2^{m+1}}\}$. Then the following hold.*

(i) *When $n \equiv 7 \pmod 8$, let $\mathcal{C}^{\maltese}$ be the Type II code of length $n+1$ over $\mathbb{Z}_{2^m}$ as constructed in Theorem 129 (i). Let $cwe_{\mathcal{C}^{\maltese}}(X_\mu : \mu \in \mathbb{Z}_{2^m})$ be as obtained in Theorem 129 (i). Then $cwe_{\mathcal{C}^{\maltese}}(\theta_{2^{m-1},\mu}(\tau,z) : \mu \in \mathbb{Z}_{2^m})$ is a Jacobi form of weight $\frac{n+1}{2}$ and index $(n+1)2^{m-1}$ on Γ.*

(ii) *When $n \equiv 5 \pmod 8$, let $\mathcal{C}^{\clubsuit}$ be the Type II code of length $n+3$ over $\mathbb{Z}_{2^m}$ as constructed in Theorem 129 (ii). Let $cwe_{\mathcal{C}^{\clubsuit}}(X_\mu : \mu \in \mathbb{Z}_{2^m})$ be as obtained in Theorem 129 (ii). Then $cwe_{\mathcal{C}^{\clubsuit}}(\theta_{2^{m-1},\mu}(\tau,z) : \mu \in \mathbb{Z}_{2^m})$ is a Jacobi form of weight $\frac{n+3}{2}$ and index $(n+3)2^{m-1}$ on Γ.*

Proof. Its proof is similar to that of Theorem 128. □

6.4 Construction of Self-dual Codes from Generalized Shadows of Self-dual Codes

In this section, self-dual codes (of higher lengths) over $\mathbb{Z}_{2^m}$ are constructed from the generalized shadow $S_g(s)$ of a self-dual code $\mathcal{C}$ of length n over $\mathbb{Z}_{2^m}$, where $s \in \mathbb{Z}_{2^m}^n \setminus \mathcal{C}$ is such that $Im(\psi_s) = \{0, 2^{m-1}\}$ and $\langle s, s\rangle \equiv 0$ or $2^{m-1} \pmod{2^m}$. Recall that with this choice of s, we have $[\mathcal{C} : \mathcal{C}_0] = 2$.

In the following theorem, we distinguish the following two cases: (i) $\langle s, s\rangle \equiv 2^{m-1} \pmod{2^m}$ and (ii) $\langle s, s\rangle \equiv 0 \pmod{2^m}$.

Theorem 132. *Let $\mathcal{C}$ be a self-dual code of length n over $\mathbb{Z}_{2^m}$. Let $S_g(s)$ be the generalized shadow of $\mathcal{C}$ with respect to a vector $s \in \mathbb{Z}_{2^m}^n \setminus \mathcal{C}$.*

(i) Let $\langle s, s\rangle \equiv 2^{m-1} \pmod{2^m}$. Let $v_1, v_2 \in \mathbb{Z}_{2^m}^2$ be as chosen in Theorem 124(i). Then the code $\mathcal{C}^{\maltese}$ is a self-dual code of length $n+2$ over $\mathbb{Z}_{2^m}$, and its complete weight enumerator is given by

$$cwe_{\mathcal{C}^{\maltese}}(X_\mu : \mu \in \mathbb{Z}_{2^m}) = \begin{cases} \displaystyle\sum_{h=1}^{2^{\frac{k}{2}+1}} \sum_{j=1}^{2^{\frac{k}{2}}} X_{(h+2j)2^{\frac{k}{2}-1}} X_{h2^{\frac{k}{2}-1}} cwe_{\mathcal{C}_{\eta(h,j)}}(X_\mu : \mu \in \mathbb{Z}_{2^m}) & \text{if } m \text{ is even;} \\ \displaystyle\sum_{h=1}^{2^{\frac{k+1}{2}}} \sum_{j=1}^{2^{\frac{k+1}{2}}} X_{(h+j)2^{\frac{k-1}{2}}} X_{j2^{\frac{k-1}{2}}} cwe_{\mathcal{C}_{\eta(h,j)}}(X_\mu : \mu \in \mathbb{Z}_{2^m}) & \text{if } m \text{ is odd.} \end{cases}$$

(ii) Let $\langle s, s\rangle \equiv 0 \pmod{2^m}$. Let $v_1, v_2 \in \mathbb{Z}_{2^m}^4$ and $w_1, w_2 \in \mathbb{Z}_{2^m}^{n+4}$ be as chosen in Theorem 124 (ii). Then the code $\mathcal{C}^{\clubsuit} = \langle \mathcal{C}^{\maltese} \cup \{w_1, w_2\}\rangle$ is a

self-dual code of length $n+4$ over $\mathbb{Z}_{2^m}$. Moreover, the complete weight enumerator of $\mathcal{C}^{\clubsuit}$ is given by

$$cwe_{\mathcal{C}^{\clubsuit}}(X_\mu : \mu \in \mathbb{Z}_{2^m})$$

$$= \begin{cases} \sum_1 X_{(h+2j+2k_1)2^{\frac{k}{2}-1}} X_{(h+2k_1+2k_2)2^{\frac{k}{2}-1}} \\ X_{(h+2k_2)2^{\frac{k}{2}-1}} X_{h2^{\frac{k}{2}-1}} cwe_{\mathcal{C}_{\eta(h,j)}}(X_\mu : \mu \in \mathbb{Z}_{2^m}) & \text{if } m \text{ is even;} \\ \sum_2 X_{(h+j+k_1+2k_2)2^{\frac{k-1}{2}}} X_{(j+k_1)2^{\frac{k-1}{2}}} X_{(h+k_1)2^{\frac{k-1}{2}}} \\ X_{k_1 2^{\frac{k-1}{2}}} cwe_{\mathcal{C}_{\eta(h,j)}}(X_\mu : \mu \in \mathbb{Z}_{2^m}) & \text{if } m \text{ is odd,} \end{cases}$$

where the summation $\sum_1$ runs over all integral 4-tuples (h, j, k_1, k_2) satisfying $1 \le j, k_1, k_2 \le 2^{\frac{k}{2}}$ and $1 \le h \le 2^{\frac{k}{2}+1}$, whereas the summation $\sum_2$ runs over all integral 4-tuples (h, j, k_1, k_2) satisfying $1 \le h, j, k_1 \le 2^{\frac{k+1}{2}}$ and $1 \le k_2 \le 2^{\frac{k-1}{2}}$.

Proof. Working in a similar way as in Theorem 124, the desired result follows. □

In the following corollary, the main theorem of Tsai [14] is deduced from the above theorem.

Corollary 133. *Let $\mathcal{C}$ be a self-dual code of length n over $\mathbb{Z}_2$. Let $\mathcal{S}_g(s)$ be the generalized shadow of $\mathcal{C}$ with respect to a vector $s \in \mathbb{Z}_2^n \setminus \mathcal{C}$.*

(i) If $\langle s, s\rangle \equiv 1 \pmod 2$, let $v_1 = (1, 0)$ and $v_2 = (1, 1)$. Then the code $\mathcal{C}^{\maltese}$ is a self-dual code of length $n+2$ over $\mathbb{Z}_2$ and its complete weight enumerator is given by

$$X_0^2 cwe_{\mathcal{C}_0}(X_0, X_1) + X_1^2 cwe_{\mathcal{C}_2}(X_0, X_1)$$
$$+ X_0 X_1 \{cwe_{\mathcal{C}_1}(X_0, X_1) + cwe_{\mathcal{C}_3}(X_0, X_1)\}.$$

(ii) If $\langle s, s\rangle \equiv 0 \pmod 2$, let $v_1, v_2 \in \mathbb{Z}_2^4$ and $w_1 \in \mathbb{Z}_2^{n+4}$ be chosen as $v_1 = (1, 0, 1, 0)$, $v_2 = (1, 1, 0, 0)$ and $w_1 = (1, 1, 1, 1, 0, \cdots, 0)$. Then the code $\mathcal{C}^{\clubsuit} = \langle \mathcal{C}^{\maltese} \cup \{w_1\}\rangle$ is a self-dual code of length $n+4$ over $\mathbb{Z}_2$ and its complete weight enumerator is given by

$$(X_0^4 + X_1^4) cwe_{\mathcal{C}_0}(X_0, X_1) + 2X_0^2 X_1^2 \{cwe_{\mathcal{C}_2}(X_0, X_1)$$
$$+ cwe_{\mathcal{C}_1}(X_0, X_1) + cwe_{\mathcal{C}_3}(X_0, X_1)\}.$$

In the following corollary, self-dual codes of higher lengths over $\mathbb{Z}_4$ are determined using the generalized shadow $\mathcal{S}_g(s)$ of a self-dual code $\mathcal{C}$ over $\mathbb{Z}_4$ with respect to a vector $s \in \mathbb{Z}_4^n \setminus \mathcal{C}$ satisfying either $\langle s, s\rangle \equiv 2 \pmod 4$ or $\langle s, s\rangle \equiv 0 \pmod 4$.

Corollary 134. *Let $\mathcal{C}$ be a self-dual code of length n over $\mathbb{Z}_4$. Let $\mathcal{S}_g(s)$ be the generalized shadow of $\mathcal{C}$ with respect to a vector $s \in \mathbb{Z}_4^n \setminus \mathcal{C}$.*

(i) If $\langle s, s\rangle \equiv 2 \pmod 4$, let $v_1 = (1,1)$ and $v_2 = (2,0)$. Then the code $\mathcal{C}^{✠}$ is a self-dual code of length $n+2$ over $\mathbb{Z}_4$ and its complete weight enumerator is given by

$$(X_0^2+X_2^2)cwe_{\mathcal{C}_0}(X_0,X_1,X_2,X_3)+2X_0X_2cwe_{\mathcal{C}_2}(X_0,X_1,X_2,X_3) \\ +(X_1^2+X_3^2)cwe_{\mathcal{C}_1}(X_0,X_1,X_2,X_3)+2X_1X_3cwe_{\mathcal{C}_3}(X_0,X_1,X_2,X_3).$$

(ii) If $\langle s, s\rangle \equiv 0 \pmod 4$, let $v_1, v_2 \in \mathbb{Z}_4^4$ and $w_1, w_2 \in \mathbb{Z}_4^{n+4}$ be chosen as $v_1 = (1,1,1,1)$, $v_2 = (2,0,0,0)$, $w_1 = (2,2,0,0,0,\cdots,0)$ and $w_2 = (0,2,2,0,0,\cdots,0)$. Then the code $\mathcal{C}^{♣} = \langle \mathcal{C}^{✠} \cup \{w_1, w_2\}\rangle$ is a self-dual code of length $n+4$ over $\mathbb{Z}_4$ and its complete weight enumerator is given by

$$\begin{aligned}&(X_0^4+X_2^4+6X_0^2X_2^2)cwe_{\mathcal{C}_0}(X_0,X_1,X_2,X_3)\\ &+4(X_0X_2^3+X_0^3X_2)cwe_{\mathcal{C}_2}(X_0,X_1,X_2,X_3)\\ &+(X_1^4+X_3^4+6X_1^2X_3^2)cwe_{\mathcal{C}_1}(X_0,X_1,X_2,X_3)\\ &+4(X_1X_3^3+X_1^3X_3)cwe_{\mathcal{C}_3}(X_0,X_1,X_2,X_3).\end{aligned}$$

6.5 Some Examples

To illustrate the above results, Type II codes of length 8 over $\mathbb{Z}_8$ and $\mathbb{Z}_{16}$ are constructed from Type I codes of lengths 4 and 7 over $\mathbb{Z}_8$ and $\mathbb{Z}_{16}$ respectively, using the shadow code. In addition to this, a self-dual code of length 8 over $\mathbb{Z}_8$ is constructed from a self-dual code of length 6 over $\mathbb{Z}_8$ using the generalized shadow of the code with respect to a vector $s = (2,2,2,0,0,0) \in \mathbb{Z}_8^6$.

a) Let $\mathcal{C}$ be the Type I code of length 4 over $\mathbb{Z}_8$ generated by $\{(2,0,2,0), (0,2,0,2), (0,0,4,0), (0,0,0,4)\}$. If we choose $v_1 = (2,0,2,0)$, $v_2 = (2,2,0,0)$, $w_1 = (2,2,2,2,0,\cdots,0)$ and $w_2 = (4,0,\cdots,0)$, then by Theorem 124(b), the code $\mathcal{C}^{♣} = \langle \mathcal{C}^{✠} \cup \{w_1, w_2\}\rangle$

is a Type II code of length 8 over $\mathbb{Z}_8$, and its complete weight enumerator is given by

$$\begin{aligned} &X_0^8 + X_2^8 + 56\,X_4X_0^3X_6^4 + 56\,X_4^3X_0X_6^4 + 84\,X_4^2X_0^2X_6^4 + 56\,X_2X_6^3X_0^4 \\ &+ 56\,X_2X_6^3X_4^4 + 8\,X_2X_6^7 + 56\,X_2^4X_4X_0^3 + 56\,X_4^3X_0^5 + X_6^8 + X_4^8 + 8\,X_4^7X_0 \\ &+ 56\,X_2^4X_4^3X_0 + 84\,X_2^4X_4^2X_0^2 + 56\,X_2^3X_6X_0^4 + 56\,X_2^3X_6X_4^4 + 84\,X_2^2X_6^2X_0^4 \\ &+ 84\,X_2^2X_6^2X_4^4 + 28\,X_4^6X_0^2 + 504\,X_2^2X_6^2X_4^2X_0^2 + 224\,X_2^3X_6X_4X_0^3 \\ &+ 224\,X_2^3X_6X_4^3X_0 + 336\,X_2^3X_6X_4^2X_0^2 + 336\,X_2^2X_6^2X_4X_0^3 + 56\,X_4^5X_0^3 \\ &+ 14\,X_4^4X_6^4 + 28\,X_4^2X_0^6 + 336\,X_2^2X_6^2X_4^3X_0 + 224\,X_2X_6^3X_4X_0^3 \\ &+ 224\,X_2X_6^3X_4^3X_0 + 336\,X_2X_6^3X_4^2X_0^2 + 8\,X_2^7X_6 + 28\,X_2^2X_6^6 \\ &+ 8\,X_4X_0^7 + 56\,X_2^3X_6^5 + 28\,X_2^6X_6^2 + 56\,X_2^5X_6^3 + 70\,X_2^4X_6^4 + 14\,X_2^4X_4^4 \\ &+ 14\,X_2^4X_0^4 + 70\,X_4^4X_0^4 + 14\,X_6^4X_0^4. \end{aligned}$$

b) Let $\mathcal{C}$ be the self-dual code of length 6 over $\mathbb{Z}_8$ generated by {(2, 0, 0, 2, 0, 0), (0, 2, 0, 0, 2, 0), (0, 0, 2, 0, 0, 2), (0, 0, 0, 4, 0, 0), (0, 0, 0, 0, 4, 0), (0, 0, 0, 0, 0, 4)}. Let

$$s = (2, 2, 2, 0, 0, 0) \in \mathbb{Z}_8^6 \setminus \mathcal{C}.$$

Note that s satisfies $\langle s, s\rangle \equiv 4 \pmod 8$ and $Im(\psi_s) = \{0, 4\}$. If we choose $v_1 = (2, 0)$ and $v_2 = (2, 2)$, then by Theorem 132(a), the code $\mathcal{C}^{\maltese}$ is a self-dual code of length 8 over $\mathbb{Z}_8$, and its complete weight enumerator is given by

$$\begin{aligned} &84X_0^2X_4^2X_2^4 + 56X_4^4X_2^3X_6 + 56X_0X_4^3X_2^4 + 56X_4^4X_2X_6^3 + 84X_0^2X_4^2X_6^4 \\ &+ 56X_6^3X_2^5 + 28X_6^2X_2^6 + 8X_6X_2^7 + 56X_6^5X_2^3 + 28X_6^6X_2^2 + 8X_6^7X_2 \\ &+ 14X_0^4X_2^4 + 14X_0^4X_6^4 + 70X_6^4X_2^4 + 84X_4^4X_2^2X_6^2 + 56X_0^5X_4^3 \\ &+ 56X_0^3X_4X_2^4 + 504X_0^2X_4^2X_2^2X_6^2 + 336X_0X_4^3X_2^2X_6^2 \\ &+ 336X_0^2X_4^2X_2^3X_6 + 224X_0X_4^3X_2^3X_6 + 336X_0^3X_4X_2^2X_6^2 \\ &+ 224X_0^3X_4X_2^3X_6 + 336X_0^2X_4^2X_2X_6^3 + 224X_0X_4^3X_2X_6^3 \\ &+ 224X_0^3X_4X_2X_6^3 + 14X_4^4X_6^4 + 14X_4^4X_2^4 + 8X_0^7X_4 + 28X_0^6X_4^2 \\ &+ 70X_0^4X_4^4 + 56X_0^3X_4^5 + 28X_0^2X_4^6 + 8X_0X_4^7 \\ &+ 56X_0X_4^3X_6^4 + X_4^8 + X_0^8 + X_2^8 + X_6^8 + 56X_0^4X_2X_6^3 \\ &+ 84X_0^4X_2^2X_6^2 + 56X_0^4X_2^3X_6 + 56X_0^3X_4X_6^4. \end{aligned}$$

c) Let $\mathcal{C}$ be the Type I code of length 7 over $\mathbb{Z}_{16}$ generated by $\{(4,0,0,0,0,0,0), (0,4,0,0,0,0,0), (0,0,4,0,0,0,0), (0,0,0,4,0,0,0), (0,0,0,0,4,0,0), (0,0,0,0,0,4,0), (0,0,0,0,0,0,4)\}$. If we choose $v_1 =$

(2) and $v_2 = (4)$, then by Theorem 129(a), the code $\mathcal{C}^{\maltese}$ is a Type II code of length 8 over $\mathbb{Z}_{16}$, and its complete weight enumerator is given by

$$\begin{aligned}
&(X_0 + X_8)cwe_{\mathcal{C}_0}(X_\mu : \mu \in \mathbb{Z}_{16}) + (X_2 + X_{10})cwe_{\mathcal{C}_1}(X_\mu : \mu \in \mathbb{Z}_{16})\\
&\quad + (X_4 + X_{12})cwe_{\mathcal{C}_2}(X_\mu : \mu \in \mathbb{Z}_{16}) + (X_6 + X_{14})cwe_{\mathcal{C}_3}(X_\mu : \mu \in \mathbb{Z}_{16}),
\end{aligned}$$

where

$$\begin{aligned}
&cwe_{\mathcal{C}_0}(X_\mu : \mu \in \mathbb{Z}_{16}) = 21X_8^2X_0^5 + 7X_4^6X_0 + 210X_8^2X_{12}^2X_0^3 + 210X_4^2X_8^3X_0^2\\
&\quad + 7X_4^6X_8 + 21X_4^2X_8^5 + 630X_4^2X_8X_{12}^2X_0^2 + 420X_4^3X_8X_{12}X_0^2\\
&\quad + 140X_4X_8^3X_{12}^3 + 210X_8^3X_{12}^2X_0^2 + 210X_4^2X_8^3X_{12}^2 + 140X_4^3X_8X_{12}^3\\
&\quad + 42X_4X_{12}X_0^5 + 105X_4^4X_8^2X_0 + 35X_{12}^4X_0^3 + 140X_4^3X_{12}X_0^3\\
&\quad + 42X_4^5X_8X_{12} + 35X_8^3X_0^4 + 140X_4^3X_8^3X_{12} + 42X_4X_{12}^5X_0 + 42X_4X_8^5X_{12}\\
&\quad + 105X_4^4X_8X_{12}^2 + 105X_4^4X_8X_0^2 + 420X_4X_8^3X_{12}X_0^2 + 105X_4^2X_8^4X_0\\
&\quad + 7X_{12}^6X_0 + 140X_4^3X_{12}^3X_0 + 7X_8X_{12}^6 + 420X_4X_8^2X_{12}X_0^3 + 210X_4X_8^4X_{12}X_0\\
&\quad + 420X_4X_8X_{12}^3X_0^2 + 105X_8^4X_{12}^2X_0 + 42X_4X_8X_{12}^5 + 7X_8X_0^6 + 7X_8^6X_0\\
&\quad + 105X_8X_{12}^4X_0^2 + 35X_4^4X_0^3 + 105X_4^2X_8X_{12}^4 + 21X_{12}^2X_0^5 + 35X_8^4X_0^3\\
&\quad + 35X_8^3X_{12}^4 + 105X_4^2X_8X_0^4 + 105X_4^2X_{12}^4X_0 + 35X_4^4X_8^3 + 140X_4X_{12}^3X_0^3\\
&\quad + 210X_4X_8X_{12}X_0^4 + 105X_4^4X_{12}^2X_0 + 210X_4^2X_{12}^2X_0^3 + 420X_4X_8^2X_{12}^3X_0\\
&\quad + 21X_4^2X_0^5 + 630X_4^2X_8^2X_{12}^2X_0 + X_8^7 + X_0^7 + 420X_4^3X_8^2X_{12}X_0 + 105X_8^2X_{12}^4X_0\\
&\quad + 105X_8X_{12}^2X_0^4 + 21X_8^5X_0^2 + 21X_8^5X_{12}^2 + 42X_4^5X_{12}X_0 + 210X_4^2X_8^2X_0^3,
\end{aligned}$$

$$\begin{aligned}
&cwe_{\mathcal{C}_1}(X_\mu : \mu \in \mathbb{Z}_{16}) = 42X_6X_{14}X_2^5 + 140X_6X_{10}^3X_{14}^3 + 42X_6^5X_{10}X_{14} + X_2^7\\
&\quad + 210X_6^2X_{10}^3X_2^2 + 630X_6^2X_{10}X_{14}^2X_2^2 + 420X_6X_{10}^2X_{14}X_2^3 + 210X_{10}^2X_{14}^2X_2^3\\
&\quad + 21X_6^2X_{10}^5 + 420X_6X_{10}^3X_{14}X_2^2 + 420X_6X_{10}^2X_{14}^3X_2 + 105X_{10}X_{14}^4X_2^2\\
&\quad + 42X_6X_{10}X_{14}^5 + 21X_6^2X_2^5 + 420X_6X_{10}X_{14}^3X_2^2 + 7X_{10}X_2^6 + 7X_{10}^6X_2\\
&\quad + 21X_{10}^5X_2^2 + 140X_6^3X_{10}X_{14}^3 + 140X_6^3X_{14}^3X_2 + 105X_{10}^2X_{14}^4X_2 + 35X_{10}^3X_2^4\\
&\quad + 105X_6^4X_{10}X_{14}^2 + 210X_6^2X_{10}^3X_{14}^2 + 105X_6^2X_{10}^4X_2 + 35X_6^4X_2^3\\
&\quad + 210X_{10}^3X_{14}^2X_2^2 + 7X_6^6X_2 + 140X_6^3X_{14}X_2^3 + 35X_6^4X_{10}^3 + 105X_6^2X_{10}X_{14}^4\\
&\quad + 140X_6^3X_{10}^3X_{14} + 105X_6^2X_{14}^4X_2 + 630X_6^2X_{10}^2X_{14}^2X_2 + 420X_6^3X_{10}X_{14}X_2^2\\
&\quad + 42X_6X_{14}^5X_2 + 7X_{14}^6X_2 + 140X_6X_{14}^3X_2^3 + 105X_{10}^4X_{14}^2X_2\\
&\quad + 210X_6X_{10}^4X_{14}X_2 + 7X_{10}X_{14}^6 + 105X_6^4X_{14}^2X_2 + 210X_6X_{10}X_{14}X_2^4\\
&\quad + 105X_6^4X_{10}^2X_2 + 210X_6^2X_{10}^2X_2^3 + X_{10}^7 + 35X_{10}^4X_2^3 + 420X_6^3X_{10}^2X_{14}X_2\\
&\quad + 105X_6^2X_{10}X_2^4 + 21X_{14}^2X_2^5 + 105X_{10}X_{14}^2X_2^4 + 42X_6X_{10}^5X_{14}\\
&\quad + 105X_6^4X_{10}X_2^2 + 210X_6^2X_{14}^2X_2^3 + 7X_6^6X_{10} + 42X_6^5X_{14}X_2 + 35X_{10}^3X_{14}^4\\
&\quad + 21X_{10}^2X_2^5 + 35X_{14}^4X_2^3 + 21X_{10}^5X_{14}^2,
\end{aligned}$$

$$
\begin{aligned}
cwe_{\mathcal{C}_2}(X_\mu : \mu \in \mathbb{Z}_{16}) &= 7X_4^6X_{12} + 105X_4X_{12}^2X_0^4 + 42X_4X_8X_0^5 \\
&\quad + 210X_4X_8X_{12}^4X_0 + 7X_4X_8^6 + 140X_4^3X_8^3X_0 + 420X_4^2X_8X_{12}^3X_0 \\
&\quad + 140X_8X_{12}^3X_0^3 + 140X_4X_8^3X_0^3 + 42X_8^5X_{12}X_0 + 21X_{12}^5X_0^2 \\
&\quad + 420X_4X_8X_{12}^2X_0^3 + 630X_4X_8^2X_{12}^2X_0^2 + 105X_4^2X_8^4X_{12} \\
&\quad + 7X_4X_{12}^6 + 105X_4^4X_8^2X_{12} + 35X_4^4X_{12}^3 + 420X_4X_8^3X_{12}^2X_0 \\
&\quad + 420X_4^2X_8X_{12}X_0^3 + 35X_8^4X_{12}^3 + 7X_4X_0^6 + 42X_8X_{12}X_0^5 \\
&\quad + 105X_4^2X_{12}X_0^4 + 210X_4^2X_8^2X_{12}^3 + 21X_4^5X_{12}^2 + 105X_4X_8^4X_{12}^2 \\
&\quad + 140X_4^3X_8X_0^3 + 630X_4^2X_8^2X_{12}X_0^2 + 42X_4^5X_8X_0 \\
&\quad + 420X_4^3X_8X_{12}^2X_0 + 210X_4^3X_8^2X_{12}^2 + 210X_4^3X_8^2X_0^2 \\
&\quad + 42X_8X_{12}^5X_0 + 35X_4^3X_{12}^4 + 105X_4X_8^4X_0^2 + X_4^7 \\
&\quad + 140X_8^3X_{12}X_0^3 + 21X_8^2X_{12}^5 + 105X_4X_8^2X_{12}^4 + 21X_4^5X_0^2 \\
&\quad + 210X_8^2X_{12}^3X_0^2 + X_{12}^7 + 7X_{12}X_0^6 + 210X_4^3X_{12}^2X_0^2 \\
&\quad + 210X_4^4X_8X_{12}X_0 + 420X_4^2X_8^3X_{12}X_0 + 105X_8^4X_{12}X_0^2 + 21X_4^2X_{12}^5 \\
&\quad + 210X_4^2X_{12}^3X_0^2 + 35X_4^3X_0^4 + 105X_4X_{12}^4X_0^2 + 105X_4^4X_{12}X_0^2 \\
&\quad + 140X_8^3X_{12}^3X_0 + 35X_4^3X_8^4 + 35X_{12}^3X_0^4 + 21X_4^5X_8^2, \text{ and}
\end{aligned}
$$

$$
\begin{aligned}
cwe_{\mathcal{C}_3}(X_\mu : \mu \in \mathbb{Z}_{16}) &= 7X_6^6X_{14} + 210X_6^4X_{10}X_{14}X_2 + 21X_6^5X_2^2 + 210X_{10}^2X_{14}^3X_2^2 \\
&\quad + 21X_{14}^5X_2^2 + 420X_6X_{10}X_{14}^2X_2^3 + 35X_6^3X_2^4 + 35X_6^3X_{14}^4 + 210X_6^3X_{10}^2X_2^2 \\
&\quad + 105X_6^2X_{14}X_2^4 + 7X_{10}^6X_{14} + 140X_6^3X_{10}^3X_2 + 35X_6^3X_{10}^4 + 21X_6^5X_{10}^2 \\
&\quad + 210X_6^2X_{14}^3X_2^2 + 420X_6^3X_{10}X_{14}^2X_2 + 140X_6X_{10}^3X_2^3 + 105X_6X_{10}^4X_{14}^2 \\
&\quad + 21X_{10}^2X_{14}^5 + 7X_6X_2^6 + 7X_6X_{14}^6 + 42X_6^5X_{10}X_2 + 35X_{10}^4X_{14}^3 + 7X_6X_{10}^6 \\
&\quad + 630X_6X_{10}^2X_{14}^2X_2^2 + 420X_6X_{10}^3X_{14}^2X_2 + 42X_6X_{10}^5X_2 + 42X_6X_{10}X_2^5 \\
&\quad + 105X_6X_{14}^4X_2^2 + 105X_6^2X_{10}^4X_{14} + 210X_6^3X_{10}^2X_{14}^2 + 210X_6^2X_{10}^2X_{14}^3 \\
&\quad + 21X_6^2X_{14}^5 + 105X_{10}^2X_{14}X_2^4 + 35X_{14}^3X_2^4 + 420X_6^2X_{10}^3X_{14}X_2 \\
&\quad + 140X_{10}^3X_{14}X_2^3 + 420X_6^2X_{10}X_{14}^3X_2 + 105X_6X_{10}^2X_{14}^4 + 105X_6X_{14}^2X_2^4 + X_6^7 \\
&\quad + 210X_6X_{10}X_{14}^4X_2 + 105X_6X_{10}^2X_2^4 + 140X_{10}X_{14}^3X_2^3 + 210X_6^3X_{14}^2X_2^2 \\
&\quad + 140X_6^3X_{10}X_2^3 + 105X_6^4X_{14}X_2^2 + 42X_{10}X_{14}^5X_2 + 630X_6^2X_{10}^2X_{14}X_2^2 \\
&\quad + 7X_{14}X_2^6 + 420X_6^2X_{10}X_{14}X_2^3 + 105X_6X_{10}^4X_2^2 + 105X_6^4X_{10}^2X_{14} \\
&\quad + 140X_{10}^3X_{14}^3X_2 + 35X_6^4X_{14}^3 + 42X_{10}X_{14}X_2^5 + 21X_6^5X_{14}^2 \\
&\quad + 105X_{10}^4X_{14}X_2^2 + 42X_{10}^5X_{14}X_2 + X_{14}^7.
\end{aligned}
$$

References

[1] E. Bannai, S. T. Dougherty, M. Haradam, Ouram Type II codes, even unimodular lattices, and invariant rings, *IEEE Trans. Inform. Theory* 45(4), pp. 1194–1205 (1999).

[2] R. A. Brualdi and V. S. Plessm Weight enumerators of self-dual codes, *IEEE Trans. Inform. Theory* 37(4), pp. 1222–1225 (1991).

[3] Y. Choie, S. T. Dougherty, H. Kim, Complete joint weight enumerators and self-dual codes, IEEE Trans. Inform. Theory 49 (2003), no. 5, 1275–1282.

[4] Y. Choie and N. Kim, The complete weight enumerator of Type II Codes over $\mathbb{Z}_{2m}$ and Jacobi forms, *IEEE Trans. Inform. Theory* 47(1), pp. 396–399 (2001).

[5] Y. Choie, H. Kim, Codes over $\mathbb{Z}_{2m}$ and Jacobi forms of genus n, J. Combin. Theory Ser. A 95 (2001), no. 2, 335–348.

[6] Y. Choie, P. Solé, Self-dual codes over $\mathbb{Z}_4$ and half-integral weight modular forms. Proc. Amer. Math. Soc. 130 (2002), no. 11, 3125–3131.

[7] J. H. Conway and N. J. A. Sloane, A new upper bound on the minimal distance of self-dual codes, *IEEE Trans. Inform. Theory* 36(6), pp. 1319–1333 (1990).

[8] S. T. Dougherty, T. A. Gulliver, M. Harada, Type II self-dual codes over finite rings and even unimodular lattices, *Journal of Algebraic Combinatorics* 9, pp. 233–250 (1999).

[9] S. T. Dougherty, M. Harada, P. Solé, Shadow lattices and shadow codes, *Discrete Mathematics* 219, pp. 49-064 (2000).

[10] S. T. Dougherty, M. Harada, P. Solé, Shadow codes over $\mathbb{Z}_4$, *Finite Fields and Their Applications* 7(4), pp. 507–529 (2001).

[11] M. Eichler, D. Zagier, *The theory of Jacobi forms*, Boston, MA: Birkäuser, 1985.

[12] M. Klemm, Über die Identität von MacWilliams für die Gewichtsfunktion von Codes, *Arch. Math.* 49, pp. 400–406 (1987).

[13] A. Sharma, A. K. Sharma: Construction of self-dual codes over $\mathbb{Z}_{2^m}$, *Cryptography and Communications* 8(1), pp. 83–101 (2016).

[14] Han-Ping Tsai, Existence of some extremal self-dual codes, *IEEE Trans. Inform. Theory* 38(6), pp. 1829–1833 (1992).

Chapter 7

Byte Weight Enumerators of Codes and Modular Forms of Genus g

7.1 Introduction

Let R be either the ring $\mathbb{Z}_{2m}$ of integers modulo $2m$ or the quaternionic ring $\Sigma_{2m} = \mathbb{Z}_{2m} + \alpha\mathbb{Z}_{2m} + \beta\mathbb{Z}_{2m} + \gamma\mathbb{Z}_{2m}$ with $\alpha = 1 + \hat{\imath}$, $\beta = 1 + \hat{\jmath}$ and $\gamma = 1 + \hat{k}$ and $m \geq 1$ an integer, where $\hat{\imath}, \hat{\jmath}, \hat{k}$ are elements of the ring $\mathbb{H}$ of real quaternions satisfying $\hat{\imath}^2 = \hat{\jmath}^2 = \hat{k}^2 = -1$, $\hat{\imath}\hat{\jmath} = -\hat{\jmath}\hat{\imath} = \hat{k}$, $\hat{\jmath}\hat{k} = -\hat{k}\hat{\jmath} = \hat{\imath}$ and $\hat{k}\hat{\imath} = -\hat{\imath}\hat{k} = \hat{\jmath}$. This chapter follows the paper [15], which generalizes the work of Choie and Dougherty [3], Choie and Kim [3–5] and Suzuki [17], and explains connections between Jacobi forms (Siegel modular forms) of genus g and byte weight enumerators (symmetrized byte weight enumerators) in genus g of Type I and Type II codes over R. A functional equation is also derived for partial Epstein zeta functions, which are summands of classical Epstein zeta functions associated with quadratic forms. It has been shown that the coefficient matrix in this functional equation is the same as the transformation matrix in the MacWilliams identity for byte weight enumerators in genus g of byte error-control codes over R.

In Section 7.2, byte error-control codes over R, their Euclidean weights, byte weight enumerators and symmetrized weight enumerators in genus g are defined. Apart from this, MacWilliams identities for byte weight enumerators and symmetrized byte weight enumerators in genus g are also derived. In Section 7.3, Jacobi forms and Siegel modular forms in genus g are defined. In Section 7.4, connections between lattices and byte weight enumerators are established by defining a certain theta series for a lattice. In Section 7.5, invariant spaces of polynomials containing byte weight enumerators and symmetrized byte weight enumerators in genus g of Type

II codes over R are determined (Theorems 149 and 151). In Section 7.6, Jacobi forms and Siegel modular forms of genus g are obtained by substituting certain theta series into byte weight enumerators and symmetrized byte weight enumerators in genus g of Type II codes over R (Theorems 154 and 156-158). In Section 7.7, Jacobi forms and Siegel modular forms of genus g respectively are determined from byte weight enumerators and symmetrized byte weight enumerators in genus g of Type I codes over R by studying their shadow codes (Theorems 159 and 160). In Section 7.8, certain partial Epstein zeta functions are discussed. A functional equation for partial Epstein zeta functions is also derived using Mellin transforms of the theta series defined by (7.6) (Theorem 165). It is also observed that the coefficient matrix in this functional equation is the same as the transformation matrix in the MacWilliams identity for the byte weight enumerator in genus g of a byte error-control code over R.

7.2 Byte Error-control Codes over R and their Byte Weight Enumerators in Genus g

Let R be either the ring $\mathbb{Z}_{2m}$ or the quaternionic ring Σ_{2m}. For an integer $N \geq 1$, let R^N be the left R-module consisting of all N-tuples over R. For a positive divisor b of N, a byte error-control code $\mathcal{C}$ of length N and byte length b over R is defined as a left R-submodule of R^N. In particular, when $b = 1$, the code $\mathcal{C}$ is called a linear code of length N over R. Byte error-control codes are used to detect or correct errors in high density data storage systems (see [14]-[16]).

The dual code of $\mathcal{C}$, denoted by $\mathcal{C}^{\perp}$, is defined as

$$\mathcal{C}^{\perp} = \Big\{ v = (v_1, v_2, \cdots, v_N) \in R^N : [u, v] = \sum_{\alpha=1}^{N} \overline{u_\alpha} v_i = 0$$
$$\text{for all } u = (u_1, u_2, \cdots, u_N) \in \mathcal{C} \Big\},$$

where

$$\overline{x} = \begin{cases} x & \text{if } x \in \mathbb{Z}_{2m}; \\ a + b\overline{\alpha} + c\overline{\beta} + d\overline{\gamma} & \text{if } x = a + b\alpha + c\beta + d\gamma \in \Sigma_{2m} \end{cases} \tag{7.1}$$

with $\overline{\alpha} = 1 - \hat{\imath}$, $\overline{\beta} = 1 - \hat{\jmath}$ and $\overline{\gamma} = 1 - \hat{k}$. Note that the dual code $\mathcal{C}^{\perp}$ is also a byte error-control code of length N and byte length b over R. Further, the code $\mathcal{C}$ is said to be self-orthogonal if $\mathcal{C} \subseteq \mathcal{C}^{\perp}$, whereas the code $\mathcal{C}$ is said to be self-dual if $\mathcal{C} = \mathcal{C}^{\perp}$.

Now to define Type I and Type II codes over R, we first recall the definition of the Euclidean weight of a vector in $\mathbb{Z}_{2m}^N$.

Recall that the Euclidean weight of the element $u = (u_1, u_2, \cdots, u_N) \in \mathbb{Z}_{2m}^N$, denoted by $wt_E(u)$, is defined as

$$wt_E(u) = \sum_{\alpha=1}^{N} \min\left\{u_i^2, (2m - u_i)^2\right\}.$$

Next to define the Euclidean weight in Σ_{2m}^N, let us define a Gray map $\psi_{2m} : \Sigma_{2m} \to \mathbb{Z}_{2m}^4$ as

$$\psi_{2m}(a + b\alpha + c\beta + d\gamma) = (b, c, d, a + b + c + d).$$

The Gray map ψ_{2m} can be further extended coordinate wise to a linear, bijective map Ψ_{2m} from Σ_{2m}^N onto $\mathbb{Z}_{2m}^{4N}$. Then the Euclidean weight of a vector $v \in \Sigma_{2m}^N$, denoted by $wt_E(v)$, is defined as the Euclidean weight of $\Psi_{2m}(v) \in \mathbb{Z}_{2m}^{4N}$.

A self-dual code over R is called a Type II code if the Euclidean weight of each of its codewords is divisible by $4m$, otherwise it is called a Type I code.

Throughout this chapter, for each $x \in R$, let us denote the real part of x by $Re(x)$, which is given by $Re(x) = x_0 + x_1 + x_2 + x_3$ if $x = x_0 + x_1\alpha + x_2\beta + x_3\gamma \in \Sigma_{2m}$ and by $Re(x) = x$ if $x \in \mathbb{Z}_{2m}$. Let $N = nb$ for some integer $n > 1$. Assume that the elements of R^b are arranged in a lexicographical ordering $\mathfrak{O}$ and any matrix indexed by elements of R^b follows the ordering $\mathfrak{O}$. For positive integers p and q, let $\mathcal{A}^{(p,q)}$ denote the set of all $p \times q$ matrices with entries from a non-empty set $\mathcal{A}$. Let $(a_1; a_2; \cdots; a_g)$ denote a $b \times g$ matrix having $a_1, a_2, \cdots, a_g \in R^b$ as its columns, where b and g are positive integers.

Now the byte weight enumerator in genus g of a byte error-control code over R is defined as follows:

Definition 135. Let $\mathcal{C}$ be a byte error-control code of length bn and byte length b over R. Then the byte weight enumerator in genus g of $\mathcal{C}$ is defined as

$$BW_{\mathcal{C},g}\left(X_a : a \in R^{(b,g)}\right) = \sum_{c_1, c_2, \cdots, c_g} \prod_{j=1}^{n} X_{\left(c_j^{(1)}; c_j^{(2)}; \cdots; c_j^{(g)}\right)}$$

where the summation $\sum\limits_{c_1,c_2,\cdots,c_g}$ runs over all codewords $c_k = \left(c_1^{(k)}, c_2^{(k)}, \cdots, c_n^{(k)}\right) \in \mathcal{C}$ with $c_j^{(k)} \in R^b$ for $1 \le k \le g$ and $1 \le j \le n$.

Note that the byte weight enumerator in genus g of $\mathcal{C}$ can be rewritten as

$$BW_{\mathcal{C},g}\left(X_a : a \in R^{(b,g)}\right) = \sum_{c_1,c_2,\cdots,c_g \in \mathcal{C}} \prod_{a \in R^{(b,g)}} X_a^{n_a(c_1,c_2,\cdots,c_g)}, \qquad (7.2)$$

where $c_k = \left(c_1^{(k)}, c_2^{(k)}, \cdots, c_n^{(k)}\right)$ with $c_j^{(k)} \in R^b$ for $1 \le k \le g$ and $1 \le j \le n$, and

$$n_a(c_1, c_2, \cdots, c_g) = |\{j : 1 \le j \le n, (c_j^{(1)}; c_j^{(2)}; \cdots; c_j^{(g)}) = a\}|$$

for each $a \in R^{(b,g)}$.

Remark 136.

(i) When $g = 1$, the byte weight enumerator in genus g of a byte error-control code $\mathcal{C}$ over R is called the byte weight enumerator of $\mathcal{C}$ (see Sharma et al. [11] and Suzuki [17]).

(ii) When $b = g = 1$, the byte weight enumerator in genus g of a byte error-control code $\mathcal{C}$ over R coincides with the complete weight enumerator of $\mathcal{C}$.

If L is any $\ell \times \ell$ matrix over $\mathbb{C}$ and $f(X) = f(X_1, X_2, \cdots, X_\ell)$ is any polynomial over $\mathbb{C}$ in ℓ variables, then the action of the matrix L on the polynomial $f(X)$ is defined as

$$L \cdot f(X) := f(LX^t),$$

where X^t denotes the matrix transpose of X.

Now the following theorem provides MacWilliams identity for the byte weight enumerator in genus g of a byte error-control code over R.

Theorem 137. *Let $\mathcal{C}$ be a byte error-control code of length bn and byte length b over R. Then*

$$BW_{\mathcal{C}^\perp,g}\left(X_a : a \in R^{(b,g)}\right) = \frac{1}{|\mathcal{C}|^g} M_R^{(2m,g)} \cdot BW_{\mathcal{C},g}\left(X_a : a \in R^{(b,g)}\right)$$

with $M_R^{(2m,g)} = \left(\otimes_{j=1}^g H_R^{(2m)}\right)$, *where* $H_R^{(2m)}$ *is a* $|R|^b \times |R|^b$ *matrix indexed by the elements of* R^b *whose* (u, v)*th entry is given by* $\left(H_R^{(2m)}\right)_{u,v} = e^{\frac{\pi i}{m} Re([u,v])}$ *for all* $u, v \in R^b$.

Proof. To prove the result, let $\eta_m = e^{\frac{\pi i}{m}}$. Further, for $v \in R^{nb}$, let us define

$$\delta_{\mathcal{C}^\perp}(v) = \begin{cases} 1 \text{ if } v \in \mathcal{C}^\perp; \\ 0 \text{ otherwise.} \end{cases}$$

We also note that $\sum_{c\in\mathcal{C}} \eta_{2m}^{Re([v,c])} = |\mathcal{C}|\delta_{\mathcal{C}^\perp}(v)$ for each $v \in R^{nb}$.

By (8.1), we have

$$BW_{\mathcal{C}^\perp,g}\left(X_a : a \in R^{(b,g)}\right) = \sum_{d_1,d_2,\cdots,d_g\in\mathcal{C}^\perp} \prod_{a\in R^{(b,g)}} X_a^{n_a(d_1,d_2,\cdots,d_g)},$$

where $d_k = \left(d_1^{(k)}, d_2^{(k)}, \cdots, d_n^{(k)}\right)$ with $d_j^{(k)} \in R^b$ for $1 \le k \le g$ and $1 \le j \le n$, and

$$n_a(d_1, d_2, \cdots, d_g) = |\{j : 1 \le j \le n, (d_j^{(1)}; d_j^{(2)}; \cdots; d_j^{(g)}) = a\}|$$

for each $a \in R^{(b,g)}$.

Now let us consider

$$\begin{aligned}
&|\mathcal{C}|^g BW_{\mathcal{C}^\perp,g}\left(X_a : a \in R^{(b,g)}\right) \\
&= |\mathcal{C}|^g \sum_{d_1,d_2,\cdots,d_g\in\mathcal{C}^\perp} \prod_{a\in R^{(b,g)}} X_a^{n_a(d_1,d_2,\cdots,d_g)} \\
&= |\mathcal{C}|^g \sum_{v_1,v_2,\cdots,v_g\in R^{nb}} \prod_{\alpha=1}^{g} \delta_{\mathcal{C}^\perp}(v_\alpha) \prod_{a\in R^{(b,g)}} X_a^{n_a(v_1,v_2,\cdots,v_g)} \\
&= \sum_{v_1,v_2,\cdots,v_g\in R^{nb}} \sum_{c_1,c_2,\cdots,c_g\in\mathcal{C}} \eta_{2m}^{\sum_{\alpha=1}^{g} Re([v_\alpha,c_\imath])} \prod_{a\in R^{(b,g)}} X_a^{n_a(v_1,v_2,\cdots,v_g)} \\
&= \sum_{c_1,c_2,\cdots,c_g\in\mathcal{C}} \prod_{j=1}^{n} \left(\sum_{(v_j^{(1)};v_j^{(2)};\cdots;v_j^{(g)})\in R^{(b,g)}} \eta_{2m}^{\sum_{i=1}^{g} Re([v_j^{(\alpha)},c_j^{(\alpha)}])} X_{(v_j^{(1)};v_j^{(2)};\cdots;v_j^{(g)})} \right) \\
&= \sum_{c_1,c_2,\cdots,c_g\in\mathcal{C}} \left(\sum_{a\in R^{(b,g)}} \eta_{2m}^{Re\left(\left[a,(c_j^{(1)},c_j^{(2)},\cdots,c_j^{(g)}\right]\right)} \right)^{n_a(c_1,c_2,\cdots,c_g)} \\
&= M_R^{(2m,g)} \cdot BW_{\mathcal{C},g}\left(X_a : a \in R^{(b,g)}\right).
\end{aligned}$$

This completes the proof of the theorem. □

Remark 138.

(i) When $g = 1$, Theorem 137 provides MacWilliams identity for the byte weight enumerator of a byte error-control code over R.

(ii) When $b = g = 1$, Theorem 137 provides MacWilliams identity for the complete weight enumerator of a linear code over R.

Next to define the symmetrized byte weight enumerator in genus g of a byte error-control code over R, let us first define an equivalence relation on $R^{(b,g)}$ as follows:

For $c, d \in R^{(b,g)}$ with their respective (j,k)th entries as c_{jk}, d_{jk}, let us define $c \sim d$ if and only if $c_{jk} = \epsilon_{jk} d_{jk}$ for $1 \le j \le b$ and $1 \le k \le g$, where each ϵ_{jk} is an element of R satisfying $\overline{\epsilon_{jk}}\epsilon_{jk} = 1$. It is easy to see that $\sim$ is an equivalence relation on $R^{(b,g)}$. Let $\widehat{R^{(b,g)}}$ denote the set of all equivalence classes of $R^{(b,g)}$ with respect to $\sim$. In particular, when $g = 1$, $\sim$ is an equivalence relation on R^b. Let $\widehat{R^b}$ denote the set of all equivalence classes of R^b with respect to $\sim$. Further, for any matrix $B = (B_1; B_2; \cdots; B_g) \in R^{(b,g)}$ with each $B_j \in R^b$, it is easy to see that $\widehat{B} = (\widehat{B_1}; \widehat{B_2}; \cdots; \widehat{B_g}) \in \widehat{R^{(b,g)}}$, where $\widehat{B_j} \in \widehat{R^b}$ for each j.

Definition 139. Let $\mathcal{C}$ be a byte error-control code of length bn and byte length b over R. Then the symmetrized byte weight enumerator in genus g of $\mathcal{C}$ is defined as

$$SW_{\mathcal{C},g}\left(X_{\widehat{a}} : \widehat{a} \in \widehat{R^{(b,g)}}\right) = \sum_{c_1, c_2, \cdots, c_g} \prod_{j=1}^{n} X_{\left(\widehat{c_j^{(1)}}; \widehat{c_j^{(2)}}; \cdots; \widehat{c_j^{(g)}}\right)},$$

where the summation $\sum\limits_{c_1, c_2, \cdots, c_g}$ runs over all codewords $c_k = \left(c_1^{(k)}, c_2^{(k)}, \cdots, c_n^{(k)}\right) \in \mathcal{C}$ with $c_j^{(k)} \in R^b$ for $1 \le k \le g$ and $1 \le j \le n$.

Remark 140.

(i) The symmetrized byte weight enumerator in genus g of a byte error-control code $\mathcal{C}$ over R can be computed from the byte weight enumerator in genus g of the code $\mathcal{C}$ on taking $X_c = X_{\widehat{d}}$ for all $c \in \widehat{d}$, where $c, d \in R^{(b,g)}$.

(ii) When $g = 1$, the symmetrized byte weight enumerator in genus g of a byte error-control code $\mathcal{C}$ over R is called the symmetrized byte weight enumerator of $\mathcal{C}$.

From now on, elements of $\widehat{R^b}$ are arranged in a lexicographical ordering $\widehat{\mathfrak{O}}$ and any matrix indexed by elements of $\widehat{R^b}$ follows the ordering $\widehat{\mathfrak{O}}$. In the following theorem, MacWilliams identity for the symmetrized byte weight enumerator in genus g of a byte error-control code over R is stated.

Theorem 141. *Let $\mathcal{C}$ be a byte error-control code of length bn and byte length b over R. Then*

$$SW_{\mathcal{C}^{\perp},g}\left(X_{\widehat{a}} : \widehat{a} \in \widehat{R^{(b,g)}}\right) = \frac{1}{|\mathcal{C}|^g}\widehat{M_R^{(2m,g)}} \cdot SW_{\mathcal{C},g}\left(X_{\widehat{a}} : \widehat{a} \in \widehat{R^{(b,g)}}\right)$$

with $\widehat{M_R^{(2m,g)}} = \left(\otimes_{j=1}^{g} \widehat{H_R^{(2m)}}\right)$, *where* $\widehat{H_R^{(2m)}}$ *is a* $|\widehat{R^b}| \times |\widehat{R^b}|$ *matrix indexed by the elements of* $\widehat{R^b}$ *whose* $(\widehat{u}, \widehat{v})$*th entry is given by* $\left(\widehat{H_R^{(2m)}}\right)_{\widehat{u},\widehat{v}} = \sum_{w \in \widehat{v}} e^{\frac{\pi i}{m} Re([u,w])}$ *for all* $\widehat{u}, \widehat{v} \in \widehat{R^b}$.

Proof. It follows immediately from Theorem 137 and Remark 140 (i). □

Remark 142. When $g = 1$, Theorem 141 provides MacWilliams identity for the symmetrized byte weight enumerator of a byte error-control code over R.

Byte weight enumerators in genus g and symmetrized byte weight enumerators in genus g of byte error-control codes over R have connections with Jacobi forms and Siegel modular forms of genus g, which are as discussed below.

7.3 Jacobi Forms of Genus g

Let $\mathcal{G}$ be either the ring $\mathbb{Z}$ of integers or the ring $\mathcal{O} = \{a + b\alpha + c\beta + d\gamma : a, b, c, d \in \mathbb{Z}\}$. Then the modular group of genus g, denoted by $\Gamma_g(\mathcal{G})$, is defined as

$$\Gamma_g(\mathcal{G}) = \left\{M \in \mathcal{G}^{(2g,2g)} : M^* J M = J\right\},$$

where $M^* = \overline{M}^t$ is the transpose of the matrix $\overline{M}$ whose (α, j)th entry is $\overline{m_{\alpha j}}$ if $m_{\alpha j}$ is the (α, j)th entry of M and $J = \begin{pmatrix} 0 & -I_g \\ I_g & 0 \end{pmatrix}$ with I_g being the $g \times g$ identity matrix. The modular group $\Gamma_g(\mathbb{Z})$ is called the Siegel modular group, while the modular group $\Gamma_g(\mathcal{O})$ is called the quaternionic modular group. Further, it is easy to note that the set

$$H_{\mathcal{G}}^{(g)} = \left\{[(\lambda,\mu),x] : \lambda,\mu \in \mathcal{G}^{(1,g)}, x \in \mathcal{G}, (x+\mu\lambda^*) = (x+\mu\lambda^*)^*\right\}$$

is a group, called the Heisenberg group, with respect to the binary operation $\circ$, defined as follows:

$$[(\lambda,\mu),x] \circ [(\lambda',\mu'),x'] = [(\lambda+\lambda',\mu+\mu'),x+x'+\lambda\mu'^* - \mu\lambda'^*]$$

for all $[(\lambda,\mu),x], [(\lambda',\mu'),x'] \in H_{\mathcal{G}}^{(g)}$.

Next the Siegel upper half-space of genus g, denoted by $\mathcal{H}_g(\mathcal{S})$, is defined as

$$\mathcal{H}_g(\mathcal{S}) = \{Z \in \mathcal{S}^{(g,g)} : Z^* = Z,\ Im(Z) > 0\}$$

with $Z^* = \overline{Z}^t$, where $\mathcal{S}$ is either $\mathbb{C}$ or $\mathbb{H}$. Let

$$\mathfrak{T} = \mathbb{C} \text{ and } \mathcal{F} = \mathbb{R} \text{ when } \mathcal{S} = \mathbb{C},$$

and let

$$\mathfrak{T} = \mathbb{H}_{\mathbb{C}} = \mathbb{C} + \mathbb{C}\alpha + \mathbb{C}\beta + \mathbb{C}\gamma \text{ and } \mathcal{F} = \mathbb{H} \text{ when } \mathcal{S} = \mathbb{H}.$$

Now let $f : \mathcal{H}_g(\mathcal{S}) \times \mathfrak{T}^{(1,g)} \to \mathbb{C}$ be a holomorphic function, and let $k, \mathfrak{d}$ be positive integers. The slash operators on f are defined as

$$\begin{aligned}(f\,|_{k,\mathfrak{d}}\, M(\tau,z) &= det(C\tau+D)^{-k} e^{-2\pi i\sigma(\mathfrak{d}z(C\tau+D)^{-1}Cz^*)}\\ &\quad \times f((A\tau+B)(C\tau+D)^{-1}, z(C\tau+D)^{-1}),\\ (f\,|_{\mathfrak{d}}\, \zeta)(\tau,z) &= e^{2\pi i\mathfrak{d}\sigma(\lambda\tau\lambda^*+\lambda z^*+z\lambda^*+(x+\mu\lambda^*))} f(\tau, z+\lambda\tau+\mu),\end{aligned}$$

where $M = \begin{pmatrix} A & B \\ C & D \end{pmatrix} \in \Gamma_g(\mathcal{G})$, $\zeta = [(\lambda,\mu),x] \in H_{\mathcal{G}}^{(g)}$, $det(A)$ denotes the determinant of matrix A and $\sigma(A)$ denotes the trace of matrix A.

Now a Jacobi form of genus g on $\Gamma_g(\mathcal{G})$ is defined as follows:

Definition 143. [5,11] Let $k, \mathfrak{d}$ be two positive integers. Then a holomorphic function $f : \mathcal{H}_g(\mathcal{S}) \times \mathfrak{T}^{(1,g)} \to \mathbb{C}$ is said to be a Jacobi form of genus g with weight k and index $\mathfrak{d}$ on $\Gamma_g(\mathcal{G})$ if it satisfies the following:

$$(f\ |_{k,\mathfrak{d}}\ M)(\tau,z) = f(\tau,z) \text{ for all } M \in \Gamma_g(\mathcal{G}), \tag{7.3}$$

$$(f\ |_{\mathfrak{d}}\ \zeta)(\tau,z) = f(\tau,z) \text{ for all } \zeta \in H_{\mathcal{G}}^{(g)} \tag{7.4}$$

and in addition, when $g = 1$, $f(\tau, z)$ has the Fourier series expansion of the form

$$f(\tau, z) = \sum_{\substack{\ell \in \mathbb{N}, x \in \mathcal{G} \\ \overline{x}x \leq 4\frac{\ell}{s}\mathfrak{d}}} c(\ell, x) e^{2\pi i \frac{\ell}{s}\tau + \frac{2\pi i}{2}(xz^* + zx^*)} \text{ with } s \in \mathbb{N}. \quad (7.5)$$

Remark 144.

(i) The set of all Jacobi forms of genus g with weight k and index $\mathfrak{d}$ on $\Gamma_g(\mathcal{G})$ is an algebra over $\mathbb{C}$ (see Eichler and Zagier [9] and Krieg [11]) and is denoted by $\mathcal{J}_{k,\mathfrak{d}}(\Gamma_g(\mathcal{G}))$.

(ii) The form $f(\tau, 0)$ is called a Siegel modular form of genus g with weight k on $\Gamma_g(\mathcal{G})$ and it satisfies the following transformation formula:

$$f((A\tau+B)(C\tau+D)^{-1}, 0) = det(C\tau+D)^k f(\tau, 0) \text{ for all } \begin{pmatrix} A & B \\ C & D \end{pmatrix} \in \Gamma_g(\mathcal{G})$$

and when $g = 1$, the function $f(\tau, 0)$ has a Fourier series expansion of the form

$$f(\tau, 0) = \sum_{\ell \in \mathbb{N}} c(\ell) e^{2\pi i \frac{\ell}{s}\tau} \text{ with } s \in \mathbb{N}.$$

(iii) The set of all Siegel modular forms of genus g with weight k on $\Gamma_g(\mathcal{G})$ is an algebra over $\mathbb{C}$ and is denoted by $\widehat{\mathcal{J}_k}(\Gamma_g(\mathcal{G}))$.

The next section relates the byte weight enumerator and the symmetrized byte weight enumerator in genus g of a byte error-control code over R with the theta series of genus g of a certain lattice induced by the code. This relation gives rise to a method to determine modular forms of genus g.

7.4 Lattices and Byte Weight Enumerators in Genus g

To define a lattice over $\mathcal{F} = \mathbb{R}$ or $\mathbb{H}$ and a theta series of genus g for the lattice, let us take $\mathcal{G} = \mathbb{Z}$ when $\mathcal{F} = \mathbb{R}$ and $\mathcal{G} = \mathcal{O}$ when $\mathcal{F} = \mathbb{H}$. A lattice Λ in $\mathcal{F}^N$ is a free $\mathcal{G}$-module spanned by N linearly independent vectors in $\mathcal{F}^N$. The dual of the lattice Λ, denoted by Λ^*, is defined as the set

$$\Lambda^* = \{u \in \mathcal{F}^N : [u, v] \in \mathcal{G} \text{ for all } v \in \Lambda\}.$$

Further, the lattice Λ is called an integral lattice if it satisfies $\Lambda \subseteq \Lambda^*$, whereas the lattice Λ is called a unimodular lattice if it satisfies $\Lambda = \Lambda^*$. A unimodular lattice Λ is called a Type II lattice (or even lattice) if $[v, v] \in 2\mathbb{Z}$ for all $v \in \Lambda$, otherwise Λ is called a Type I lattice.

Now to construct a lattice from a byte error-control code over R, let us define a map $h : R \to \mathcal{G}$ as

$$h(r) = \begin{cases} \rho(r) & \text{if } r \in \mathbb{Z}_{2m}; \\ \rho(a) + \rho(b)\alpha + \rho(c)\beta + \rho(d)\gamma & \text{if } r = a + b\alpha + c\beta + d\gamma \in \Sigma_{2m}, \end{cases}$$

where the map $\rho : \mathbb{Z}_{2m} \to \mathbb{Z}$ is defined as

$$\rho(a) = \begin{cases} a & \text{if } 0 \le a \le m; \\ a - 2m & \text{if } m + 1 \le a \le 2m - 1. \end{cases}$$

Note that the map h can be further extended to a map from R^{bn} into $\mathcal{G}^{bn}$ by applying it coordinate wise. It is easy to show that the map $h : R^{bn} \to \mathcal{G}^{bn}$ is a ring homomorphism. Further, if $\mathcal{C}$ is a byte error-control code of length bn and byte length b over R, then the set

$$\Lambda(\mathcal{C}) = \frac{1}{\sqrt{2m}} \left\{ h(\mathcal{C}) + 2m\mathbb{Z}^{nb} \right\}$$

is a free $\mathcal{G}$-module in $\mathcal{F}^{bn}$ and is called the lattice induced from $\mathcal{C}$ (see Bannai et al. [1, Theorem 3.1] and Choie and Dougherty [3, Theorem 4.2]).

Definition 145. Let Λ be a lattice in $\mathcal{F}^N$, and let Y be a fixed element in Λ. Then the theta series of genus g for the lattice Λ is a function $\theta^{(g)}_{\Lambda,Y}$: $\mathcal{H}_g(\mathcal{S}) \times \mathfrak{T}^{(1,g)} \to \mathbb{C}$, defined as

$$\theta^{(g)}_{\Lambda,Y}(\tau, z) = \sum_{x \in \Lambda^g} e^{\frac{2\pi i}{2}[\sigma(x\tau x^*) + \sigma((Y^*x)z^* + z(Y^*x)^*)]}.$$

(Throughout this chapter, if A is any non-empty set consisting of (bn)-tuples and k is any positive integer, then elements of A^k are represented as $bn \times k$ matrices. Also for each positive integer k, let $\mathbf{1}_k$ denote the all-one vector of length k.)

The following theorem relates the byte weight enumerator in genus g of a byte error-control code $\mathcal{C}$ over R with the theta series of genus g for the induced lattice $\Lambda(\mathcal{C})$.

Theorem 146. *Let $\mathcal{C}$ be a byte error-control code of length bn and byte length b over R with the byte weight enumerator in genus g as $BW_{\mathcal{C},g}\left(X_a : a \in R^{(b,g)}\right)$. Then for $Y = \sqrt{2m}\mathbf{1}^t_{bn} \in \Lambda(\mathcal{C})$, we have*

$$\theta^{(g)}_{\Lambda(\mathcal{C}),Y}(\tau, z) = BW_{\mathcal{C},g}\left(\theta^{(g)}_{2m,a}(\tau, z) : a \in R^{(b,g)}\right),$$

where

$$\theta^{(g)}_{2m,a}(\tau, z) = \sum_{\substack{x\in\mathcal{G}^{(b,g)}\\ x\equiv a(\mathrm{mod}\ 2m)}} e^{\frac{2\pi i}{4m}\sigma(x\tau x^*)}e^{\frac{2\pi i}{2}\sigma(\mathbf{1}_b xz^*+zx^*\mathbf{1}_b^t)} \text{ for each } a \in R^{(b,g)}. \tag{7.6}$$

Proof. To prove the result, let us consider

$$\begin{aligned}
&\theta^{(g)}_{\Lambda(\mathcal{C}),Y}(\tau, z)\\
&= \sum_{v\in\mathcal{C}^g}\sum_{x\in\frac{1}{\sqrt{2m}}h^{-1}(v)} e^{\frac{2\pi i}{2}[\sigma(x\tau x^*)+\sigma((Y^*x)z^*+z(x^*Y))]}\\
&= \sum_{v\in\mathcal{C}^g}\sum_{x\in h^{-1}(v)} e^{2\pi i\sigma\left(\frac{x\tau x^*}{4m}\right)+\frac{2\pi i}{2}\sigma((\mathbf{1}_{bn}x)z^*+z(x^*\mathbf{1}^t_{bn}))}\\
&= \sum_{v\in\mathcal{C}^g}\sum_{x\in h^{-1}(0)} e^{2\pi i\sigma\left(\frac{(x+v)\tau(x+v)^*}{4m}\right)+\frac{2\pi i}{2}\sigma((\mathbf{1}_{bn}(x+v))z^*+z((x+v)^*\mathbf{1}^t_{bn}))}\\
&= \sum_{v=(v_1,v_2,\cdots,v_n)^t\in\mathcal{C}^g}\prod_{j=1}^{n}\\
&\quad\times \sum_{x_j\in 2m\mathcal{G}^{(b,g)}} e^{\frac{2\pi i}{4m}\sigma((x_j+v_j)\tau(x_j+v_j)^*)+\frac{2\pi i}{2}\sigma((\mathbf{1}_b(x_j+v_j))z^*+z((x_j+v_j)^*\mathbf{1}^t_b))}\\
&= \sum_{v=(v_1,v_2,\cdots,v_n)^t\in\mathcal{C}^g}\prod_{j=1}^{n}\theta^{(g)}_{2m,v_j}(\tau, z) = BW_{\mathcal{C},g}\left(\theta^{(g)}_{2m,a}(\tau, z) : a \in R^{(b,g)}\right),
\end{aligned}$$

which proves the theorem. □

Further, one can observe that $\theta^{(g)}_{2m,a}(\tau, 0) = \theta^{(g)}_{2m,c}(\tau, 0)$ for all $c \in \widehat{a}$, where $a, c \in R^{(b,g)}$. Let us define $\theta^{(g)}_{2m,\widehat{a}}(\tau, 0) = \theta^{(g)}_{2m,a}(\tau, 0)$ for every $a \in R^{(b,g)}$. In view of this, the following theorem relates theta series of genus g for the induced lattice $\Lambda(\mathcal{C})$ of a byte error-control code $\mathcal{C}$ over R with the symmetrized byte weight enumerator in genus g of the code $\mathcal{C}$.

Theorem 147. *Let $\mathcal{C}$ be a byte error-control code of length bn and byte length b over R with the symmetrized byte weight enumerator in genus g as $SW_{\mathcal{C},g}\left(X_{\widehat{a}} : \widehat{a} \in \widehat{R^{(b,g)}}\right)$. Then we have*

$$\theta^{(g)}_{\Lambda(\mathcal{C}),\sqrt{2m}\mathbf{1}^t_{bn}}(\tau, 0) = SW_{\mathcal{C},g}\left(\theta^{(g)}_{2m,\widehat{a}}(\tau, 0) : \widehat{a} \in \widehat{R^{(b,g)}}\right).$$

Proof. It follows from Theorem 146, Remark 140(i) and using the fact that $\theta^{(g)}_{2m,\widehat{a}}(\tau, 0) = \theta^{(g)}_{2m,a}(\tau, 0) = \theta^{(g)}_{2m,c}(\tau, 0)$ for all $c \in \widehat{a}$ with $a, c \in R^{(b,g)}$. □

In the following section, invariant spaces of polynomials containing byte weight enumerators and symmetrized byte weight enumerators in genus g of Type II codes over R are determined.

7.5 Invariant Space Containing Byte Weight Enumerators in Genus g of Type II Codes over R

To determine the space of polynomials invariant under a certain matrix group and containing byte weight enumerators (symmetrized byte weight enumerators) in genus g of Type II codes over R, let $\mathfrak{S} = 1$, $\mathcal{G} = \mathbb{Z}$ when $R = \mathbb{Z}_{2m}$ and $\mathfrak{S} = 4$, $\mathcal{G} = \mathcal{O}$ when $R = \Sigma_{2m}$. Let $Sym(g; \mathcal{G})$ denote the set of all $g \times g$ matrices $A = (a_{\alpha j})$ over $\mathcal{G}$ satisfying $A = A^* = (\overline{A})^t$, where the (α, j)th entry of A^* is $\overline{a_{j\alpha}}$ (which is as defined by (7.1)) for $1 \leq \alpha, j \leq g$.

Definition 148. Let $G_R^{(2m,g)}$ be the group generated by the matrices

$$\mathfrak{M}_R^{(2m,g)} = \left(\frac{1}{\sqrt{2m}}\right)^{\mathfrak{S}bg} e^{-\frac{\pi i \mathfrak{S} bg}{4}} M_R^{(2m,g)} \quad \text{(as defined in Theorem 137)}$$

and

$$N_R^{(2m,g)}(T) \ \text{ for all } \ T \in Sym(g; \mathcal{G}),$$

where for each $T \in Sym(g; \mathcal{G})$ and $a, c \in R^{(b,g)}$, the (a, c)th entry of the matrix $N_R^{(2m,g)}(T)$ is given by

$$\left(N_R^{(2m,g)}(T)\right)_{a,c} = \begin{cases} e^{2\pi i \sigma(\frac{1}{4m} a T a^*)} & \text{if } a = c; \\ 0 & \text{otherwise.} \end{cases}$$

A homogeneous polynomial $f(X)$ with $X = \left(X_a : a \in R^{(b,g)}\right)$ is said to be invariant under the group $G_R^{(2m,g)}$ if $L \cdot f(X) = f(LX^t) = f(X)$ for all $L \in G_R^{(2m,g)}$ and the space of all such homogeneous polynomials over $\mathbb{C}$ is denoted by $\mathbb{C}[X]^{G_R^{(2m,g)}}$.

The following theorem determines the invariant space of polynomials containing byte weight enumerators in genus g of Type II codes over R.

Theorem 149. *The byte weight enumerator* $BW_{\mathcal{C},g}\left(X_a : a \in R^{(b,g)}\right)$ *in genus* g *of a Type II code* $\mathcal{C}$ *over* R *belongs to* $\mathbb{C}[X]^{G_R^{(2m,g)}}$.

Proof. To prove the result, it is enough to prove that the byte weight enumerator $BW_{\mathcal{C},g}\left(X_a : a \in R^{(b,g)}\right)$ of a Type II code $\mathcal{C}$ is invariant under the matrices $\mathfrak{M}_R^{(2m,g)} = \left(\frac{1}{\sqrt{2m}}\right)^{\mathfrak{S}bg} e^{-\frac{\pi i \mathfrak{S} bg}{4}} M_R^{(2m,g)}$ and $N_R^{(2m,g)}(T)$ for each $T \in Sym(g;\mathcal{G})$.

To prove the invariance of the byte weight enumerator in genus g under the matrix $\mathfrak{M}_R^{(2m,g)}$, by Corollary 2.5 of Choie and Dougherty [3] and by applying Corollary 3.3 of Bannai et al. [1], we note that $bn\mathfrak{S} \equiv 0 \pmod 8$, which gives $\frac{bn\mathfrak{S}}{4} \equiv 0 \pmod 2$. Now by applying Theorem 137, one can easily see that the byte weight enumerator $BW_{\mathcal{C},g}\left(X_a : a \in R^{(b,g)}\right)$ is invariant under the matrix $\mathfrak{M}_R^{(2m,g)}$.

Next to prove the invariance of the byte weight enumerator in genus g under matrices of the type $N_R^{(2m,g)}(T)$ with $T = (t_{pq}) \in Sym(g;\mathcal{G})$, we observe that

$$N_R^{(2m,g)}(T) \cdot BW_{\mathcal{C},g}\left(X_a : a \in R^{(b,g)}\right) = \sum_{c_1,c_2,\cdots,c_g \in \mathcal{C}} \prod_{j=1}^{n} e^{\frac{2\pi i}{4m}\sigma(a_j T a_j^*)} X_{a_j},$$

where for $1 \le j \le n$, $a_j = \left(c_j^{(1)}; c_j^{(2)}; \cdots; c_j^{(g)}\right)$ if $c_k = (c_1^{(k)}, c_2^{(k)}, \cdots, c_n^{(k)})$ with $c_j^{(k)} \in R^b$ for $1 \le k \le g$. Now it suffices to show that

$$\sum_{j=1}^{n} \sigma(a_j T a_j^*) \equiv 0 \pmod{4m}. \tag{7.7}$$

For this, let us write $c_j^{(p)} = (c_{j1}^{(p)}, c_{j2}^{(p)}, \cdots, c_{jb}^{(p)})^t$ for each j and p, and let us consider

$$\begin{aligned}
&\sum_{j=1}^{n} \sigma(a_j T a_j^*) \\
&= \sum_{j=1}^{n} \left\{ \sum_{p=1}^{g} \sum_{\ell=1}^{b} \overline{c_{j\ell}^{(p)}} t_{pp} c_{j\ell}^{(p)} + \sum_{1 \le q < s \le g} \sum_{\ell=1}^{b} \left(c_{j\ell}^{(q)} t_{qs} \overline{c_{j\ell}^{(s)}} + c_{j\ell}^{(s)} \overline{t_{qs}} \overline{c_{j\ell}^{(q)}} \right) \right\} \\
&= \left\{ \sum_{p=1}^{g} \sum_{j=1}^{n} \sum_{\ell=1}^{b} \overline{c_{j\ell}^{(p)}} c_{j\ell}^{(p)} t_{pp} + \sum_{1 \le q < s \le g} \sum_{j=1}^{n} \sum_{\ell=1}^{b} 2Re\left(c_{j\ell}^{(q)} t_{qs} \overline{c_{j\ell}^{(s)}} \right) \right\},
\end{aligned}$$

where $Re(x) = x_0 + x_1 + x_2 + x_3$ for each $x = x_0 + x_1\alpha + x_2\beta + x_3\gamma \in R$ (assuming $x_1 = x_2 = x_3 = 0$ if $x \in \mathbb{Z}_{2m}$).

Now as $\mathcal{C}$ is a Type II code, we have $\sum_{j=1}^{n}\sum_{\ell=1}^{b}\overline{c_{j\ell}^{(p)}}c_{j\ell}^{(p)} \equiv wt_E(c_p) \equiv 0 \pmod{4m}$ for each p. Thus it remains to show that

$$\sum_{j=1}^{n}\sum_{\ell=1}^{b} 2Re\left(c_{j\ell}^{(q)}t_{qs}\overline{c_{j\ell}^{(s)}}\right) \equiv 0 \pmod{4m}$$

for each q and s. For this, we write $c_{j\ell}^{(q)}, \overline{c_{j\ell}^{(s)}} \in R$ in the form $x_0+x_1\alpha+x_2\beta+x_3\gamma$ with $x_0,x_1,x_2,x_3 \in \mathbb{Z}_{2m}$ and each $t_{qs} \in \mathcal{G}$ in the form $y_0+y_1\alpha+y_2\beta+y_3\gamma$ with $y_0,y_1,y_2,y_3 \in \mathbb{Z}$ (assuming $x_1 = x_2 = x_3 = y_1 = y_2 = y_3 = 0$ in the corresponding representations when $R = \mathbb{Z}_{2m}$ and $\mathcal{G} = \mathbb{Z}$). Now for each q and s, on substituting the corresponding representations in the sum

$$\sum_{j=1}^{n}\sum_{\ell=1}^{b} 2Re\left(c_{j\ell}^{(q)}t_{qs}\overline{c_{j\ell}^{(s)}}\right)$$

and using the fact that

$$[c_q, c_s] \equiv 0 \pmod{2m},$$

we see that

$$\sum_{j=1}^{n}\sum_{\ell=1}^{b} Re\left(c_{j\ell}^{(q)}t_{qs}\overline{c_{j\ell}^{(s)}}\right) \equiv 0 \pmod{2m},$$

from which the assertion (7.7) follows immediately. □

Next to determine the invariant space containing symmetrized byte weight enumerators in genus g of a Type II code over R, let us define the following:

Definition 150. Let $H_R^{(2m,g)}$ be the group generated by the matrices $\left(\frac{1}{\sqrt{2m}}\right)^{\mathfrak{S}bg} e^{-\frac{\pi i \mathfrak{S}bg}{4}}\widehat{M_R^{(2m,g)}}$ (as defined in Theorem 141) and $\widehat{N_R^{(2m,g)}}(T)$ for all $T \in Sym(g;\mathcal{G})$, where for each $T \in Sym(g;\mathcal{G})$ and $\widehat{a},\widehat{c} \in \widehat{R^{(b,g)}}$, the $(\widehat{a},\widehat{c})$th entry of $\widehat{N_R^{(2m,g)}}(T)$ is given by

$$\left(\widehat{N_R^{(2m,g)}}(T)\right)_{\widehat{a},\widehat{c}} = \begin{cases} e^{2\pi i\sigma(\frac{1}{4m}aTa^*)} & \text{if } \widehat{a}=\widehat{c};\\ 0 & \text{otherwise}\end{cases}$$

(note that $\sigma(aTa^*) = \sigma(cTc^*)$ for every $c \in \widehat{a}$, where $a, c \in R^{(b,g)}$).

A homogeneous polynomial $f(\widehat{X})$ with $\widehat{X} = \left(X_{\widehat{a}} : \widehat{a} \in \widehat{R^{(b,g)}}\right)$ is said to be invariant under the group $H_R^{(2m,g)}$ if $L \cdot f(\widehat{X}) = f(L\widehat{X}^t) = f(\widehat{X})$ for all $L \in H_R^{(2m,g)}$, and the space of all such homogeneous polynomials over $\mathbb{C}$ is denoted by $\mathbb{C}[X]^{H_R^{(2m,g)}}$.

Now the following theorem determines the invariant space of polynomials containing symmetrized byte weight enumerators in genus g of Type II codes over R.

Theorem 151. *The symmetrized byte weight enumerator $SW_{\mathcal{C},g}\left(X_{\widehat{a}} : \widehat{a} \in \widehat{R^{(b,g)}}\right)$ in genus g of a Type II code $\mathcal{C}$ over R belongs to $\mathbb{C}[X]^{H_R^{(2m,g)}}$.*

Proof. It follows immediately from Theorem 149 and Remark 140(i). □

7.6 Determination of Jacobi and Siegel Modular Forms of Genus g from Type II Codes over R

The following lemma plays an important role in the determination of Jacobi and Siegel modular forms of genus g.

Lemma 152. *For each $a \in R^{(b,g)}$, let the theta series $\theta_{2m,a}^{(g)}(\tau, z)$ be as defined by (7.6). Then for each $T \in Sym(g;\mathcal{G})$ and $\lambda, \kappa \in \mathcal{G}^{(1,g)}$, we have the following transformation formulae:*

(i) $\theta_{2m,a}^{(g)}(\tau + T, z) = e^{\frac{2\pi i}{4m}\sigma(aTa^*)}\theta_{2m,a}^{(g)}(\tau, z)$.

(ii) $\theta_{2m,a}^{(g)}(-\tau^{-1}, z\tau^{-1}) = (det\ \tau)^{\mathfrak{S}b/2}\left(\frac{1}{2mi}\right)^{\mathfrak{S}bg/2} e^{2\pi imbz\tau^{-1}z^*}$
$\sum_{\nu \in R^{(b,g)}} e^{-\frac{\pi i}{m}\sigma(Re(\nu a^*))}\theta_{2m,\nu}^{(g)}(\tau, z)$.

(iii) $\theta_{2m,a}^{(g)}(\tau, z + \lambda\tau + \kappa) = e^{-2\pi imb(\lambda\tau\lambda^* + \lambda z^* + z\lambda^*)}\theta_{2m,a}^{(g)}(\tau, z)$.

Proof. For $R = \mathbb{Z}_{2m}$, working in a similar way as in Lemma 5.4 of Choie and Kim [5], the result follows. Now let us assume, throughout the proof, that $R = \Sigma_{2m}$.

(i) By the definition of theta series, for each $T \in Sym(g;\mathcal{G})$, we have

$$\begin{aligned}\theta_{2m,a}^{(g)}(\tau + T, z) &= \sum_{\substack{x \in \mathcal{O}^{(b,g)} \\ x \equiv a \pmod{2m}}} e^{\frac{2\pi i}{4m}\sigma(x(\tau+T)x^*)} e^{\frac{2\pi i}{2}\sigma(\mathbf{1}_b x z^* + z x^* \mathbf{1}_b^t)} \\ &= \sum_{\substack{x \in \mathcal{O}^{(b,g)} \\ x \equiv a \pmod{2m}}} e^{\frac{2\pi i}{4m}\sigma(xTx^*)} e^{\frac{2\pi i}{4m}\sigma(x\tau x^*)} e^{\frac{2\pi i}{2}\sigma(\mathbf{1}_b x z^* + z x^* \mathbf{1}_b^t)}.\end{aligned}$$

Now it suffices to show that $e^{\frac{2\pi i}{4m}\sigma(xTx^*)} = e^{\frac{2\pi i}{4m}\sigma(aTa^*)}$ for each $x \in \mathcal{O}^{(b,g)}$ satisfying $x \equiv a \pmod{2m}$.

Note that if $x \in \mathcal{O}^{(b,g)}$ satisfies $x \equiv a \pmod{2m}$, then we can write $x = a - 2mu$ for some $u \in \mathcal{O}^{(b,g)}$. This gives

$$e^{\frac{2\pi i}{4m}\sigma(xTx^*)} = e^{\frac{2\pi i}{4m}\sigma(aTa^*)-\frac{2\pi i}{2}\sigma(uTa^*+aTu^*)+2\pi i m\sigma(uTu^*)} = e^{\frac{2\pi i}{4m}\sigma(aTa^*)},$$

which follows from the fact that $\sigma(uTu^*) \in \mathbb{Z}$ and $\sigma(uTa^* + aTu^*) = \sigma(2Re(uTa^*)) \in 2\mathbb{Z}$.

(ii) Again, by the definition of theta series, we have

$$\begin{aligned}\theta^{(g)}_{2m,a}(\tau, z) &= \sum_{\substack{x\in\mathcal{O}^{(b,g)}\\ x\equiv a \pmod{2m}}} e^{\frac{2\pi i}{4m}\sigma(x\tau x^*)}e^{\frac{2\pi i}{2}\sigma(\mathbf{1}_b xz^*+zx^*\mathbf{1}_b^t)}\\ &= \sum_{u\in\mathcal{O}^{(b,g)}} e^{\frac{2\pi i}{4m}\sigma(a\tau a^*-2m(u\tau a^*+a\tau u^*)+4m^2u\tau u^*)}\\ &\quad\times e^{\frac{2\pi i}{2}\sigma(az^*\mathbf{1}_b+\mathbf{1}_b^t za^*-2m(uz^*\mathbf{1}_b+\mathbf{1}_b^t zu^*))}.\end{aligned}$$

This gives

$$\begin{aligned}&\theta^{(g)}_{2m,a}(-\tau^{-1}, z\tau^{-1})\\ &\quad= \sum_{u\in\mathcal{O}^{(b,g)}} e^{\frac{2\pi i}{4m}\sigma\left(-a\tau^{-1}a^*+2m(u\tau^{-1}a^*+a\tau^{-1}u^*)-4m^2u\tau^{-1}u^*\right)}\\ &\qquad\times e^{\frac{2\pi i}{2}\sigma\left(a(z\tau^{-1})^*\mathbf{1}_b+\mathbf{1}_b^t(z\tau^{-1})a^*-2m(u(z\tau^{-1})^*\mathbf{1}_b+\mathbf{1}_b^t(z\tau^{-1})u^*)\right)}.\end{aligned}$$

Now by Lemma I 3.3 of Krieg [11], we see that $\tau = P^*DP$, where D is a $g \times g$ real diagonal matrix and P is a $g \times g$ matrix satisfying $P^*P = I_g$. On substituting this in the above equation, we obtain

$$\begin{aligned}&\theta^{(g)}_{2m,a}(-\tau^{-1}, z\tau^{-1})\\ &= \sum_{u\in\mathcal{O}^{(b,g)}} e^{\frac{2\pi i}{4m}\sigma\left(-aP^*D^{-1}Pa^*+2m(uP^*D^{-1}Pa^*+aP^*D^{-1}Pu^*)-4m^2uP^*D^{-1}Pu^*\right)}\\ &\quad\times e^{\frac{2\pi i}{2}\sigma\left(a(zP^*D^{-1}P)^*\mathbf{1}_b+\mathbf{1}_b^t(zP^*D^{-1}P)a^*-2m(u(zP^*D^{-1}P)^*\mathbf{1}_b+\mathbf{1}_b^t(zP^*D^{-1}P)u^*)\right)}.\end{aligned}$$

Further, on taking $\tilde{z} = \mathbf{1}_b^t z$, $z' = P\tilde{z}^*$, $a' = Pa^*$, $u' = Pu^*$ and on expanding the trace terms, we get

$$\begin{aligned}&\theta^{(g)}_{2m,a}(-\tau^{-1}, z\tau^{-1})\\ &\quad= \prod_{j=1}^{g}\prod_{k=1}^{b}\sum_{u'_{jk}\in\mathcal{O}} e^{-\frac{2\pi i}{4md_j}a'^*_{jk}a'_{jk}+\frac{2\pi i}{2d_j}(u'^*_{jk}a'_{jk}+a'^*_{jk}u'_{jk})-2\pi i m u'^*_{jk}d_j^{-1}u'_{jk}}\\ &\qquad\times e^{\frac{2\pi i}{2d_j}(a'^*_{jk}z'_{jk}+z'^*_{jk}a'_{jk})-\frac{2\pi i m}{d_j}(u'^*_{jk}z'_{jk}+z'^*_{jk}u'_{jk})},\end{aligned}$$

where $a' = (a'_{jk})$, $u' = (u'_{jk})$, $z' = (z'_{jk})$ and $D = diag(d_1, d_2, \cdots, d_g)$, a diagonal matrix. Now let us write $a'_{jk} = a^{(1)}_{jk} + a^{(2)}_{jk}\hat{\imath} + a^{(3)}_{jk}\hat{\jmath} + a^{(4)}_{jk}\hat{k}$, $u'_{jk} = u^{(1)}_{jk} + u^{(2)}_{jk}\hat{\imath} + u^{(3)}_{jk}\hat{\jmath} + u^{(4)}_{jk}\hat{k}$ and $z'_{jk} = z^{(1)}_{jk} + z^{(2)}_{jk}\hat{\imath} + z^{(3)}_{jk}\hat{\jmath} + z^{(4)}_{jk}\hat{k}$, where $a^{(p)}_{jk}, u^{(p)}_{jk}, z^{(p)}_{jk}$ are integers for each j, k and p. On substitution, we obtain

$$\theta^{(g)}_{2m,a}(-\tau^{-1}, z\tau^{-1})$$

$$= \prod_{j=1}^{g}\prod_{k=1}^{b}\prod_{\ell=1}^{4}\sum_{u^{(\ell)}_{jk}\in\mathbb{Z}} e^{-\frac{2\pi i}{4md_j}\left(a^{(\ell)}_{jk}\right)^2 + \frac{2\pi i}{d_j}u^{(\ell)}_{jk}a^{(\ell)}_{jk} - \frac{2\pi i m}{d_j}\left(u^{(\ell)}_{jk}\right)^2 + \frac{2\pi i}{d_j}a^{(\ell)}_{jk}z^{(\ell)}_{jk} - \frac{4\pi i m}{d_j}u^{(\ell)}_{jk}z^{(\ell)}_{jk}}.$$

Now by applying the Poisson's summation formula (see Ebeling [9, Theorem 2.3]), we get

$$\theta^{(g)}_{2m,a}(-\tau^{-1}, z\tau^{-1})$$

$$= \prod_{j=1}^{g}\prod_{k=1}^{b}\prod_{\ell=1}^{4}\sum_{y^{(\ell)}_{jk}\in\mathbb{Z}} \sqrt{\frac{d_j}{2m\alpha}}\, e^{\frac{\pi i}{2m}d_j\left(y^{(\ell)}_{jk}\right)^2 + \frac{2\pi i m}{d_j}\left(z^{(\ell)}_{jk}\right)^2 - \frac{\pi i}{m}y^{(\ell)}_{jk}a^{(\ell)}_{jk} + 2\pi i y^{(\ell)}_{jk}z^{(\ell)}_{jk}}.$$

On recombining the terms and using the fact that $det\ \tau = det\ D$, we obtain

$$\theta^{(g)}_{2m,a}(-\tau^{-1}, z\tau^{-1})$$

$$= (det\ \tau)^{2b}\left(\frac{1}{2m\alpha}\right)^{2bg}\sum_{y\in\mathcal{O}^{(b,g)}}$$

$$\times\, e^{\frac{\pi i}{2m}\sigma(y\tau y^*) + 2\pi i m\sigma(\mathbf{1}^t_b z\tau^{-1}z^*\mathbf{1}_b) - \frac{\pi i}{2m}\sigma(ya^* + ay^*) + \pi i\sigma(yz^*\mathbf{1}_b + \mathbf{1}^t_b zy^*)}$$

$$= (det\ \tau)^{2b}\left(\frac{1}{2m\alpha}\right)^{2bg} e^{2\pi i m b z\tau^{-1}z^*}\sum_{\nu\in\Sigma^{(b,g)}_{2m}}\sum_{u\in\mathcal{O}^{(b,g)}} e^{\frac{\pi i}{2m}\sigma((\nu-2mu)\tau(\nu-2mu)^*)}$$

$$\times\, e^{-\frac{\pi i}{2m}\sigma((\nu-2mu)a^* + a(\nu-2mu)^*)}e^{\pi i\sigma((\nu-2mu)z^*\mathbf{1}_b + \mathbf{1}^t_b z(\nu-2mu)^*)}$$

$$= (det\ \tau)^{2b}\left(\frac{1}{2m\alpha}\right)^{2bg} e^{2\pi i m b z\tau^{-1}z^*}\sum_{\nu\in\Sigma^{(b,g)}_{2m}} e^{-\frac{\pi i}{m}\sigma(Re(\nu a^*))}\theta^{(g)}_{2m,\nu}(\tau, z).$$

(iii) Again by definition, for each $\lambda, \kappa \in \mathcal{O}^{(1,g)}$, we have

$$e^{2\pi i m b(\lambda\tau\lambda^* + \lambda z^* + z\lambda^*)}\theta^{(g)}_{2m,a}(\tau, z + \lambda\tau + \kappa)$$

$$= e^{2\pi i m b(\lambda\tau\lambda^* + \lambda z^* + z\lambda^*)} \sum_{\substack{x\in\mathcal{O}^{(b,g)} \\ x\equiv a \pmod{2m}}}$$

$$\times\, e^{\frac{2\pi i}{4m}\sigma(x\tau x^*)} e^{\frac{2\pi i}{2}\sigma(\mathbf{1}_b x(z+\lambda\tau+\kappa)^* + (z+\lambda\tau+\kappa)x^*\mathbf{1}_b^t)}$$

$$= \sum_{\substack{x\in\mathcal{O}^{(b,g)} \\ x\equiv a \pmod{2m}}} e^{\frac{2\pi i}{4m}\sigma((x+2m\mathbf{1}_b^t\lambda)\tau(x+2m\mathbf{1}_b^t\lambda)^*)}$$

$$\times\, e^{\frac{2\pi i}{2}\sigma(\mathbf{1}_b(x+2m\mathbf{1}_b^t\lambda)(z+\lambda\tau+\kappa)^* + (z+\lambda\tau+\kappa)(x+2m\mathbf{1}_b^t\lambda)^*\mathbf{1}_b^t)},$$

which equals $\theta^{(g)}_{2m,a}(\tau, z)$, as $x + 2m\mathbf{1}_b^t\lambda \in \mathcal{O}^{(b,g)}$ satisfies $x + 2m\mathbf{1}_b^t\lambda \equiv a \pmod{2m}$.

This completes the proof of the lemma. □

From now on, we shall distinguish the cases $R = \mathbb{Z}_{2m}$ and $R = \Sigma_{2m}$.

7.6.1 *The case* $R = \mathbb{Z}_{2m}$

Let $R = \mathbb{Z}_{2m}$ so that we have $\mathcal{G} = \mathbb{Z}$ and $\mathfrak{S} = 1$. Here Jacobi forms of genus g are determined from byte weight enumerators in genus g of Type II codes over $\mathbb{Z}_{2m}$ by relating the invariant space $\mathbb{C}[X]^{G_R^{(2m,g)}}$ with a space of Jacobi forms of genus g as follows:

Theorem 153. *For an integer $\ell \geq 1$, let $\mathbb{C}[X]_{8\ell}^{G_R^{(2m,g)}}$ denote the space of all homogeneous polynomials of degree $bn = 8\ell$ invariant under the group $G_R^{(2m,g)}$. Then the map*

$$\varphi_m : \bigoplus_{\ell\geq 1} \mathbb{C}[X]_{8\ell}^{G_R^{(2m,g)}} \to \bigoplus_{\ell\geq 1} \mathcal{J}_{4\ell b, 8m\ell b}(\Gamma_g(\mathbb{Z})),$$

defined as

$$\varphi_m\left(f_n\left(X_a : a \in \mathbb{Z}_{2m}^{(b,g)}\right)\right) = f_n\left(\theta^{(g)}_{2m,a}(\tau, z) : a \in \mathbb{Z}_{2m}^{(b,g)}\right),$$

is an algebra homomorphism.

Proof. To prove the result, we first assert that if $f_n\left(X_a : a \in \mathbb{Z}_{2m}^{(b,g)}\right) \in \mathbb{C}[X]_{8\ell}^{G_R^{(2m,g)}}$, then $f_n\left(\theta_{2m,a}^{(g)}(\tau,z) : a \in \mathbb{Z}_{2m}^{(b,g)}\right)$ lies in $\mathcal{J}_{4\ell b, 8m\ell b}(\Gamma_g(\mathbb{Z}))$. To prove this, by Theorem II 2.3 of Krieg [11], we see that the modular group $\Gamma_g(\mathbb{Z})$ is generated by the matrices $\begin{pmatrix} 0 & -I_g \\ I_g & 0 \end{pmatrix}$ and $\begin{pmatrix} I_g & T \\ 0 & I_g \end{pmatrix}$, $T \in Sym(g;\mathbb{Z})$. From this, it is enough to show that the polynomial $f_n\left(\theta_{2m,a}^{(g)}(\tau,z) : a \in \mathbb{Z}_{2m}^{(b,g)}\right)$ satisfies (7.3) and (7.4) for the generator $\begin{pmatrix} 0 & -I_g \\ I_g & 0 \end{pmatrix}$ and generators of the type $\begin{pmatrix} I_g & T \\ 0 & I_g \end{pmatrix}$ with $T \in Sym(g;\mathbb{Z})$. Besides this, when $g = 1$, we need to show that $f_n\left(\theta_{2m,a}^{(g)}(\tau,z) : a \in \mathbb{Z}_{2m}^{(b,g)}\right)$ has a Fourier representation of the form (7.5).

For the generator $\begin{pmatrix} 0 & -I_g \\ I_g & 0 \end{pmatrix}$, we have

$$
\begin{aligned}
& f_n\left(\theta_{2m,a}^{(g)}(-\tau^{-1}, z\tau^{-1}) : a \in \mathbb{Z}_{2m}^{(b,g)}\right) \\
&= f_n\left((2m)^{-bg/2}\left(det\frac{\tau}{\alpha}\right)^{b/2} e^{2\pi i m b z \tau^{-1} z^t} M_R^{(2m,g)} \cdot \left(\theta_{2m,a}^{(g)}(\tau,z) : a \in \mathbb{Z}_{2m}^{(b,g)}\right)\right) \\
&= (det\ \tau)^{4\ell b} e^{2\pi i 8 m \ell b z \tau^{-1} z^t} f_n\left((2m)^{-bg/2} e^{-\frac{\pi i b r}{4}} M_R^{(2m,g)} \right. \\
&\quad \left. \cdot \left(\theta_{2m,a}^{(g)}(\tau,z) : a \in \mathbb{Z}_{2m}^{(b,g)}\right)\right) \\
&= (det\ \tau)^{4\ell b} e^{2\pi i 8 m \ell b z \tau^{-1} z^t} f_n\left(\theta_{2m,a}^{(g)}(\tau,z) : a \in \mathbb{Z}_{2m}^{(b,g)}\right),
\end{aligned}
$$

as f_n is invariant under $\mathfrak{M}_R^{(2m,g)} = (2m)^{-bg/2} e^{-\frac{\pi i b g}{4}} M_R^{(2m,g)}$, by Theorem 149. For a generator of the type $\begin{pmatrix} I_g & T \\ 0 & I_g \end{pmatrix}$ with $T \in Sym(g;\mathbb{Z})$, we have

$$
\begin{aligned}
& f_n\left(\theta_{2m,a}^{(g)}(\tau+T, z) : a \in \mathbb{Z}_{2m}^{(b,g)}\right) \\
&\quad = f_n\left(N_R^{(2m,g)}(T) \cdot \left(\theta_{2m,a}^{(g)}(\tau,z) : a \in \mathbb{Z}_{2m}^{(b,g)}\right)\right) \\
&\quad = f_n\left(\theta_{2m,a}^{(g)}(\tau,z) : a \in \mathbb{Z}_{2m}^{(b,g)}\right),
\end{aligned}
$$

as f_n is invariant under $N_R^{(2m,g)}(T)$, by Theorem 149. Moreover, for all

$\lambda, \mu \in \mathbb{Z}^{(1,g)}$, we note that

$$\begin{aligned} f_n &\left(\theta^{(g)}_{2m,a}(\tau, z + \lambda\tau + \mu) : a \in \mathbb{Z}^{(b,g)}_{2m}\right) \\ &= f_n\left(e^{-2\pi i b(m\lambda\tau\lambda^t + 2m\lambda z^t)}\left(\theta^{(g)}_{2m,a}(\tau, z) : a \in \mathbb{Z}^{(b,g)}_{2m}\right)\right) \\ &= e^{-2\pi i 8\ell b(m\lambda\tau\lambda^t + 2m\lambda z^t)} f_n\left(\theta^{(g)}_{2m,a}(\tau, z) : a \in \mathbb{Z}^{(b,g)}_{2m}\right). \end{aligned}$$

When $g = 1$, it is easy to see that $f_n\left(\theta^{(g)}_{2m,a}(\tau, z) : a \in \mathbb{Z}^{(b,g)}_{2m}\right)$ has a Fourier series expansion of the form (7.5). From this, it follows that $f_n\left(\theta^{(g)}_{2m,a}(\tau, z) : a \in \mathbb{Z}^{(b,g)}_{2m}\right)$ is a Jacobi form of genus g with weight $4\ell b$ and index $8m\ell b$ on $\Gamma_g(\mathbb{Z})$. Moreover, it is easy to show that the map φ_m is a well-defined algebra homomorphism. □

Now the following theorem determines Jacobi forms of genus g from byte weight enumerators in genus g of Type II codes over $\mathbb{Z}_{2m}$ as a consequence of the above theorem.

Theorem 154. *Let C be a Type II code of length bn and byte length b over $\mathbb{Z}_{2m}$. Then*

$$BW_{C,g}\left(\theta^{(g)}_{2m,a}(\tau, z) : a \in \mathbb{Z}^{(b,g)}_{2m}\right)$$

is a Jacobi form of genus g having weight $bn/2$ and index mbn on $\Gamma_g(\mathbb{Z})$.

Proof. It follows immediately from Theorems 149 and 153. □

Next to obtain Siegel modular forms of genus g from symmetrized byte weight enumerators in genus g of Type II codes over $\mathbb{Z}_{2m}$, we first relate the invariant space $\mathbb{C}[X]^{H_R^{(2m,g)}}$ with a space of Siegel modular forms of genus g in the following theorem.

Theorem 155. *For an integer $\ell \geq 1$, let $\mathbb{C}[X]_{8\ell}^{H_R^{(2m,g)}}$ be the space of all homogeneous polynomials of degree $bn = 8\ell$ invariant under the group $H_R^{(2m,g)}$. Then the map*

$$\Phi_m : \bigoplus_{\ell \geq 1} \mathbb{C}[X]_{8\ell}^{H_R^{(2m,g)}} \to \bigoplus_{\ell \geq 1} \widehat{\mathcal{J}}_{4\ell b}(\Gamma_g(\mathbb{Z})),$$

defined as

$$\Phi_m\left(f_n\left(X_{\widehat{a}} : \widehat{a} \in \widehat{\mathbb{Z}^{(b,g)}_{2m}}\right)\right) = f_n\left(\theta^{(g)}_{m,\widehat{a}}(\tau, 0) : \widehat{a} \in \widehat{\mathbb{Z}^{(b,g)}_{2m}}\right),$$

is an algebra homomorphism.

Proof. Its proof is similar to that of Theorem 153. □

As a consequence of the above theorem, Siegel modular forms of genus g are obtained from symmetrized byte weight enumerators in genus g of Type II codes over $\mathbb{Z}_{2m}$ in the following theorem.

Theorem 156. *Let C be a Type II code of length bn and byte length b over $\mathbb{Z}_{2m}$. Then*

$$SW_{C,g}\left(\theta_{2m,\widehat{a}}^{(g)}(\tau,0) : \widehat{a} \in \widehat{\mathbb{Z}_{2m}^{(b,g)}}\right)$$

is a Siegel modular form of genus g with weight $bn/2$ on $\Gamma_g(\mathbb{Z})$.

Proof. It follows immediately from Theorems 151 and 155. □

7.6.2 *The case $R = \Sigma_{2m}$*

Let $R = \Sigma_{2m}$ so that we have $\mathcal{G} = \mathcal{O}$ and $\mathfrak{S} = 4$. The following theorem determines a Jacobi form of genus g from the byte weight enumerator in genus g of a Type II code over Σ_{2m}.

Theorem 157. *Let C be a Type II code of length bn and byte length b over Σ_{2m}. Then*

$$BW_{C,g}\left(\theta_{2m,a}^{(g)}(\tau,z) : a \in \Sigma_{2m}^{(b,g)}\right)$$

is a Jacobi form of genus g with weight $2bn$ and index mbn on $\Gamma_g(\mathcal{O})$.

Proof. By Theorem II 2.3 of Krieg [11], we see that the modular group $\Gamma_g(\mathcal{O})$ is generated by the following matrices:

$$\begin{pmatrix} 0 & -I_g \\ I_g & 0 \end{pmatrix}, \begin{pmatrix} U^* & 0 \\ 0 & U^{-1} \end{pmatrix}, \ U \in GL(g;\mathcal{O}) \text{ and } \begin{pmatrix} I_g & T \\ 0 & I_g \end{pmatrix}, \ T \in Sym(g;\mathcal{O}). \tag{7.8}$$

Therefore to prove this theorem, it is enough to show that $BW_{C,g}\left(\theta_{2m,a}^{(g)}(\tau,z) : a \in \Sigma_{2m}^{(b,g)}\right)$ satisfies (7.3) and (7.4) for the three types of generators (as stated in (7.8)) of $\Gamma_g(\mathcal{O})$ and has a Fourier series representation of the form (7.5) when $g = 1$.

For a generator of the type $\begin{pmatrix} I_g & T \\ 0 & I_g \end{pmatrix}$ with $T \in Sym(g; \mathcal{O})$, we note that

$$\begin{aligned}
&BW_{\mathcal{C},g}\left(\theta^{(g)}_{2m,a}(\tau+T,z) : a \in \Sigma^{(b,g)}_{2m}\right)\\
&= BW_{\mathcal{C},g}\left(e^{\frac{2\pi i}{4m}\sigma(aTa^*)}\theta^{(g)}_{2m,a}(\tau,z) : a \in \Sigma^{(b,g)}_{2m}\right)\\
&= BW_{\mathcal{C},g}\left(N_R^{(2m,g)}(T)\cdot\left(\theta^{(g)}_{2m,a}(\tau,z) : a \in \Sigma^{(b,g)}_{2m}\right)\right)\\
&= BW_{\mathcal{C},g}\left(\theta^{(g)}_{2m,a}(\tau,z) : a \in \Sigma^{(b,g)}_{2m}\right),
\end{aligned}$$

as the byte weight enumerator of genus g is invariant under $N_R^{(2m,g)}(T)$ by Theorem 149. For the generator $\begin{pmatrix} 0 & -I_g \\ I_g & 0 \end{pmatrix}$, we have

$$\begin{aligned}
&(det\ \tau)^{-2bn}e^{-2\pi imbnz\tau^{-1}z^*}BW_{\mathcal{C},g}\left(\theta^{(g)}_{2m,a}(-\tau^{-1},z\tau^{-1}) : a \in \Sigma^{(b,g)}_{2m}\right)\\
&= (det\ \tau)^{-2bn}e^{-2\pi imbnz\tau^{-1}z^*}BW_{\mathcal{C},g}\Bigg((det\ \tau)^{2b}\left(\frac{1}{2m\alpha}\right)^{2bg}e^{2\pi imbz\tau^{-1}z^*}\\
&\quad\times\sum_{\nu\in\Sigma^{(b,g)}_{2m}} e^{-\frac{\pi i}{m}\sigma(Re(\nu a^*))}\theta^{(g)}_{2m,\nu}(\tau,z) : a \in \Sigma^{(b,g)}_{2m}\Bigg)\\
&= BW_{\mathcal{C},g}\left(\left(\frac{1}{2m}\right)^{2bg}M_R^{(2m,g)}\cdot\left(\theta^{(g)}_{2m,a}(\tau,z) : a \in \Sigma^{(b,g)}_{2m}\right)\right)\\
&= e^{\pi ibng}BW_{\mathcal{C},g}\left(\left(\frac{1}{2m}\right)^{2bg}e^{-\pi ibg}M_R^{(2m,g)}\cdot\left(\theta^{(g)}_{2m,a}(\tau,z) : a \in \Sigma^{(b,g)}_{2m}\right)\right)\\
&= BW_{\mathcal{C},g}\left(\theta^{(g)}_{2m,a}(\tau,z) : a \in \Sigma^{(b,g)}_{2m}\right),
\end{aligned}$$

as $e^{\pi ibng}=1$ and the byte weight enumerator of genus g is invariant under $\mathfrak{M}_R^{(2m,g)} = \left(\frac{1}{2m}\right)^{2bg}e^{-\pi ibg}M_R^{(2m,g)}$ by Theorem 149. Further, for a generator of the type $\begin{pmatrix} U^* & 0 \\ 0 & U^{-1} \end{pmatrix}$ with $U \in GL(g; \mathcal{O})$, we have

$$\begin{aligned}
&BW_{\mathcal{C},g}\left(\theta^{(g)}_{2m,a}(U^*\tau U, zU) : a \in \Sigma^{(b,g)}_{2m}\right) \\
&\quad = \theta^{(g)}_{\Lambda(\mathcal{C}),\sqrt{2m}\mathbf{1}^t_{bn}}(U^*\tau U, zU) \\
&\quad = \sum_{x \in \Lambda(\mathcal{C})^g} e^{\frac{2\pi i}{2}\sigma(xU^*\tau Ux^*) + \frac{2\pi i}{2}\sigma(((\sqrt{2m}\mathbf{1}^t_{bn})^*x)(zU)^* + (zU)((\sqrt{2m}\mathbf{1}^t_{bn})^*x)^*)} \\
&\quad = \theta^{(g)}_{\Lambda(\mathcal{C}),\sqrt{2m}\mathbf{1}^t_{bn}}(\tau, z),
\end{aligned}$$

by Corollary IV 1.7 of Krieg [11]. This, by Theorem 146, implies that

$$BW_{\mathcal{C},g}\left(\theta^{(g)}_{2m,a}(U^*\tau U, zU) : a \in \Sigma^{(b,g)}_{2m}\right) = BW_{\mathcal{C},g}\left(\theta^{(g)}_{2m,a}(\tau, z) : a \in \Sigma^{(b,g)}_{2m}\right).$$

Moreover, for every $\lambda, \kappa \in \mathcal{O}^{(1,g)}$, we have

$$\begin{aligned}
&e^{2\pi imbn(\lambda\tau\lambda^* + \lambda z^* + z\lambda^*)} BW_{\mathcal{C},g}\left(\theta^{(g)}_{2m,a}(\tau, z + \lambda\tau + \kappa) : a \in \Sigma^{(b,g)}_{2m}\right) \\
&\quad = e^{2\pi imbn(\lambda\tau\lambda^* + \lambda z^* + z\lambda^*)} BW_{\mathcal{C},g} \\
&\qquad \times \left(e^{-2\pi imb(\lambda\tau\lambda^* + \lambda z^* + z\lambda^*)}\theta^{(g)}_{2m,a}(\tau, z) : a \in \Sigma^{(b,g)}_{2m}\right) \\
&\quad = BW_{\mathcal{C},g}\left(\theta^{(g)}_{2m,a}(\tau, z) : a \in \Sigma^{(b,g)}_{2m}\right).
\end{aligned}$$

When $g = 1$, one can easily prove that $BW_{\mathcal{C},g}\left(\theta^{(g)}_{2m,a}(\tau, z) : a \in \Sigma^{(b,g)}_{2m}\right)$ has a Fourier series expansion of the form (7.5), which completes the proof of the theorem. □

The following theorem determines Siegel modular forms of genus g from symmetrized byte weight enumerators in genus g of Type II codes over Σ_{2m}.

Theorem 158. *Let $\mathcal{C}$ be a Type II code of length bn and byte length b over Σ_{2m}. Then*

$$SW_{\mathcal{C},g}\left(\theta^{(g)}_{2m,\widehat{a}}(\tau, 0) : \widehat{a} \in \widehat{\Sigma^{(b,g)}_{2m}}\right)$$

is a Siegel modular form of genus g with weight $2bn$ on $\Gamma_g(\mathcal{O})$.

Proof. It follows immediately from Theorem 157, and Remarks 140(i) and 144(ii). □

The next section discusses the shadow of a Type I code over R, and determines Jacobi forms and Siegel modular forms of genus g respectively from byte weight enumerators and symmetrized byte weight enumerators in genus g of Type I codes over R.

7.7 Determination of Jacobi Forms and Siegel Modular Forms of Genus g from Type I Codes over R

Throughout this section, let $\mathcal{C}$ be a Type I code of length bn and byte length b over R. Let $\mathcal{C}_0 = \{c \in \mathcal{C} : wt_E(c) \equiv 0 \pmod{4m}\}$. By Lemma 1.4 of Chapter 6, we see that $\mathcal{C}_0$ is a linear subcode of index 2 in $\mathcal{C}$ and the code $\mathcal{C}$ is of index 2 in $\mathcal{C}_0^\perp$. So we can write $\mathcal{C} = \mathcal{C}_0 \cup \mathcal{C}_2$ and $\mathcal{C}_0^\perp = \mathcal{C}_0 \cup \mathcal{C}_1 \cup \mathcal{C}_2 \cup \mathcal{C}_3$, where $\mathcal{C}_1 = s + \mathcal{C}_0$, $\mathcal{C}_2 = t + \mathcal{C}_0$ and $\mathcal{C}_3 = s + t + \mathcal{C}_0$ for some $s \in \mathcal{C}_0^\perp \setminus \mathcal{C}$ and $t \in \mathcal{C} \setminus \mathcal{C}_0$. Recall that the shadow $\mathcal{S}$ of $\mathcal{C}$ is defined as

$$\mathcal{S} = \mathcal{C}_0^\perp \setminus \mathcal{C} = \mathcal{C}_1 \cup \mathcal{C}_3.$$

Now to determine Jacobi forms and Siegel modular forms of genus g, let $\mathbb{T}$ be a $bn \times g$ matrix with each column as t, and let $\mathbb{S}$ be a $bn \times g$ matrix with each column as s. Further, for any $bn \times g$ matrix $A = (A_1, A_2, \cdots, A_n)^t$ over R with each $A_j \in R^{(b,g)}$, let us define

$$A \odot BW_{\mathcal{C},g}(X_a : a \in R^{(b,g)}) = \sum_{v=(v_1,v_2,\cdots,v_n)^t \in \mathcal{C}^g} \prod_{j=1}^{n} X_{v_j + h^{-1}(A_j)},$$

$$\widehat{A} \odot SW_{\mathcal{C},g}(X_{\widehat{a}} : \widehat{a} \in \widehat{R^{(b,g)}}) = \sum_{v=(v_1,v_2,\cdots,v_n)^t \in \mathcal{C}^g} \prod_{j=1}^{n} X_{\widehat{v_j + h^{-1}(A_j)}},$$

where $\widehat{A} = (\widehat{A_1}, \widehat{A_2}, \cdots, \widehat{A_n})^t$ with $\widehat{A_j} \in \widehat{R^{(b,g)}}$ for $1 \le j \le n$. Moreover, for any rational number $\mathfrak{q}$, let us denote $\mathbb{Z} + \mathfrak{q} = \{n + \mathfrak{q} : n \in \mathbb{Z}\}$. We recall that $\mathcal{G} = \mathbb{Z}$, $\mathcal{F} = \mathbb{R}$, $\mathfrak{S} = 1$ when $R = \mathbb{Z}_{2m}$ and $\mathcal{G} = \mathcal{O}$, $\mathcal{F} = \mathbb{H}$, $\mathfrak{S} = 4$ when $R = \Sigma_{2m}$.

Now the following theorem determines Jacobi forms of genus g from byte weight enumerators in genus g of Type I codes over R.

Theorem 159. *If $\mathcal{C}$ is a Type I code of length bn and byte length b over R, then $L = \Lambda(\mathcal{C})$ is a Type I lattice in $\mathcal{F}^{bn}$. Further, for any positive integer* *k satisfying* $bn + kb \equiv 0 \pmod{\frac{4}{\mathfrak{S}}}$, *let* $\mathcal{G}_0^{kb} = \{v \in \mathcal{G}^{kb} : [v, v] \in 2\mathbb{Z}\}$, *and let* $\mathcal{G}_\alpha^{kb}$ $(0 \le \alpha \le 3)$ *be the cosets of* $\mathcal{G}_0^{kb}$ *in the dual lattice* $(\mathcal{G}_0^{kb})^*$. *Then*

$$\tilde{L} = (\mathcal{G}_0^{kb} \oplus L_0) \cup (\mathcal{G}_1^{kb} \oplus L_1) \cup (\mathcal{G}_2^{kb} \oplus L_2) \cup (\mathcal{G}_3^{kb} \oplus L_3)$$

is a unimodular lattice in $\mathcal{F}^{bn+kb}$, *where* $L_\alpha = \Lambda(\mathcal{C}_\alpha)$ *for* $0 \le \alpha \le 3$. *When* $bn + kb \equiv 0 \pmod{\frac{8}{\mathfrak{S}}}$, $\tilde{L}$ *is a Type II lattice in* $\mathcal{F}^{bn+kb}$. *Moreover, the theta series of genus g for the lattice* $\tilde{L}$, *given by*

$$\theta^{(g)}_{\tilde{L},\sqrt{2m}\mathbf{1}^t_{(n+k)b}}(\tau, z) = \sum_{\alpha=0}^{3} \theta^{(g)}_{\mathcal{G}_\alpha^{kb},\sqrt{2m}\mathbf{1}^t_{kb}}(\tau, z)\, \theta^{(g)}_{L_\alpha,\sqrt{2m}\mathbf{1}^t_{bn}}(\tau, z), \tag{7.9}$$

is a Jacobi form of genus g with weight $\mathfrak{S}(n + k)b/2$ *and index* $m(n + k)b$ *on* $\Gamma_g(\mathcal{G})$, *where*

$$\theta^{(g)}_{L_0,\sqrt{2m}\mathbf{1}^t_{bn}}(\tau, z) = BW_{\mathcal{C}_0,g}\left(\theta^{(g)}_{2m,a}(\tau, z) : a \in R^{(b,g)}\right),$$

$$\theta^{(g)}_{L_1,\sqrt{2m}\mathbf{1}^t_{bn}}(\tau, z) = h^{-1}(\mathbb{S}) \odot BW_{\mathcal{C}_0,g}\left(\theta^{(g)}_{2m,a}(\tau, z) : a \in R^{(b,g)}\right),$$

$$\theta^{(g)}_{L_2,\sqrt{2m}\mathbf{1}^t_{bn}}(\tau, z) = h^{-1}(\mathbb{T}) \odot BW_{\mathcal{C}_0,g}\left(\theta^{(g)}_{2m,a}(\tau, z) : a \in R^{(b,g)}\right),$$

$$\theta^{(g)}_{L_3,\sqrt{2m}\mathbf{1}^t_{bn}}(\tau, z) = h^{-1}(\mathbb{S} + \mathbb{T}) \odot BW_{\mathcal{C}_0,g}\left(\theta^{(g)}_{2m,a}(\tau, z) : a \in R^{(b,g)}\right).$$

Proof. To prove the result, we will distinguish the following two cases: **I.** $R = \mathbb{Z}_{2m}$ and **II.** $R = \Sigma_{2m}$.

Case I. First let $R = \mathbb{Z}_{2m}$. Here we have $\mathcal{F} = \mathbb{R}$, $\mathcal{G} = \mathbb{Z}$ and $\mathfrak{S} = 1$ so that k is a positive integer satisfying $bn + kb \equiv 0 \pmod 4$. Further, we see that

$$\mathcal{G}_0^{kb} = \{v \in \mathbb{Z}^{kb} : O_v \text{ is even}\}$$

and

$$\mathcal{G}_2^{kb} = \{v \in \mathbb{Z}^{kb} : O_v \text{ is odd}\},$$

where for $v \in \mathbb{Z}^{kb}$, the number O_v equals the total number of odd components in v. From this, one can observe that

$$(\mathcal{G}_0^{kb})^* = \mathbb{Z}^{kb} \cup (\mathbb{Z} + 1/2)^{kb}.$$

This gives

$$\mathcal{G}_1^{kb} = \left\{x + (\frac{1}{2}, \frac{1}{2}, \cdots, \frac{1}{2}) : x \in \mathcal{G}_0^{kb}\right\}$$

and

$$\mathcal{G}_3^{kb} = \left\{x + (\frac{1}{2}, \frac{1}{2}, \cdots, \frac{1}{2}) : x \in \mathcal{G}_2^{kb}\right\}.$$

Now consider the set $\tilde{L} = (\mathcal{G}_0^{kb} \oplus L_0) \cup (\mathcal{G}_1^{kb} \oplus L_1) \cup (\mathcal{G}_2^{kb} \oplus L_2) \cup (\mathcal{G}_3^{kb} \oplus L_3)$. It is clear that $\tilde{L}$ is a lattice in $\mathbb{R}^{bn+kb}$. To show that $\tilde{L}$ is a unimodular lattice, it suffices to show that $(\tilde{L})^* = \tilde{L}$. To prove this, we will consider the case $kb \equiv 1 \pmod 4$, while proofs for the remaining cases are similar and are left to the reader.

Let $kb \equiv 1 \pmod 4$. In this case, we first observe that the cosets $\mathcal{G}_i^{kb}$, $0 \le \alpha \le 3$, satisfy Table 7.1, where the rational number δ at the $(\alpha+1, j+1)$th position $(0 \le \alpha, j \le 3,\ \alpha \neq j)$ of the table means $[x, y] \in \mathbb{Z} + \delta$ for each $x \in \mathcal{G}_\alpha^{kb}$ and $y \in \mathcal{G}_j^{kb}$, while the rational number δ at the $(\alpha+1, \alpha+1)$th position $(0 \le \alpha \le 3)$ of the table means $[x, y] \in 2\mathbb{Z} + \delta$ for each $x, y \in \mathcal{G}_\alpha^{kb}$.

Table 7.1 Values of δ.

$[\cdot,\cdot]$	$\mathcal{G}_0^{kb}$	$\mathcal{G}_1^{kb}$	$\mathcal{G}_2^{kb}$	$\mathcal{G}_3^{kb}$
$\mathcal{G}_0^{kb}$	0	0	0	0
$\mathcal{G}_1^{kb}$	0	$\frac{kb}{4}$	$\frac{1}{2}$	$\frac{3}{4}$
$\mathcal{G}_2^{kb}$	0	$\frac{1}{2}$	1	$\frac{1}{2}$
$\mathcal{G}_3^{kb}$	0	$\frac{3}{4}$	$\frac{1}{2}$	$\frac{kb}{4}$

As $bn \equiv 3 \pmod 4$ in this case, by Lemma 3 of Dougherty et al. [7] and Theorem 1 of Dougherty et al. [7], we see that the cosets $L_\alpha, 0 \le \alpha \le 3$, satisfy Table 7.2, where the rational number ν at the $(\alpha+1, j+1)$th position

Table 7.2 Values of ν.

$[\cdot,\cdot]$	L_0	L_1	L_2	L_3
L_0	0	0	0	0
L_1	0	$\frac{bn}{4}$	$\frac{1}{2}$	$-\frac{3}{4}$
L_2	0	$\frac{1}{2}$	1	$\frac{1}{2}$
L_3	0	$-\frac{3}{4}$	$\frac{1}{2}$	$\frac{bn}{4}$

$(0 \le \alpha, j \le 3,\ \alpha \neq j)$ of the table means $[x, y] \in \mathbb{Z} + \nu$ for each $x \in L_\alpha$ and $y \in L_j$, while the rational number ν at the $(\alpha+1, \alpha+1)$th position $(0 \le \alpha \le 3)$ of the table means $[x, y] \in 2\mathbb{Z} + \nu$ for each $x, y \in L_\alpha$.

Next from Tables 7.1 and 7.2, we note that $[X, Y] \in \mathbb{Z}$ for every $X, Y \in \tilde{L}$, as $bn + kb \equiv 0 \pmod 4$. From this, it follows that $\tilde{L} \subseteq (\tilde{L})^*$. Therefore to show that $\tilde{L}$ is a unimodular lattice, it remains to show that $(\tilde{L})^* \subseteq \tilde{L}$. For this, let $X = (X_1, X_2) \in (\tilde{L})^*$, where $X_1 \in \mathbb{R}^{kb}$ and $X_2 \in \mathbb{R}^{bn}$. This implies that $[X, Y] \in \mathbb{Z}$ for all $Y = (Y_1, Y_2) \in \tilde{L}$. In particular, we have $[X, Y] \in \mathbb{Z}$ for all $Y = (Y_1, Y_2) \in \mathcal{G}_0^{kb} \oplus L_0$. Now as $\mathbf{0} \in \mathcal{G}_0^{kb}$, we have $[X_2, Y_2] \in \mathbb{Z}$ for all $Y_2 \in L_0$. This gives $X_2 \in L_0^* = \Lambda(\mathcal{C}_0^{\perp}) = L_0 \cup L_1 \cup L_2 \cup L_3$. Further from Tables 7.1 and 7.2, we see that if $X_2 \in L_\alpha$ for some α, then $X_1 \in \mathcal{G}_\alpha^{kb}$, which implies that $X = (X_1, X_2) \in \tilde{L}$.

From now onwards, we suppose that $bn + kb \equiv 0 \pmod 8$. Here we will show that $\tilde{L}$ is a Type II lattice. To prove this, we assert that $[v, v] \in 2\mathbb{Z}$ for all $v = (v_1, v_2) \in \tilde{L}$.

To prove this assertion, first let $v_1 \in \mathcal{G}_0^{kb}$ and $v_2 \in L_0$. Here using Tables 7.1 and 7.2, we see that $[v_1, v_1], [v_2, v_2] \in 2\mathbb{Z}$, which implies that $[v, v] = [v_1, v_1] + [v_2, v_2] \in 2\mathbb{Z}$.

Now let $v_1 \in \mathcal{G}_2^{kb}$ and $v_2 \in L_2$. Here from Tables 7.1 and 7.2, we see that $[v_1, v_1], [v_2, v_2] \in 2\mathbb{Z} + 1$, which gives $[v, v] = [v_1, v_1] + [v_2, v_2] \in 2\mathbb{Z}$.

Next let $v_1 \in \mathcal{G}_\alpha^{kb}$ and $v_2 \in L_\alpha$ for some $\alpha \in \{1, 3\}$. Here using Tables 7.1 and 7.2 again, we see that $[v_2, v_2] \in 2\mathbb{Z} + \frac{bn}{4}$ and $[v_1, v_1] \in 2\mathbb{Z} + \frac{kb}{4}$. This gives $[v, v] = [v_1, v_1] + [v_2, v_2] \in 2\mathbb{Z}$, as $bn + kb \equiv 0 \pmod 8$. From this, it follows that $\tilde{L}$ is a Type II lattice when $bn + kb \equiv 0 \pmod 8$.

Further, by definition, the theta series of genus g for the lattice $\tilde{L}$ is given by

$$\theta^{(g)}_{\tilde{L}, \sqrt{2m}\mathbf{1}^t_{(n+k)b}}(\tau, z) = \sum_{\alpha=0}^{3} \theta^{(g)}_{\mathcal{G}_i^{kb}, \sqrt{2m}\mathbf{1}^t_{kb}}(\tau, z)\, \theta^{(g)}_{L_\alpha, \sqrt{2m}\mathbf{1}^t_{bn}}(\tau, z),$$

which is a Jacobi form of genus g with weight $(n+k)b/2$ and index $m(n+k)b$ on $\Gamma_g(\mathbb{Z})$ by Theorem 3.2 of Choie and Kim [5].

We next proceed to compute theta series $\theta^{(g)}_{L_\alpha, \sqrt{2m}\mathbf{1}^t_{bn}}(\tau, z)$, $0 \le \alpha \le 3$, in terms of the byte weight enumerator in genus g of the code $\mathcal{C}_0$. For this, we first observe that as $\mathcal{C}_0$ is a byte error-control code of length bn and byte length b over R, by Theorem 146, we have $\theta^{(g)}_{L_0, \sqrt{2m}\mathbf{1}^t_{bn}}(\tau, z) = BW_{\mathcal{C}_0, g}\left(\theta^{(g)}_{2m, a}(\tau, z) : a \in R^{(b,g)}\right)$.

Moreover, by the definition of theta series of genus g for the lattice $L_1 = \Lambda(\mathcal{C}_1)$, we have

$$\begin{aligned}
&\theta^{(g)}_{L_1,\sqrt{2m}\mathbf{1}^t_{bn}}(\tau,z)\\
&= \sum_{v\in\mathcal{C}_1^g}\ \sum_{x\in\frac{1}{\sqrt{2m}}h^{-1}(v)} e^{\frac{2\pi i}{2}[\sigma(x\tau x^*)+\sigma((\sqrt{2m}\mathbf{1}_{bn}x)z^*+z(x^*\sqrt{2m}\mathbf{1}^t_{bn}))]}\\
&= \sum_{v\in\mathbb{S}+\mathcal{C}_0^g}\ \sum_{x\in h^{-1}(v)} e^{2\pi i\sigma\left(\frac{x\tau x^*}{4m}\right)+\frac{2\pi i}{2}\sigma((\mathbf{1}_{bn}x)z^*+z(x^*\mathbf{1}^t_{bn}))}\\
&= \sum_{v\in\mathcal{C}_0^g}\ \sum_{x\in h^{-1}(\mathbb{S})+h^{-1}(v)} e^{2\pi i\sigma\left(\frac{x\tau x^*}{4m}\right)+\frac{2\pi i}{2}\sigma((\mathbf{1}_{bn}x)z^*+z(x^*\mathbf{1}^t_{bn}))}\\
&= \sum_{v\in\mathcal{C}_0^g}\ \sum_{x\in h^{-1}(0)} e^{2\pi i\sigma\left(\frac{(x+v+h^{-1}(\mathbb{S}))\tau(x+v+h^{-1}(\mathbb{S}))^*}{4m}\right)}\\
&\quad\times e^{\frac{2\pi i}{2}\sigma\left((\mathbf{1}_{bn}(x+v+h^{-1}(\mathbb{S})))z^*+z((x+v+h^{-1}(\mathbb{S}))^*\mathbf{1}^t_{bn})\right)}\\
&= \sum_{v=(v_1,v_2,\cdots,v_n)^t\in\mathcal{C}_0^g}\ \prod_{j=1}^{n}\ \sum_{x_j\in 2m\mathcal{G}^{(b,g)}} e^{\frac{2\pi i}{4m}\sigma\left((x_j+v_j+h^{-1}(\mathbb{S}_j))\tau(x_j+v_j+h^{-1}(\mathbb{S}_j))^*\right)}\\
&\quad\times e^{\frac{2\pi i}{2}\sigma\left((\mathbf{1}_b(x_j+v_j+h^{-1}(\mathbb{S}_j)))z^*+z((x_j+v_j+h^{-1}(\mathbb{S}_j))^*\mathbf{1}^t_b)\right)},
\end{aligned}$$

where $\mathbb{S}=(\mathbb{S}_1,\mathbb{S}_2,\cdots,\mathbb{S}_n)^t$ with each $\mathbb{S}_j\in R^{(b,g)}$. From this, we obtain

$$\begin{aligned}
&\theta^{(g)}_{L_1,\sqrt{2m}\mathbf{1}^t_{bn}}(\tau,z)\\
&\quad= \sum_{v=(v_1,v_2,\cdots,v_n)^t\in\mathcal{C}_0^g}\ \prod_{j=1}^{n}\theta^{(g)}_{2m,v_j+h^{-1}(\mathbb{S}_j)}(\tau,z)\\
&\quad= h^{-1}(\mathbb{S})\odot BW_{\mathcal{C}_0,g}\left(\theta^{(g)}_{2m,a}(\tau,z): a\in R^{(b,g)}\right).
\end{aligned}$$

Working in a similar way as above, one can show that

$$\theta^{(g)}_{L_2,\sqrt{2m}\mathbf{1}^t_{bn}}(\tau,z) = h^{-1}(\mathbb{T})\odot BW_{\mathcal{C}_0,g}\left(\theta^{(g)}_{2m,a}(\tau,z): a\in R^{(b,g)}\right)\ \text{and}$$
$$\theta^{(g)}_{L_3,\sqrt{2m}\mathbf{1}^t_{bn}}(\tau,z) = h^{-1}(\mathbb{S}+\mathbb{T})\odot BW_{\mathcal{C}_0,g}\left(\theta^{(g)}_{2m,a}(\tau,z): a\in R^{(b,g)}\right).$$

Case II. Let $R=\Sigma_{2m}$. Here we have $\mathcal{G}=\mathcal{O}$, $\mathcal{F}=\mathbb{H}$ and $\mathfrak{S}=4$ so that k is any arbitrary positive integer. Further by Corollary 2.5 of Choie and Dougherty [3], we see that if $\mathcal{C}$ is a Type I code of length bn over Σ_{2m}, then $\psi_{2m}(\mathcal{C})$ is a Type I code of length $4bn$ over $\mathbb{Z}_{2m}$. Now working in a similar way as in the previous case with $L=\Lambda(\psi_{2m}(\mathcal{C}))$ and $\mathcal{G}^{kb}=\psi_{2m}(\mathcal{O}^{kb})=\mathbb{Z}^{4kb}$, the desired result follows immediately. □

By applying the above theorem, one can determine Siegel modular forms of genus g from symmetrized byte weight enumerators in genus g of Type I codes over R as follows:

Theorem 160. *Let $\mathcal{C}$ be a Type I code of length bn and byte length b over R. Then $L = \Lambda(\mathcal{C})$ is a Type I lattice in $\mathcal{F}^{bn}$. For any positive integer k satisfying $bn + kb \equiv 0 \pmod{\frac{8}{\mathfrak{S}}}$, let $\tilde{L}$ be the Type II lattice in $\mathcal{F}^{bn+kb}$ as constructed in Theorem 159. Then*

$$\theta^{(g)}_{\tilde{L},\sqrt{2m}\mathbf{1}^t_{(n+k)b}}(\tau,0) = \sum_{\alpha=0}^{3} \theta^{(g)}_{\mathcal{G}_i^{kb},\sqrt{2m}\mathbf{1}^t_{kb}}(\tau,0)\, \theta^{(g)}_{L_\alpha,\sqrt{2m}\mathbf{1}^t_{bn}}(\tau,0)$$

is a Siegel modular form of genus g with weight $\mathfrak{S}(n+k)b/2$ on $\Gamma_g(\mathcal{G})$, where

$$\theta^{(g)}_{L_0,\sqrt{2m}\mathbf{1}^t_{bn}}(\tau,0) = SW_{\mathcal{C}_0,g}\left(\theta^{(g)}_{2m,\widehat{a}}(\tau,0) : \widehat{a} \in \widehat{R^{(b,g)}}\right),$$
$$\theta^{(g)}_{L_1,\sqrt{2m}\mathbf{1}^t_{bn}}(\tau,0) = \widehat{h^{-1}(\mathbb{S})} \odot SW_{\mathcal{C}_0,g}\left(\theta^{(g)}_{2m,\widehat{a}}(\tau,0) : \widehat{a} \in \widehat{R^{(b,g)}}\right),$$
$$\theta^{(g)}_{L_2,\sqrt{2m}\mathbf{1}^t_{bn}}(\tau,0) = \widehat{h^{-1}(\mathbb{T})} \odot SW_{\mathcal{C}_0,g}\left(\theta^{(g)}_{2m,\widehat{a}}(\tau,0) : \widehat{a} \in \widehat{R^{(b,g)}}\right),$$
$$\theta^{(g)}_{L_3,\sqrt{2m}\mathbf{1}^t_{bn}}(\tau,0) = \widehat{h^{-1}(\mathbb{S}+\mathbb{T})} \odot SW_{\mathcal{C}_0,g}\left(\theta^{(g)}_{2m,\widehat{a}}(\tau,0) : \widehat{a} \in \widehat{R^{(b,g)}}\right).$$

Proof. It follows immediately from Theorem 159, and Remarks 140(i) and 144(ii). □

The following theorem shows that one can compute the byte weight enumerator and the symmetrized byte weight enumerator in genus g for the code $\mathcal{C}_0$ provided the same are known for the code $\mathcal{C}$.

Theorem 161. *Let $\mathcal{C}$ be a Type I code of length bn and byte length b over R and $\mathcal{C}_0 = \{c \in \mathcal{C} : wt_E(c) \equiv 0 \pmod{4m}\}$. Then we have*

(i) $BW_{\mathcal{C}_0,g}(X)$

$$= \frac{1}{2^g}\left\{ BW_{\mathcal{C},g}(X) + \sum_{j=1}^{g} BW_{\mathcal{C},g}\left(\eta^{wt_E(a_j)} X_a : a = (a_1; a_2; \cdots; a_g) \in R^{(b,g)}\right)\right.$$
$$+ \sum_{1\le j_1<j_2\le g} BW_{\mathcal{C},g}\left(\eta^{\sum_{k=1}^{2} wt_E(a_{j_k})} X_a : a = (a_1; a_2; \cdots; a_g) \in R^{(b,g)}\right) + \cdots\cdots$$
$$\left. + \sum_{1\le j_1<j_2<\cdots<j_g\le g} BW_{\mathcal{C},g}\left(\eta^{\sum_{k=1}^{g} wt_E(a_{j_k})} X_a : a = (a_1; a_2; \cdots; a_g) \in R^{(b,g)}\right)\right\},$$

(ii) $SW_{\mathcal{C}_0,g}(\widehat{X})$

$$= \frac{1}{2^g}\left\{SW_{\mathcal{C},g}(\widehat{X}) + \sum_{j=1}^{g} SW_{\mathcal{C},g}\left(\eta^{wt_E(a_j)}X_{\widehat{a}} : \widehat{a} = (\widehat{a_1};\widehat{a_2};\cdots;\widehat{a_g}) \in \widehat{R^{(b,g)}}\right)\right.$$

$$+ \sum_{1\le j_1<j_2\le g} SW_{\mathcal{C},g}\left(\eta^{\sum_{k=1}^{2} wt_E(a_{j_k})} X_{\widehat{a}} : \widehat{a} = (\widehat{a_1};\widehat{a_2};\cdots;\widehat{a_g}) \in \widehat{R^{(b,g)}}\right) + \cdots\cdots$$

$$\left. + \sum_{1\le j_1<j_2<\cdots<j_g\le g} SW_{\mathcal{C},g}\left(\eta^{\sum_{k=1}^{g} wt_E(a_{j_k})} X_{\widehat{a}} : \widehat{a} = (\widehat{a_1};\widehat{a_2};\cdots;\widehat{a_g}) \in \widehat{R^{(b,g)}}\right)\right\},$$

where $X = \left(X_a : a = (a_1;a_2;\cdots;a_g) \in R^{(b,g)}\right)$, $\widehat{X} = \left(X_{\widehat{a}} : \widehat{a} = (\widehat{a_1};\widehat{a_2};\cdots;\widehat{a_g}) \in \widehat{R^{(b,g)}}\right)$ *and* η *is a primitive* $(4m)$*th root of unity in* $\mathbb{C}$. *(Note that* $\eta^{wt_E(u)} = \eta^{wt_E(w)}$ *for every* $w \in \widehat{u}$, *where* $u, w \in R^b$.*)*

Proof. Proof is a straightforward exercise and is left to the reader. □

Remark 162. In view of Theorems 159-161, we see that one can determine a Jacobi form and a Siegel modular form of genus g knowing the byte weight enumerator and the symmetrized byte weight enumerator in genus g of a Type I code over R.

The next section discusses an interesting application of the theta series (as defined by (7.6)) in deriving a functional equation for partial Epstein zeta functions.

7.8 Partial Epstein Zeta Functions and their Functional Equation

In this section, partial Epstein zeta functions for each element in $R^{(b,g)}$ are defined and their functional equation is derived using the Mellin transform of the theta series defined by (7.6).

Definition 163. Let Y be a $b \times b$ matrix of a positive definite quadratic form, and let r, h be $b \times g$ real matrices. Then for any $s \in \mathbb{C}$ with $Re(s) > \mathfrak{S}bg$, the Epstein zeta function associated with (Y, r, h) is defined as

$$Z_b(Y, r, h, s) = \sum_{\substack{a\in\mathcal{G}^{(b,g)}\\ a+g\neq 0}} \frac{e^{2\pi i\sigma(ha^*)}}{(\sigma((a+r)^*Y(a+r)))^{s/2}}.$$

(Recall that $\mathfrak{S} = 1$, $\mathcal{G} = \mathbb{Z}$ when $R = \mathbb{Z}_{2m}$ and $\mathfrak{S} = 4$, $\mathcal{G} = \mathcal{O}$ when $R = \Sigma_{2m}$.)

In particular, when $r = h = 0$ and $Y = I_b$, we have

$$Z_b(I_b, 0, 0, s) = \sum_{\substack{a \in \mathcal{G}^{(b,g)} \\ a \neq 0}} \frac{1}{(\sigma(a^* a))^{s/2}},$$

where I_b denotes the $b \times b$ identity matrix. Let us denote the function $Z_b(I_b, 0, 0, s)$ by $Z_b(s)$ and define the partial Epstein zeta functions as follows:

Definition 164. For a complex variable s with $Re(s) > \mathfrak{S}bg$, partial Epstein zeta functions $Z_{b,\mu}(s)$, $\mu \in R^{(b,g)}$, are defined as

$$\begin{aligned} Z_{b,\mu}(s) &= \sum_{\substack{a \in \mathcal{G}^{(b,g)}, a \neq 0 \\ a \equiv \mu \pmod{2m}}} \frac{1}{(\sigma(aa^*))^{s/2}} \\ &= \sum_{\substack{n \in \mathcal{G}^{(b,g)} \\ 2mn + \mu \neq 0}} \frac{1}{\sigma((\mu + 2mn)(\mu + 2mn)^*)^{s/2}}. \end{aligned}$$

In the following theorem, a functional equation for partial Epstein zeta functions is derived.

Theorem 165. *For each $\mu \in R^{(b,g)}$, the partial Epstein zeta function $Z_{b,\mu}(s)$ (defined for $Re(s) > \mathfrak{S}bg$) extends analytically to a meromorphic function on the whole complex s-plane, which are holomorphic except for a simple pole at $s = \mathfrak{S}bg$ with residue $\frac{2}{\Gamma(\mathfrak{S}gb/2)} \left(\frac{\pi}{4m^2}\right)^{\mathfrak{S}gb/2}$, where Γ is the Gamma function. For each $\mu \in R^{(b,g)}$, if we define*

$$\Lambda_{b,\mu}(s) = \left(\frac{\pi}{2m}\right)^{-s/2} \Gamma\left(\frac{s}{2}\right) Z_{b,\mu}(s),$$

then we have

$$\Lambda_{b,\mu}(s) = \left(\frac{1}{2m}\right)^{\mathfrak{S}bg/2} \sum_{v \in R^{(b,g)}} e^{\frac{\pi i}{m}\sigma(\mu v^*)} \Lambda_{b,v}(\mathfrak{S}bg - s).$$

In other words, if elements of $R^{(b,g)}$ are listed as $\mu_0 = 0, \mu_1, \cdots, \mu_{(2m)^{\mathfrak{S}bg}-1}$, then the functions $Z_{b,\mu}(s)$ ($\mu \in R^{(b,g)}$) satisfy the following functional *equation:*

$$\Gamma\left(\frac{s}{2}\right)\begin{pmatrix} Z_{b,\mu_0}(s) \\ Z_{b,\mu_1}(s) \\ \vdots \\ Z_{b,\mu_{(2m)^{\mathfrak{S}bg}-1}}(s)\end{pmatrix}$$

$$= \frac{(\pi)^{-(\mathfrak{S}bg-2s)/2}}{(2m)^s}\Gamma\left(\frac{\mathfrak{S}bg-s}{2}\right) M_R^{(2m,g)}\begin{pmatrix} Z_{b,\mu_0}(\mathfrak{S}bg-s) \\ Z_{b,\mu_1}(\mathfrak{S}bg-s) \\ \vdots \\ Z_{b,\mu_{(2m)^{\mathfrak{S}bg}-1}}(\mathfrak{S}bg-s)\end{pmatrix},$$

where $M_R^{(2m,g)}$ *is the matrix as defined in Theorem 137.*

Proof. For $\mu \in R^{(b,g)}$, consider the theta series

$$\theta_{2m,\mu}^{(g)}(\tau, z) = \sum_{\substack{x\in\mathcal{G}^{(b,g)} \\ x\equiv\mu(mod\ 2m)}} e^{\frac{2\pi i}{4m}\sigma(x\tau x^*)}e^{\frac{2\pi i}{2}\sigma(\mathbf{1}_b x z^* + z x^* \mathbf{1}_b^t)}.$$

On substituting $\tau = itI_g$ with $t > 0$ and $z = 0$, we obtain

$$\begin{aligned}\theta_{2m,\mu}^{(g)}(itI_g, 0) &= \sum_{\substack{x\in\mathcal{G}^{(b,g)} \\ x\equiv\mu(mod\ 2m)}} e^{\frac{2\pi i}{4m}\sigma(xitI_g x^*)} \\ &= \sum_{n\in 2m\mathcal{G}^{(b,g)}} e^{-\frac{\pi t}{2m}\sigma((n+\mu)(n+\mu)^*)} = \theta_{2m,\mu}(t).\end{aligned}$$

Similarly, on substituting $\tau = -\frac{1}{it}I_g$ and $z = 0$, we get

$$\theta_{2m,\mu}^{(g)}\left(-\frac{1}{it}I_g, 0\right) = \sum_{n\in 2m\mathcal{G}^{(b,g)}} e^{-\frac{\pi}{2mt}\sigma((n+\mu)(n+\mu)^*)} = \theta_{2m,\mu}(1/t).$$

As $R^{(b,g)} = \{\mu_0 = 0, \mu_1, \cdots, \mu_{(2m)^{\mathfrak{S}bg}-1}\}$, using Lemma 152, we obtain

$$\begin{pmatrix} \theta_{2m,\mu_0}(t) \\ \theta_{2m,\mu_1}(t) \\ \\ \theta_{2m,\mu_{(2m)^{\mathfrak{S}bg}-1}}(t)\end{pmatrix} = \left(\frac{1}{2mt}\right)^{\mathfrak{S}bg/2} M_R^{(2m,g)}\begin{pmatrix} \theta_{2m,\mu_0}(1/t) \\ \theta_{2m,\mu_1}(1/t) \\ \\ \theta_{2m,\mu_{(2m)^{\mathfrak{S}bg}-1}}(1/t)\end{pmatrix}. \tag{7.10}$$

From this, it follows that if t is very large, then $\theta_{2m,\mu_0}(t)$ is asymptotic to 1 and $\theta_{2m,\mu_\alpha}(t)$ is asymptotic to zero for each α, $1 \le \alpha \le (2m)^{\mathfrak{S}bg} - 1$.

Furthermore, when t tends to 0, $\theta_{2m,\mu_i}(t)$ is asymptotic to $\left(\frac{1}{2mt}\right)^{\mathfrak{S}bg/2}$ for $0 \leq \alpha \leq (2m)^{\mathfrak{S}bg} - 1$. Now for each μ_α, let us define

$$\phi_{b,\mu_0}(s) = \int_1^\infty t^{s/2-1}\left(\theta_{2m,\mu_0}(t) - 1\right) dt + \int_0^1 t^{s/2-1}\left(\theta_{2m,\mu_0}(t) - \left(\frac{1}{2mt}\right)^{\mathfrak{S}bg/2}\right) dt, \tag{7.11}$$

$$\phi_{b,\mu_\alpha}(s) = \int_1^\infty t^{s/2-1}\theta_{2m,\mu_i}(t)dt + \int_0^1 t^{s/2-1}\left(\theta_{2m,\mu_\alpha}(t) - \left(\frac{1}{2mt}\right)^{\mathfrak{S}bg/2}\right) dt \tag{7.12}$$

for $1 \leq \alpha \leq (2m)^{\mathfrak{S}bg} - 1$. From this, for each s in the half-plane with $Re(s) > \mathfrak{S}bg$, we get

$$\phi_{b,\mu_0}(s) = \int_0^\infty t^{s/2-1}\left(\sum_{\substack{n \in 2m\mathcal{G}^{(b,g)} \\ n \neq 0}} e^{-\frac{\pi i}{2m}\sigma(nn^*)}\right) dt + \frac{2}{s} + \left(\frac{1}{2m}\right)^{\mathfrak{S}bg/2} \frac{2}{\mathfrak{S}bg - s}.$$

Now on taking the Mellin transform of the function $e^{-\frac{\pi i}{2m}\sigma(nn^*)}$, we obtain

$$\begin{aligned}\phi_{b,\mu_0}(s) &= \left(\frac{\pi}{2m}\right)^{-s/2}\Gamma\left(\frac{s}{2}\right) Z_{b,\mu_0}(s) + \frac{2}{s} + \left(\frac{1}{2m}\right)^{\mathfrak{S}bg/2} \frac{2}{\mathfrak{S}bg - s} \\ &= \Lambda_{b,\mu_0}(s) + \frac{2}{s} + \left(\frac{1}{2m}\right)^{\mathfrak{S}bg/2} \frac{2}{\mathfrak{S}bg - s}.\end{aligned} \tag{7.13}$$

This yields

$$Z_{b,\mu_0}(s) = \frac{(\pi/2m)^{s/2}}{\Gamma(s/2)}\left(\phi_{b,\mu_0}(s) - \frac{2}{s} - \left(\frac{1}{2m}\right)^{\mathfrak{S}bg/2} \frac{2}{\mathfrak{S}bg - s}\right).$$

Since the integrals in (7.11) converge for every value of s, $\phi_{b,\mu_0}(s)$ is an entire function of s. Furthermore, the functions $(\pi/2m)^{s/2}$, $\frac{1}{\Gamma(s/2)}$ are entire functions and $s\Gamma(s/2) = 2\Gamma(s/2+1)$ is non-zero when s tends to zero. Thus $Z_{b,\mu_0}(s)$ has a simple pole at $s = \mathfrak{S}bg$ with residue $\frac{2}{\Gamma(\mathfrak{S}gb/2)}\left(\frac{\pi}{4m^2}\right)^{\mathfrak{S}gb/2}$.

Working in a similar way for each α, $1 \leq \alpha \leq (2m)^{\mathfrak{S}bg} - 1$, we obtain

$$\begin{aligned}
\phi_{b,\mu_i}(s) &= \int_0^\infty t^{s/2-1} \left(\sum_{n \in 2m\mathcal{G}^{(b,g)}} e^{-\frac{\pi i}{2m}\sigma((n+\mu_i)(n+\mu_i)^*)} \right) dt \\
&\quad + \left(\frac{1}{2m}\right)^{\mathfrak{S}bg/2} \frac{2}{\mathfrak{S}bg - s} \\
&= \left(\frac{\pi}{2m}\right)^{-s/2} \Gamma\left(\frac{s}{2}\right) Z_{b,\mu_\alpha}(s) + \left(\frac{1}{2m}\right)^{\mathfrak{S}bg/2} \frac{2}{\mathfrak{S}bg - s} \\
&= \Lambda_{b,\mu_\alpha}(s) + \left(\frac{1}{2m}\right)^{\mathfrak{S}bg/2} \frac{2}{\mathfrak{S}bg - s}.
\end{aligned}$$

This gives

$$Z_{b,\mu_\alpha}(s) = \frac{(\pi/2m)^{s/2}}{\Gamma(s/2)} \left(\phi_{b,\mu_i}(s) - \left(\frac{1}{2m}\right)^{\mathfrak{S}bg/2} \frac{2}{\mathfrak{S}bg - s} \right)$$

has a simple pole at $s = \mathfrak{S}bg$ with residue $\frac{2}{\Gamma(\mathfrak{S}gb/2)}\left(\frac{\pi}{4m^2}\right)^{\mathfrak{S}gb/2}$ for each α.

Next we proceed to derive a functional equation for $Z_{b,\mu_\alpha}(s)$, $0 \leq \alpha \leq (2m)^{\mathfrak{S}bg} - 1$. For this, we see, by (7.13), that

$$\begin{aligned}
\Lambda_{b,\mu_0}(s) &= \phi_{b,\mu_0}(s) - \frac{2}{s} - \left(\frac{1}{2m}\right)^{\mathfrak{S}bg/2} \frac{2}{\mathfrak{S}bg - s} \\
&= \int_1^\infty t^{s/2-1} \left(\theta_{2m,\mu_0}(t) - 1\right) dt \\
&\quad + \int_0^1 t^{s/2-1} \left(\theta_{2m,\mu_0}(t) - \left(\frac{1}{2mt}\right)^{\mathfrak{S}bg/2} \right) dt \\
&\quad - \frac{2}{s} - \left(\frac{1}{2m}\right)^{\mathfrak{S}bg/2} \frac{2}{\mathfrak{S}bg - s}.
\end{aligned}$$

On taking $t = 1/u$ and by using (7.10), we obtain

$$\begin{aligned}
\Lambda_{b,\mu_0}(s) &= \int_0^1 u^{-s/2-1} \left[\left(\frac{u}{2m}\right)^{\mathfrak{S}bg/2} \sum_{a \in R^{(b,g)}} e^{\frac{\pi}{m}\sigma(\mu_0 a^*)} \theta_{2m,a}(u) - 1 \right] du \\
&\quad + \int_1^\infty u^{-s/2-1} \left[\left(\frac{u}{2m}\right)^{\mathfrak{S}bg/2} \sum_{a \in R^{(b,g)}} e^{\frac{\pi}{m}\sigma(\mu_0 a^*)} \theta_{2m,a}(u) \right. \\
&\quad \left. - \left(\frac{1}{2mt}\right)^{\mathfrak{S}bg/2} \right] du - \frac{2}{s} - \left(\frac{1}{2m}\right)^{\mathfrak{S}bg/2} \frac{2}{\mathfrak{S}bg - s}
\end{aligned}$$

$$= \left(\frac{1}{2m}\right)^{\mathfrak{S}bg/2} e^{\frac{\pi i}{m}\sigma(\mu_0\mu_0^*)} \left\{ \int_1^\infty u^{\frac{\mathfrak{S}bg-s}{2}} (\theta_{2m,\mu_0}(u) - 1) \frac{du}{u} - \frac{2}{\mathfrak{S}bg - s} \right.$$

$$+ \int_0^1 u^{\frac{\mathfrak{S}bg-s}{2}} \left(\theta_{2m,\mu_0}(u) - \left(\frac{1}{2mu}\right)^{\mathfrak{S}bg/2}\right) \frac{du}{u} - \left.\left(\frac{1}{2m}\right)^{\mathfrak{S}bg/2} \frac{2}{s} \right\}$$

$$+ \left(\frac{1}{2m}\right)^{\mathfrak{S}bg/2} \sum_{\substack{a\in R^{(b,g)} \\ a\neq 0}} e^{\frac{\pi i}{m}\sigma(\mu_0 a^*)} \left\{ \int_1^\infty u^{\frac{\mathfrak{S}bg-s}{2}} \theta_{2m,a}(u) \frac{du}{u} \right.$$

$$+ \int_0^1 u^{\frac{\mathfrak{S}bg-s}{2}} \left(\theta_{2m,a}(u) - \left(\frac{1}{2mu}\right)^{\mathfrak{S}bg/2}\right) \frac{du}{u} - \left.\left(\frac{1}{2m}\right)^{\mathfrak{S}bg/2} \frac{2}{s} \right\}$$

$$= \left(\frac{1}{2m}\right)^{\mathfrak{S}bg/2} \sum_{a\in R^{(b,g)}} e^{\frac{\pi i}{m}\sigma(\mu_0 a^*)} \Lambda_{b,a}(\mathfrak{S}bg - s),$$

as $\sum_{a\in R^{(b,g)}} e^{\frac{\pi i}{m}\sigma(\mu_0 a^*)} = (2m)^{\mathfrak{S}bg}$. Further, working in a similar way as above, we note that

$$\Lambda_{b,\mu_\alpha}(s) = \left(\frac{1}{2m}\right)^{\mathfrak{S}bg/2} \sum_{a\in R^{(b,g)}} e^{\frac{\pi i}{m}\sigma(\mu_\alpha a^*)} \Lambda_{b,a}(\mathfrak{S}bg - s) \text{ for } 0 \le \alpha \le (2m)^{\mathfrak{S}bg} - 1,$$

which completes the proof of the theorem. □

Remark 166. Note that the functional equation derived in the above theorem can be rewritten in terms of Epstein zeta functions as follows:

$$\pi^{-\frac{s}{2}} \Gamma\left(\frac{s}{2}\right) \begin{pmatrix} Z_b(I_b, \frac{\mu_0}{2m}, 0, s) \\ Z_b(I_b, \frac{\mu_1}{2m}, 0, s) \\ \vdots \\ Z_b(I_b, \frac{\mu_{(2m)^{\mathfrak{S}bg}-1}}{2m}, 0, s) \end{pmatrix}$$

$$= \pi^{-\frac{(\mathfrak{S}bg-s)}{2}} \Gamma\left(\frac{\mathfrak{S}bg - s}{2}\right) \begin{pmatrix} Z_b(I_b, 0, \frac{\mu_0}{2m}, \mathfrak{S}bg - s) \\ Z_b(I_b, 0, \frac{\mu_1}{2m}, \mathfrak{S}bg - s) \\ \vdots \\ Z_b(I_b, 0, \frac{\mu_{(2m)^{\mathfrak{S}bg}-1}}{2m}, \mathfrak{S}bg - s) \end{pmatrix}.$$

References

[1] E. Bannai, S. T. Dougherty, M. Harada and M. Oura, Type II codes, even unimodular lattices, and invariant rings, *IEEE Trans. Inform. Theory* 45(4), pp. 1194–1205 (1999).

[2] Y. Choie and S. T. Dougherty, Codes over Σ_{2m} and Jacobi forms over the quaternions, *Appl. Algebra Eng. Commun. Comp.* 15, pp. 129–147 (2004).
[3] Y. Choie, S. T. Dougherty, H. Kim, Complete joint weight enumerators and self-dual codes, IEEE Trans. Inform. Theory 49 (2003), no. 5, 1275–1282.
[4] Y. Choie and N. Kim, The complete weight enumerator of Type II codes over $\mathbb{Z}_{2m}$ and Jacobi forms, *IEEE Trans. Inform. Theory* 47(1), pp. 396–399 (2001).
[5] Y. Choie and H. Kim, Codes over $\mathbb{Z}_{2m}$ and Jacobi forms of genus n, *J. Combin. Theory Ser. A* 95, pp. 335–348 (2001).
[6] S. T. Dougherty, M. Harada, M. Oura, Note on the g-fold joint weight enumerators of self-dual codes over $\mathbb{Z}_k$, *Appl. Algebra Eng. Commun. Comp.* 11(6), pp. 437–445 (2001).
[7] S. T. Dougherty, M. Harada and P. Solé, Shadow lattices and shadow codes, *Discrete Mathematics* 219, pp. 49–64 (2000).
[8] W. Ebeling, *Lattices and codes. A course partially based on lectures by F. Hirzebruch*, Adv. Lect. Math., Vieweg, Braunschweig (1994).
[9] M. Eichler and D. Zagier, *The theory of Jacobi forms*, Boston, MA: Birkäuser (1985).
[10] A. Krieg, *Modular forms on half-spaces of quaternions*, Lect. Notes in Math. 1143, Berlin, Heidelberg, New York, Springer (1985).
[11] A. Sharma and A. K. Sharma, MacWilliams identities for m-spotty weight enumerators of codes over rings, *Australasian Journal of Combinatorics* 58(1), pp. 67–105 (2014).
[12] A. Sharma and A. K. Sharma, Byte weight enumerators and modular forms of genus r, *Journal of Algebra and its applications* 14(6), pp. 1–25 (2015).
[13] K. Suzuki, Complete m-spotty weight enumerators of binary codes, Jacobi forms, and partial Epstein zeta functions, *Discrete Mathematics* 312, pp. 265–278 (2012).
[14] K. Suzuki, T. Kashiyama and E. Fujiwara, A General Class of m-Spotty Byte Error Control Codes, *IEICE Transactions* 90-A(7), pp. 1418–1427 (2007).
[15] K. Suzuki, T. Kashiyama and E. Fujiwara, A general class of m-spotty byte error control codes, *Proceedings of Asian-European Workshop on Information Theory*, pp. 24–26, Viareggio, Italy (2004).
[16] G. Umanesan and E. Fujiwara, A Class of Random Multiple Bits in a Byte Error Correcting and Single Byte Error Detecting ($S_{t/b}EC$-S_bED) Codes, *IEEE Trans. Computers* 52(7), pp. 835–847 (2003).

Chapter 8

Byte Weight Enumerators of Codes over $\mathbb{F}_p$ and Modular Forms over a Totally Real Field

8.1 Introduction

An injective homomorphism from an invariant space of polynomials into a space of Jacobi forms on the modular group $\Gamma = SL_2(\mathbb{Z})$ is defined by Choie and Solé [5]. Using this homomorphism, Jacobi forms on Γ are determined from complete weight enumerators (or Lee weight enumerators) of ternary self-dual codes. In [9], a connection between Lee weight enumerators of self-dual codes over $\mathbb{F}_p$ (p is an odd prime) and Hilbert modular forms over the totally real field $k_p = \mathbb{Q}(\zeta_p + \zeta_p^{-1})$ ($\zeta_p = e^{2\pi i/p}$) is established. As an application, Hilbert modular forms over k_p are determined from Lee weight enumerators of self-dual codes over $\mathbb{F}_p$ for every odd prime p. There have been further extensions to obtain Hilbert modular forms coming from codes defined over various fields [1, 2]. The results derived in [9] are further extended by Choie and Jeong [4]. More precisely, an algebra homomorphism is defined from an invariant space of polynomials into a space of Jacobi forms over the totally real field k_p by Choie and Jeong [4]. In the same work, Jacobi forms over the totally real field k_p are obtained from complete weight enumerators (or Lee weight enumerators) of self-dual codes over $\mathbb{F}_p$.

This chapter follows the paper [16], which extends the work of Choie and Jeong [4] and the results obtained in [9] to determine Jacobi forms over the totally real field k_p from byte weight enumerators of self-dual codes over $\mathbb{F}_p$, where p is an odd prime. Moreover, Siegel modular forms of genus g over k_p are defined and some new Siegel modular forms of genus g over

k_p are determined for all primes $p \in \mathfrak{P}$, where the set $\mathfrak{P}$ consists of all those odd primes p such that the ring $\mathcal{O}_{k_p}$ of algebraic integers of k_p is a Euclidean domain.

In Section 8.2, byte weight enumerators in genus g for codes over $\mathbb{F}_p$, Jacobi forms over the totally real field k_p and lattices induced by codes over $\mathbb{F}_p$ are discussed, where p is an odd prime. In Section 8.3, byte weight enumerator of a linear code $\mathcal{C}$ over $\mathbb{F}_p$ is related with the theta series of a lattice induced by the code $\mathcal{C}$ (Theorem 172). In Section 8.4, certain invariant space of polynomials is related with a space of Jacobi forms over k_p, where p is an odd prime (Theorems 175). As a consequence, Jacobi forms over k_p are obtained from byte weight enumerators of self-dual codes over $\mathbb{F}_p$ for every odd prime p (Theorem 178). In Section 8.5, Siegel modular forms of genus g ($g \geq 1$ is an integer) over k_p are defined. Further, Siegel modular forms are obtained from byte weight enumerators in genus g of self-dual codes over $\mathbb{F}_p$ for each $p \in \mathfrak{P}$ (Theorems 183-185). In Section 8.6, an invariant space of polynomials containing byte weight enumerators in genus g of self-dual codes over $\mathbb{F}_p$ is determined and is further related with a space of Siegel modular forms of genus g over k_p for each $p \in \mathfrak{P}$ (Theorems 187 and 188). In Section 8.7, a functional equation for certain partial Epstein zeta functions is derived using the Mellin transform of the theta series (Theorem 193). It is also observed that the coefficient matrix of the functional equation for partial Epstein zeta functions is the same as the transformation matrix in the MacWilliams identity for byte weight enumerator in genus g of a linear code over $\mathbb{F}_p$.

8.2 Some Preliminaries

Let p be an odd prime, and let $\mathbb{F}_p$ be the finite field of order p. For a positive integer N, let $\mathbb{F}_p^N$ be the vector space consisting of all N-tuples over the finite field $\mathbb{F}_p$. For a positive divisor b of N, a byte error-control code $\mathcal{C}$ of length N and byte length b over $\mathbb{F}_p$ is defined as an additive subgroup of $\mathbb{F}_p^N$. With every byte error-control code $\mathcal{C}$, there is associated a dual code $\mathcal{C}^\perp$, which is also a byte error-control code of length N over $\mathbb{F}_p$ and is defined as follows:

$$\mathcal{C}^\perp = \{y \in \mathbb{F}_p^N : \langle x, y\rangle = 0 \text{ for all } x \in \mathcal{C}\},$$

where

$$\langle x, y\rangle = x_1y_1 + x_2y_2 + \cdots + x_Ny_N$$

for $x = (x_1, x_2, \cdots, x_N)$ and $y = (y_1, y_2, \cdots, y_N)$ in $\mathbb{F}_p^N$. Further, the code $\mathcal{C}$ is said to be self-dual if $\mathcal{C} = \mathcal{C}^{\perp}$. By Lemma 4.1 (1) of Choie and Jeong [4], we note that if $\mathcal{C}$ is a self-dual code of length N over $\mathbb{F}_p$, then $N(p-1) \equiv 0 \pmod 8$.

Throughout this chapter, let us assume that $N = bn$ for some positive integer n, and let us represent each vector $u \in \mathbb{F}_p^{bn}$ as $u = (u_1, u_2, \cdots, u_n)$, where $u_h \in \mathbb{F}_p^b$ is called the hth byte of u for each h. Let us assume that elements of $\mathbb{F}_p^b$ are arranged in a lexicographical ordering $\mathfrak{O}$ and any matrix indexed by elements of $\mathbb{F}_p^b$ follows this ordering.

8.2.1 *Byte Weight Enumerators in Genus g*

In this section, the byte weight enumerator in genus g (g is a positive integer) is defined for a byte error-control code of length bn and byte length b over $\mathbb{F}_p$ and the corresponding MacWilliams identity is also derived. For this, let $(a_1; a_2; \cdots; a_g)$ denote a $b \times g$ matrix having $a_1, a_2, \cdots, a_g \in \mathbb{F}_p^b$ as its columns.

Definition 167. Let $\mathcal{C}$ be a byte error-control code of length bn and byte length b over $\mathbb{F}_p$. Then the byte weight enumerator in genus g of $\mathcal{C}$ is defined as

$$BW_{\mathcal{C},g}\left(X_a : a \in \mathbb{F}_p^{(b,g)}\right) = \sum_{c_1,c_2,\cdots,c_g} \prod_{j=1}^{n} X_{\left(c_j^{(1)};c_j^{(2)};\cdots;c_j^{(g)}\right)},$$

where the summation $\sum\limits_{c_1,c_2,\cdots,c_g}$ runs over all codewords $c_m = \left(c_1^{(m)}, c_2^{(m)}, \cdots, c_n^{(m)}\right) \in \mathcal{C}$ with $c_j^{(m)} \in \mathbb{F}_p^b$ for $1 \le m \le g$ and $1 \le j \le n$.

Note that the byte weight enumerator in genus g of the code $\mathcal{C}$ can be rewritten as

$$BW_{\mathcal{C},g}\left(X_a : a \in \mathbb{F}_p^{(b,g)}\right) = \sum_{c_1,c_2,\cdots,c_g \in \mathcal{C}} \prod_{a \in \mathbb{F}_p^{(b,g)}} X_a^{n_a(c_1,c_2,\cdots,c_g)}, \tag{8.1}$$

where $c_k = \left(c_1^{(k)}, c_2^{(k)}, \cdots, c_n^{(k)}\right)$ with $c_j^{(k)} \in \mathbb{F}_p^b$ for $1 \le k \le g$ and $1 \le j \le n$, and

$$n_a(c_1, c_2, \cdots, c_g) = |\{j : 1 \leq j \leq n, (c_j^{(1)}; c_j^{(2)}; \cdots; c_j^{(g)}) = a\}|$$

for each $a \in \mathbb{F}_p^{(b,g)}$.

The following theorem derives MacWilliams identity for the byte weight enumerator in genus g of a byte error-control code over $\mathbb{F}_p$.

Theorem 168. *Let $\mathcal{C}$ be a byte error-control code of length bn and byte length b over $\mathbb{F}_p$. Then the byte weight enumerator in genus g of the dual code $\mathcal{C}^\perp$ is given by*

$$BW_{\mathcal{C}^\perp, g}\left(X_a : a \in \mathbb{F}_p^{(b,g)}\right) = \frac{1}{|\mathcal{C}|^g} BW_{\mathcal{C},g}\left(M_{p,g}\left(X_a : a \in \mathbb{F}_p^{(b,g)}\right)^t\right)$$

with $M_{p,g} = \otimes_{j=1}^g M_p$, where M_p is a $(p^b \times p^b)$ matrix indexed by the elements of $\mathbb{F}_p^b$ with its (u, v)th entry as $(M_p)_{u,v} = \zeta_p^{\langle u,v \rangle}$ for each $u, v \in \mathbb{F}_p^b$. Here ζ_p is a primitive p-th root of unity in $\mathbb{C}$ and $\left(X_a : a \in \mathbb{F}_p^{(b,g)}\right)^t$ denotes the transpose of the row matrix $\left(X_a : a \in \mathbb{F}_p^{(b,g)}\right)$.

Proof. To prove the result, for $v \in \mathbb{F}_p^{nb}$, let us define

$$\delta_{\mathcal{C}^\perp}(v) = \begin{cases} 1 \text{ if } v \in \mathcal{C}^\perp; \\ 0 \text{ otherwise.} \end{cases}$$

Now we observe that $\sum_{c \in \mathcal{C}} \zeta_p^{\langle v,c \rangle} = |\mathcal{C}| \delta_{\mathcal{C}^\perp}(v)$ for each $v \in \mathbb{F}_p^{nb}$.

By (8.1), we have

$$BW_{\mathcal{C}^\perp, g}\left(X_a : a \in \mathbb{F}_p^{(b,g)}\right) = \sum_{d_1, d_2, \cdots, d_g \in \mathcal{C}^\perp} \prod_{a \in \mathbb{F}_p^{(b,g)}} X_a^{n_a(d_1, d_2, \cdots, d_g)},$$

where $d_k = \left(d_1^{(k)}, d_2^{(k)}, \cdots, d_n^{(k)}\right)$ with $d_j^{(k)} \in \mathbb{F}_p^b$ for $1 \leq k \leq g$ and $1 \leq j \leq n$, and

$$n_a(d_1, d_2, \cdots, d_g) = |\{j : 1 \leq j \leq n, (d_j^{(1)}; d_j^{(2)}; \cdots; d_j^{(g)}) = a\}|$$

for each $a \in \mathbb{F}_p^{(b,g)}$.

Now let us consider

$$
\begin{aligned}
&|\mathcal{C}|^g BW_{\mathcal{C}^\perp,g}\left(X_a : a \in \mathbb{F}_p^{(b,g)}\right) \\
&= |\mathcal{C}|^g \sum_{d_1,d_2,\cdots,d_g \in \mathcal{C}^\perp} \prod_{a\in\mathbb{F}_p^{(b,g)}} X_a^{n_a(d_1,d_2,\cdots,d_g)} \\
&= |\mathcal{C}|^g \sum_{v_1,v_2,\cdots,v_g\in\mathbb{F}_p^{nb}} \prod_{h=1}^{g} \delta_{\mathcal{C}^\perp}(v_h) \prod_{a\in\mathbb{F}_p^{(b,g)}} X_a^{n_a(v_1,v_2,\cdots,v_g)} \\
&= \sum_{v_1,v_2,\cdots,v_g\in\mathbb{F}_p^{nb}} \sum_{c_1,c_2,\cdots,c_g\in\mathcal{C}} \zeta_p^{\sum_{h=1}^{g}\langle v_h,c_h\rangle} \prod_{a\in\mathbb{F}_p^{(b,g)}} X_a^{n_a(v_1,v_2,\cdots,v_g)} \\
&= \sum_{c_1,c_2,\cdots,c_g\in\mathcal{C}} \prod_{j=1}^{n} \left(\sum_{(v_j^{(1)};v_j^{(2)};\cdots;v_j^{(g)})\in\mathbb{F}_p^{(b,g)}} \zeta_p^{\sum_{h=1}^{g}\langle v_j^{(h)},c_j^{(h)}\rangle} X_{(v_j^{(1)};v_j^{(2)};\cdots;v_j^{(g)})} \right) \\
&= \sum_{c_1,c_2,\cdots,c_g\in\mathcal{C}} \left(\sum_{a\in\mathbb{F}_p^{(b,g)}} \zeta_p^{\langle a,(c_j^{(1)},c_j^{(2)},\cdots,c_j^{(g)})\rangle} \right)^{n_a(c_1,c_2,\cdots,c_g)} \\
&= BW_{\mathcal{C},g}\left(M_{p,g}\left(X_a : a\in\mathbb{F}_p^{(b,g)}\right)^t \right).
\end{aligned}
$$

This completes the proof of the theorem. □

Note that when $g = 1$, the byte weight enumerator in genus g of a byte error-control code $\mathcal{C}$ over $\mathbb{F}_p$ coincides with the byte weight enumerator of $\mathcal{C}$.

The byte weight enumerator in genus g of a self-dual code over $\mathbb{F}_p$ can be further related to a Jacobi form over a totally real field, which is as discussed below.

8.2.2 *Jacobi Form over a Totally Real Field*

For an odd prime p, let $\zeta_p = e^{2\pi i/p}$, and let $K_p = \mathbb{Q}(\zeta_p)$ be the cyclotomic field. It is easy to see that the totally real subfield of K_p is $k_p = \mathbb{Q}(\zeta_p + \zeta_p^{-1})$, where $[K_p : k_p] = 2$ and $[k_p : \mathbb{Q}] = \frac{p-1}{2}$. Let $\sigma_1, \sigma_2, \cdots, \sigma_{\frac{p-1}{2}}$ be the different embeddings of k_p into $\mathbb{R}$ with the identity map σ_1. The trace of an element $v \in k_p$, denoted by $Tr_{k_p/\mathbb{Q}}(v)$, is defined as

$$Tr_{k_p/\mathbb{Q}}(v) = \sum_{j=1}^{\frac{p-1}{2}} \sigma_j(v).$$

Let $\mathcal{O}_{k_p}$ denote the ring of algebraic integers of k_p. Then the Jacobi group of k_p, denoted by $\Gamma^J(k_p)$, is defined as

$$\Gamma^J(k_p) = SL_2(\mathcal{O}_{k_p}) \propto \mathcal{O}_{k_p}^2,$$

where $SL_2(\mathcal{O}_{k_p})$ is the special linear group consisting of 2×2 matrices over $\mathcal{O}_{k_p}$. Note that $\Gamma^J(k_p)$ is a group with respect to the binary operation $\circ$, which is defined as

$$(M, [\lambda, \mu]) \circ (M', [\alpha, \beta]) = (MM', [\lambda, \mu]M' + [\alpha, \beta])$$

for all $M, M' \in SL_2(\mathcal{O}_{k_p})$ and $[\lambda, \mu], [\alpha, \beta] \in \mathcal{O}_{k_p}^2$.

Let $\mathcal{H}$ be the complex upper half-plane. Now to define the action of the group $\Gamma^J(k_p)$ on the space $\mathcal{H}^{\frac{p-1}{2}} \times \mathbb{C}^{\frac{p-1}{2}}$ for each conjugate of the field, let us denote the variables of the space $\mathcal{H}^{\frac{p-1}{2}} \times \mathbb{C}^{\frac{p-1}{2}}$ by $(\tau, z) = (\tau_1, \tau_2, \cdots, \tau_{\frac{p-1}{2}}, z_1, z_2, \cdots, z_{\frac{p-1}{2}})$, where $\tau_h \in \mathcal{H}$ and $z_h \in \mathbb{C}$ for $1 \le h \le \frac{p-1}{2}$. Then the action of $\Gamma^J(k_p)$ on the space $\mathcal{H}^{\frac{p-1}{2}} \times \mathbb{C}^{\frac{p-1}{2}}$ is defined as

$$\begin{aligned}\begin{pmatrix} \alpha & \beta \\ \gamma & \delta \end{pmatrix}(\tau, z) &:= \left(\frac{\alpha\tau + \beta}{\gamma\tau + \delta}, \frac{z}{\gamma\tau + \delta}\right) \\ &:= \left(\frac{\alpha^{(1)}\tau_1 + \beta^{(1)}}{\gamma^{(1)}\tau_1 + \delta^{(1)}}, \cdots, \frac{\alpha^{(\frac{p-1}{2})}\tau_{\frac{p-1}{2}} + \beta^{(\frac{p-1}{2})}}{\gamma^{(\frac{p-1}{2})}\tau_{\frac{p-1}{2}} + \delta^{(\frac{p-1}{2})}}, \frac{z_1}{\gamma^{(1)}\tau_1 + \delta^{(1)}}, \right. \\ &\qquad \left. \cdots, \frac{z_{\frac{p-1}{2}}}{\gamma^{(\frac{p-1}{2})}\tau_{\frac{p-1}{2}} + \delta^{(\frac{p-1}{2})}}\right)\end{aligned}$$

for all $\begin{pmatrix} \alpha & \beta \\ \gamma & \delta \end{pmatrix} \in SL_2(\mathcal{O}_{k_p})$ and

$$\begin{aligned}&[\lambda, \mu](\tau, z) \\ &\quad := (\tau, z + \lambda\tau + \mu) \\ &\quad := \left(\tau_1, \cdots, \tau_{\frac{p-1}{2}}, z_1 + \lambda^{(1)}\tau_1 + \mu^{(1)}, \cdots, z_{\frac{p-1}{2}} + \lambda^{(\frac{p-1}{2})}\tau_{\frac{p-1}{2}} + \mu^{(\frac{p-1}{2})}\right)\end{aligned}$$

for all $[\lambda, \mu] \in \mathcal{O}_{k_p}^2$.

Let $f : \mathcal{H}^{\frac{p-1}{2}} \times \mathbb{C}^{\frac{p-1}{2}} \to \mathbb{C}$ be a function. Then for $\ell \in \mathbb{Z}$, $m \in \mathcal{O}_{k_p}$, $\begin{pmatrix} \alpha & \beta \\ \gamma & \delta \end{pmatrix} \in SL_2(\mathcal{O}_{k_p})$ and $[\lambda, \mu] \in \mathcal{O}_{k_p}^2$, the slash operators on f are defined as

$$\left(f\mid_{\ell,m}\begin{pmatrix}\alpha & \beta\\ \gamma & \delta\end{pmatrix}\right)(\tau,z)$$

$$=\left(\prod_{j=1}^{\frac{p-1}{2}} e^{-2\pi i m^{(j)}\frac{\gamma^{(j)}z_j^2}{\gamma^{(j)}\tau_j+\delta^{(j)}}}\right)\prod_{j=1}^{\frac{p-1}{2}}\left(\gamma^{(j)}\tau_j+\delta^{(j)}\right)^{-\ell} f\left(\frac{\alpha\tau+\beta}{\gamma\tau+\delta},\frac{z}{\gamma\tau+\delta}\right)$$

and $(f\mid_m[\lambda,\mu])(\tau,z)=\left(\prod_{j=1}^{\frac{p-1}{2}} e^{2\pi i m^{(j)}\left(\lambda^{(j)^2}\tau_j+2\lambda^{(j)}z_j\right)}\right) f(\tau,z+\lambda\tau+\mu)$.

Definition 169. [4,10] Let $\ell\in\mathbb{Z}$ and $m\in\mathcal{O}_{k_p}$. Then an analytic function $f:\mathcal{H}^{\frac{p-1}{2}}\times\mathbb{C}^{\frac{p-1}{2}}\to\mathbb{C}$ is said to be a Jacobi form of weight ℓ and index m over k_p if it satisfies the following:

$$\left(f\mid_{\ell,m}\begin{pmatrix}\alpha & \beta\\ \gamma & \delta\end{pmatrix}\right)(\tau,z)=f(\tau,z)\text{ for all }\begin{pmatrix}\alpha & \beta\\ \gamma & \delta\end{pmatrix}\in SL_2(\mathcal{O}_{k_p}),\qquad(8.2)$$

$$(f\mid_m[\lambda,\mu])(\tau,z)=f(\tau,z)\text{ for all }[\lambda,\mu]\in\mathcal{O}_{k_p}^2,\qquad(8.3)$$

and $f(\tau,z)$ has a Fourier series expansion of the form

$$f(\tau,z)=\sum_{N,R\in\delta_{k_p}^{-1},N\geq 0} c(N,R)\prod_{j=1}^{\frac{p-1}{2}} e^{\frac{2\pi i}{\omega}\left(N^{(j)}\tau_j+R^{(j)}z_j\right)}\qquad(8.4)$$

with a suitable $\omega\in\mathbb{Z}$, where $N\geq 0$ means that N is either totally-positive or zero, each coefficient $c(N,R)$ is a constant and $\delta_{k_p}^{-1}=\{a\in k_p: Tr_{k_p/\mathbb{Q}}(ab)\in\mathbb{Z}\text{ for all }b\in\mathcal{O}_{k_p}\}$ is called the inverse different of k_p.

Remark 170. The set of all Jacobi forms with weight ℓ and index m over k_p forms an algebra over $\mathbb{C}$ and is denoted by $\mathcal{J}_{\ell,m}(k_p)$.

8.2.3 *Lattices Induced by Codes over $\mathbb{F}_p$*

Now to define a lattice induced by a byte error-control code over $\mathbb{F}_p$, let us first recall the following basic definitions and results.

For each $v,w\in K_p$, let us define a map $\cdot: K_p\times K_p\to k_p$ as

$$v\cdot w=v\bar{w}+\bar{v}w,$$

where $\bar{v}$ denotes the complex conjugate of v for each $v\in K_p$. Further, for each positive integer ℓ, we extend this map to $\cdot: K_p^\ell\times K_p^\ell\to k_p$ as

$$v\cdot w=\sum_{j=1}^{\ell} v_j\cdot w_j$$

for each $v = (v_1, \cdots, v_\ell), w = (w_1, \cdots, w_\ell) \in K_p^\ell$. Note that the map $\cdot$ is a totally positive-definite scalar product on K_p^ℓ, i.e., $\sigma_s(v \cdot v) = (v \cdot v)^{(s)} > 0$ for $1 \le s \le \frac{p-1}{2}$ and for each $v \in K_p^\ell \setminus \{0\}$.

A k_p-lattice Λ in K_p^ℓ is defined as a finitely generated $\mathcal{O}_{k_p}$-submodule of K_p^ℓ containing a k_p-basis of K_p^ℓ. If $\{e_1, e_2, \cdots, e_{(p-1)\ell}\}$ is a $\mathbb{Z}$-basis of the lattice Λ, then the discriminant of Λ, denoted by $\Delta(\Lambda)$, is defined as

$$\Delta(\Lambda) = \det(Tr_{k_p/\mathbb{Q}}(e_h \cdot e_j)),$$

where det denotes the matrix determinant function. The dual lattice of Λ, denoted by Λ^*, is defined as

$$\Lambda^* = \{w \in K_p^\ell : Tr_{k_p/\mathbb{Q}}(w \cdot v) \in \mathbb{Z} \text{ for all } v \in \Lambda\}.$$

Note that the dual lattice Λ^* is also a k_p-lattice in K_p^ℓ. A lattice Λ is said to be unimodular if $\Lambda^* = \Lambda$ and is said to be integral if $Tr_{k_p/\mathbb{Q}}(v \cdot w) \in \mathbb{Z}$ for all $v, w \in \Lambda$. Further, a unimodular lattice Λ is called a Type II lattice (or an even lattice) if $Tr_{k_p/\mathbb{Q}}(\epsilon(v \cdot v)) \in 2\mathbb{Z}$ for all $v \in \Lambda$ and $\epsilon \in \mathcal{O}_{k_p}$.

Let $\mathcal{O}_{K_p}$ denote the ring of algebraic integers of K_p. One can easily show that

$$\mathcal{O}_{K_p} = \left\{ \sum_{j=0}^{p-2} a_j \zeta_p^j : a_j \in \mathbb{Z} \text{ for } 0 \le j \le p-2 \right\}.$$

Next let β_p be the principal ideal of $\mathcal{O}_{K_p}$ generated by the element $1 - \zeta_p \in \mathcal{O}_{K_p}$. Then it is easy to show that β_p is a k_p-lattice in K_p with

$$\beta_p^* = \mathcal{O}_{K_p},\ \Delta(\beta_p) = p,\ \ Tr_{k_p/\mathbb{Q}}\left(\frac{x \cdot y}{p}\right) \in \mathbb{Z},\ \ Tr_{k_p/\mathbb{Q}}\left(\frac{x \cdot x}{p}\right) \in 2\mathbb{Z}$$
$$\text{for each } x \in \beta_p \text{ and } y \in \mathcal{O}_{K_p} \quad (8.5)$$

Now define a map $\rho_p : \mathcal{O}_{K_p} \to \mathbb{Z}/p\mathbb{Z}$ as

$$\rho_p(\alpha) = a_0 + a_1 + \cdots + a_{p-2} \pmod p$$

for each $\alpha = a_0 + a_1\zeta_p + \cdots + a_{p-2}\zeta_p^{p-2} \in \mathcal{O}_{K_p}$. Note that ρ_p is a ring homomorphism with kernel β_p and can be viewed as a reduction map modulo β_p. From this, it follows that $\mathcal{O}_{K_p}/\beta_p \cong \mathbb{Z}/p\mathbb{Z} \cong \mathbb{F}_p$, which implies that

$$\mathcal{O}_{K_p} = \bigcup_{a \in \mathbb{F}_p} (a + \beta_p). \quad (8.6)$$

Let us further extend this map to $\rho_p : \mathcal{O}_{K_p}^\ell \to \mathbb{F}_p^\ell$ by applying reduction modulo β_p coordinate wise. Now if $\mathcal{C}$ is a byte error-control code of length bn and byte length b over $\mathbb{F}_p$, then the set $\Lambda_\mathcal{C} = \frac{1}{\sqrt{p}}\rho_p^{-1}(\mathcal{C})$ is a k_p-lattice in K_p^{bn} and is called the lattice induced by $\mathcal{C}$. For more details, one may refer to [9, Ch. 5].

8.3 Theta Series for a Lattice and Byte Weight Enumerator of a Code over $\mathbb{F}_p$

A theta series for the lattice induced by a byte error-control code over $\mathbb{F}_p$ is defined as follows:

Definition 171. Let $\mathcal{C}$ be a byte error-control code of length bn and byte length b over $\mathbb{F}_p$, and let $\Lambda_{\mathcal{C}}$ be the lattice induced by $\mathcal{C}$. Then for $\mathcal{Y} \in \Lambda_{\mathcal{C}}$, the theta series for the induced lattice $\Lambda_{\mathcal{C}}$ is a function $\Theta_{\mathcal{C},\mathcal{Y}} : \mathcal{H}^{\frac{p-1}{2}} \times \mathbb{C}^{\frac{p-1}{2}} \to \mathbb{C}$, defined as

$$\Theta_{\mathcal{C},\mathcal{Y}}(\tau, z) = \sum_{x \in \Lambda_{\mathcal{C}}} e^{\pi i Tr_{k_p/\mathbb{Q}}((x \cdot x)\tau)} e^{2\pi i Tr_{k_p/\mathbb{Q}}((x \cdot \mathcal{Y})z)},$$

where

$$Tr_{k_p/\mathbb{Q}}((x \cdot x)\tau) = \sum_{s=1}^{\frac{p-1}{2}} (x \cdot x)^{(s)} \tau_s \quad \text{and} \quad Tr_{k_p/\mathbb{Q}}((x \cdot \mathcal{Y})z) = \sum_{s=1}^{\frac{p-1}{2}} (x \cdot \mathcal{Y})^{(s)} z_s.$$

The following theorem relates the theta series for the lattice induced by a byte error-control code $\mathcal{C}$ over $\mathbb{F}_p$ with the byte weight enumerator of $\mathcal{C}$.

Theorem 172. *Let $\mathcal{C}$ be a byte error-control code of length bn and byte length b over $\mathbb{F}_p$. Then for $\mathcal{Y} = \frac{1}{\sqrt{p}}(Y, Y, \cdots, Y) \in \Lambda_{\mathcal{C}}$ with $Y = (y, y, \cdots, y) \in \beta_p^b$, we have*

$$\Theta_{\mathcal{C},\mathcal{Y}}(\tau, z) = BW_{\mathcal{C}}\left(\theta_{a,Y}(\tau, z) : a \in \mathbb{F}_p^b\right),$$

where for each $a \in \mathbb{F}_p^b$,

$$\theta_{a,Y}(\tau, z) = \theta_{a+\beta_p^b, Y}(\tau, z) = \sum_{v \in a+\beta_p^b} e^{\pi i \, Tr_{k_p/\mathbb{Q}}\left(\frac{v \cdot v}{p}\tau\right)} e^{2\pi i \, Tr_{k_p/\mathbb{Q}}\left(\frac{v \cdot Y}{p} z\right)} \tag{8.7}$$

with

$$Tr_{k_p/\mathbb{Q}}\left(\frac{v \cdot v}{p}\tau\right) = \sum_{s=1}^{\frac{p-1}{2}} \frac{(v \cdot v)^{(s)}}{p} \tau_s$$

and

$$Tr_{k_p/\mathbb{Q}}\left(\frac{v \cdot Y}{p} z\right) = \sum_{s=1}^{\frac{p-1}{2}} \frac{(v \cdot Y)^{(s)}}{p} z_s.$$

Proof. Let $\mathcal{Y} = \frac{1}{\sqrt{p}}(Y, Y, \cdots, Y) = \frac{1}{\sqrt{p}} Y_0 \in \Lambda_{\mathcal{C}}$ with $Y = (y, y, \cdots, y) \in \beta_p^b$. Then by the definition of theta series for the induced lattice, we have

$$\begin{aligned}
\Theta_{\mathcal{C},\mathcal{Y}}(\tau, z) &= \sum_{x \in \Lambda_{\mathcal{C}}} e^{\pi i Tr_{k_p/\mathbb{Q}}((x \cdot x)\tau)} e^{2\pi i Tr_{k_p/\mathbb{Q}}((x \cdot \mathcal{Y})z)} \\
&= \sum_{c \in \mathcal{C}} \sum_{x \in \frac{1}{\sqrt{p}} \rho_p^{-1}(c)} e^{\pi i Tr_{k_p/\mathbb{Q}}((x \cdot x)\tau)} e^{2\pi i Tr_{k_p/\mathbb{Q}}((x \cdot \mathcal{Y})z)} \\
&= \sum_{c \in \mathcal{C}} \sum_{x \in \rho_p^{-1}(c)} e^{\pi i Tr_{k_p/\mathbb{Q}}\left(\frac{x \cdot x}{p}\tau\right)} e^{2\pi i Tr_{k_p/\mathbb{Q}}\left(\frac{x \cdot Y_0}{p} z\right)} \\
&= \sum_{c \in \mathcal{C}} \sum_{x \in c + \beta_p^{bn}} e^{\pi i Tr_{k_p/\mathbb{Q}}\left(\frac{x \cdot x}{p}\tau\right)} e^{2\pi i Tr_{k_p/\mathbb{Q}}\left(\frac{x \cdot Y_0}{p} z\right)} \\
&= \sum_{\substack{c \in \mathcal{C} \\ c = (c_1, c_2, \cdots, c_n)}} \prod_{j=1}^{n} \sum_{x_j \in c_j + \beta_p^b} e^{\pi i Tr_{k_p/\mathbb{Q}}\left(\frac{x_j \cdot x_j}{p}\tau\right)} e^{2\pi i Tr_{k_p/\mathbb{Q}}\left(\frac{x_j \cdot Y}{p} z\right)} \\
&= \sum_{\substack{c \in \mathcal{C} \\ c = (c_1, c_2, \cdots, c_n)}} \prod_{j=1}^{n} \theta_{c_j, Y}(\tau, z),
\end{aligned}$$

which equals $BW_{\mathcal{C}}\left(\theta_{a,Y}(\tau, z) : a \in \mathbb{F}_p^b\right)$. □

8.4 Invariant Spaces and Jacobi Forms over k_p

In this section, a space of polynomials that is invariant under a certain matrix group and contains byte weight enumerators of self-dual codes over $\mathbb{F}_p$ is determined. For this, let the elements of $\mathbb{F}_p$ be listed as $0, 1, 2, \cdots, p-1$, and let us identify each element $r \in \mathbb{F}_p$ by $r \in K_p$ for $0 \le r \le p-1$. Using this, one can identify each tuple $a \in \mathbb{F}_p^\ell$ by $a \in K_p^\ell$ for each integer $\ell \ge 1$. Furthermore, under this identification, it is clear that

$$u \cdot v = 2\langle u, v \rangle \text{ for all } u, v \in \mathbb{F}_p^\ell. \tag{8.8}$$

Now let $\mathcal{O}_{k_p}$ denote the ring of algebraic integers of k_p.

Definition 173. Let G_p be the group generated by the matrix $\frac{1}{\sqrt{p^b}} M_p$ (with M_p as defined in Theorem 168) and matrices of the form $N_p(\gamma)$ for each $\gamma \in \mathcal{O}_{k_p}$, where for $a, c \in \mathbb{F}_p^b$, the (a, c)-entry of $N_p(\gamma)$ is given by

$$(N_p(\gamma))_{a,c} = \begin{cases} e^{\pi i Tr_{k_p/\mathbb{Q}}\left(\frac{a \cdot a}{p}\gamma\right)} & \text{if } a = c; \\ 0 & \text{otherwise.} \end{cases}$$

A homogeneous polynomial $f(X)$ with $X = (X_a : a \in \mathbb{F}_p^b)$ is said to be invariant under the group G_p if $f(L \cdot X) = f(LX^t) = f(X)$ for all $L \in G_p$, and the space of all such homogeneous polynomials over $\mathbb{C}$ is denoted by $\mathbb{C}[X]^{G_p}$.

The following theorem determines an invariant space of polynomials containing byte weight enumerators of self-dual codes over $\mathbb{F}_p$.

Theorem 174. *If $\mathcal{C}$ is a self-dual code of length bn and byte length b over $\mathbb{F}_p$, then its byte weight enumerator $BW_{\mathcal{C}}(X_a : a \in \mathbb{F}_p^b)$ belongs to $\mathbb{C}[X]^{G_p}$.*

Proof. Here it is sufficient to show that the byte weight enumerator $BW_{\mathcal{C}}(X_a : a \in \mathbb{F}_p^b)$ of the code $\mathcal{C}$ is invariant under the matrices $\frac{1}{\sqrt{p^b}} M_p$ and $N_p(\gamma)$ for each $\gamma \in \mathcal{O}_{k_p}$.

To prove this, we first observe, by Theorem 168, that $BW_{\mathcal{C}}(X_a : a \in \mathbb{F}_p^b)$ is invariant under the matrix $\frac{1}{\sqrt{p^b}} M_p$. Now to prove the invariance of $BW_{\mathcal{C}}(X_a : a \in \mathbb{F}_p^b)$ under matrices $N_p(\gamma)$ with $\gamma \in \mathcal{O}_{k_p}$, we consider

$$\begin{aligned} BW_{\mathcal{C}}\left(N_p(\gamma) \cdot (X_a : a \in \mathbb{F}_p^b)\right) &= BW_{\mathcal{C}}\left(e^{\pi i Tr_{k_p/\mathbb{Q}}(\frac{a \cdot a}{p}\gamma)} X_a : a \in \mathbb{F}_p^b\right) \\ &= \sum_{(c_1, c_2, \cdots, c_n) \in \mathcal{C}} \prod_{j=1}^{n} e^{\pi i Tr_{k_p/\mathbb{Q}}\left(\frac{c_j \cdot c_j}{p}\gamma\right)} X_{c_j}. \end{aligned}$$

From this, we see that it is enough to prove the following:

$$\sum_{j=1}^{n} Tr_{k_p/\mathbb{Q}}\left(\frac{c_j \cdot c_j}{p}\gamma\right) = Tr_{k_p/\mathbb{Q}}\left(\frac{c \cdot c}{p}\gamma\right) \in 2\mathbb{Z}$$

for all $c = (c_1, c_2, \cdots, c_n) \in \mathcal{C}$.

To do this, we note that $\mathcal{C}$ is a self-dual code over $\mathbb{F}_p$, so we have $\langle c, c \rangle = 0$ in $\mathbb{F}_p$ for all $c \in \mathcal{C}$, which, by (8.8), implies that $c \cdot c = 2\langle c, c \rangle = 0$ in K_p. This gives $Tr_{k_p/\mathbb{Q}}\left(\frac{c \cdot c}{p}\gamma\right) = 0 \in 2\mathbb{Z}$, which completes the proof of the theorem. □

The following theorem relates the invariant space of polynomials containing byte weight enumerators of self-dual codes over $\mathbb{F}_p$ with a space of Jacobi forms over k_p.

Theorem 175. *For an integer $n \geq 1$ satisfying $bn(p-1) \equiv 0 \pmod{8}$, let $\mathbb{C}[X]_n^{G_p}$ be the space of all homogeneous polynomials in $\mathbb{C}[X]^{G_p}$ of degree* *n. Let* $Y = (y, y, \cdots, y) \in \beta_p^b$ *be such that* $bny\bar{y}/p \in \mathcal{O}_{k_p}$. *Then the map*

$$\varphi : \bigoplus_{n\geq 1} \mathbb{C}[X]_n^{G_p} \to \bigoplus_{n\geq 1} \mathcal{J}_{nb,nby\bar{y}/p}(k_p)$$

defined as

$$\varphi\left(f_n\left(X_a : a \in \mathbb{F}_p^b\right)\right) = f_n\left(\theta_{a,Y}(\tau, z) : a \in \mathbb{F}_p^b\right)$$

is an algebra homomorphism.

Remark 176. Note that for each odd prime p, there exists a vector $Y = (y, y, \cdots, y) \in \beta_p^b$ satisfying $\frac{bny\bar{y}}{p} \in \mathcal{O}_{k_p}$, i.e., there exists $y \in \beta_p$ satisfying $\frac{bny\bar{y}}{p} \in \mathcal{O}_{k_p}$. For this, we first observe that $p = (1-\zeta_p)(1-\zeta_p^2)\cdots(1-\zeta_p^{p-1})$, which implies that $p \in \beta_p$. Further, it is easy to see that $y = p \in \beta_p$ satisfies $\frac{bny\bar{y}}{p} = bnp \in \mathcal{O}_{k_p}$.

Now to prove Theorem 175, we need to prove the following lemma:

Lemma 177. *Let* β_p *be the principal ideal of* $\mathcal{O}_{K_p}$ *generated by* $(1-\zeta_p) \in \mathcal{O}_{K_p}$. *For each* $a \in \mathbb{F}_p^b$ *and a fixed* $Y = (y, y, \cdots, y) \in \beta_p^b$ *with* $y \in \beta_p$, *let* $\theta_{a,Y}$ $(a \in \mathbb{F}_p^b)$ *be as defined in* (8.7). *Then we have the following:*

(i) $\theta_{a,Y}(\tau + \gamma, z) = e^{\pi i Tr_{k_p/\mathbb{Q}}(\frac{a\cdot a}{p}\gamma)}\theta_{a,Y}(\tau, z)$ *for each* $\gamma \in \mathcal{O}_{k_p}$.

(ii) $\theta_{a,Y}\left(-\frac{1}{\tau}, \frac{z}{\tau}\right) = \left(\frac{1}{\sqrt{p}}\right)^b e^{\pi i Tr_{k_p/\mathbb{Q}}\left(\frac{Y\cdot Y}{p}\frac{z^2}{\tau}\right)} \prod_{s=1}^{\frac{p-1}{2}} \left(\frac{\tau_s}{h}\right)^b$ $\sum_{\ell\in\mathbb{F}_p^b} e^{2\pi i\frac{\langle a,\ell\rangle}{p}}\theta_{\ell,Y}(\tau, z)$.

(iii) $\theta_{a,Y}(\tau, z+\lambda\tau+\mu) = e^{-\pi i Tr_{k_p/\mathbb{Q}}\left(\frac{Y\cdot Y}{p}(\lambda^2\tau+2\lambda z)\right)}\theta_{a,Y}(\tau, z)$ *for each* $[\lambda, \mu] \in \mathcal{O}_{k_p}^2$.

Proof. One can easily prove (i) and (iii) using (8.5), while the proof of (ii) is similar to that of Lemma 3.3 of Choie and Jeong [4]. □

Proof of Theorem 175. First of all, we will show that if $f_n\left(X_a : a \in \mathbb{F}_p^b\right) \in \mathbb{C}[X]_n^{G_p}$, then $f_n\left(\theta_{a,Y}(\tau, z) : a \in \mathbb{F}_p^b\right) \in \mathcal{J}_{bn,bny\bar{y}/p}(k_p)$. For this, we know that the group $SL_2(\mathcal{O}_{k_p})$ is generated by the matrices $\begin{pmatrix} 0 & -1 \\ 1 & 0 \end{pmatrix}$ and $\begin{pmatrix} 1 & \gamma \\ 0 & 1 \end{pmatrix}$ for $\gamma \in \mathcal{O}_{k_p}$ (see [6, 7, 13, 18]). Therefore to show that $f_n\left(\theta_{a,Y}(\tau, z) : a \in \mathbb{F}_p^b\right)$ satisfies (8.2) for each matrix in $SL_2(\mathcal{O}_{k_p})$, it is enough to show that the polynomial $f_n\left(\theta_{a,Y}(\tau, z) : a \in \mathbb{F}_p^b\right)$ satisfies (8.2) for the matrix $\begin{pmatrix} 0 & -1 \\ 1 & 0 \end{pmatrix}$ and matrices of the type $\begin{pmatrix} 1 & \gamma \\ 0 & 1 \end{pmatrix}$, $\gamma \in \mathcal{O}_{k_p}$, which follows from Lemma 177 and the fact that f_n is invariant under G_p. Further, by Lemma 177(iii), one can show that $f_n\left(\theta_{a,Y}(\tau, z) : a \in \mathbb{F}_p^b\right)$ satisfies (8.3) for each $[\lambda, \mu] \in \mathcal{O}_{k_p}^2$.

Next by definition of the theta series, we see that

$$\theta_{a,Y}(\tau, z) = \sum_{\substack{v \in a+\beta_p^b \\ v=(v_1,\cdots,v_b)}} \prod_{s=1}^{\frac{p-1}{2}} e^{\frac{2\pi i}{p}\left(\sum_{\ell=1}^{b}(v_\ell\overline{v_\ell})^{(s)}\right)\tau_s} e^{\frac{2\pi i}{p}\left(\sum_{\ell=1}^{b}(v_\ell\overline{y}+\overline{v_\ell}y)^{(s)}\right)z_s}. \quad (8.9)$$

Note that for each $v = (v_1, \cdots, v_b) \in a + \beta_p^b$, one can easily show that $\sum_{\ell=1}^{b} v_\ell\overline{v_\ell} \geq 0$ and $\sum_{\ell=1}^{b} v_\ell\overline{v_\ell}$, $\sum_{\ell=1}^{b} (v_\ell\overline{y} + \overline{v_\ell}y) \in \delta_{k_p}^{-1}$. Now on comparing equation (8.9) with (8.4) and on taking $N = \sum_{\ell=1}^{b} v_\ell\overline{v_\ell}$, $R = \sum_{\ell=1}^{b} (v_\ell\overline{y} + \overline{v_\ell}y)$ and $\omega = p$, we see that $f_n\left(\theta_{a,Y}(\tau, z) : a \in \mathbb{F}_p^b\right)$ has a Fourier series expansion of the form (8.4). From this, it follows that $f_n\left(\theta_{a,Y}(\tau, z) : a \in \mathbb{F}_p^b\right)$ is a Jacobi form of weight bn and index $bny\overline{y}/p$ over k_p. Further, it is easy to show that the map φ is an algebra homomorphism. □

In the following theorem, Jacobi forms over k_p are constructed from byte weight enumerators of self-dual codes over $\mathbb{F}_p$ by applying the above theorem.

Theorem 178. *Let $\mathcal{C}$ be a self-dual code of length bn and byte length b over $\mathbb{F}_p$. Let $Y = (y, y, \cdots, y) \in \beta_p^b$ be such that $bny\overline{y}/p \in \mathcal{O}_{k_p}$. Then $BW_{\mathcal{C}}\left(\theta_{a,Y}(\tau, z) : a \in \mathbb{F}_p^b\right)$ is a Jacobi form of weight bn and index $bny\overline{y}/p$ over k_p.*

Proof. It follows immediately from Theorems 174 and 175. □

8.5 Siegel Modular Forms of Genus g over k_p

In this section, Siegel modular forms of genus g over k_p are defined, where p is an odd prime and $g \geq 1$ is an integer. For this, let $\mathcal{A}^{(c,d)}$ denote the set of all $c \times d$ matrices over the non-empty set $\mathcal{A}$, where c and d are positive integers. Further, for each matrix $B = (B_{\ell m})$ with entries $B_{\ell m} \in k_p$, let us define $B^{(s)} = (B_{\ell m}^{(s)})$ for $1 \leq s \leq \frac{p-1}{2}$.

Now the modular group of genus g over the totally real field k_p, denoted by $Sp_{2g}(\mathcal{O}_{k_p})$, is defined as

$$Sp_{2g}(\mathcal{O}_{k_p}) = \left\{M \in \mathcal{O}_{k_p}^{(2g,2g)} : M^* J M = J\right\},$$

where M^* is the conjugate transpose of the matrix M and $J = \begin{pmatrix} 0 & -I_g \\ I_g & 0 \end{pmatrix}$ with I_g as the $g \times g$ identity matrix. The Siegel upper half-space of genus g, denoted by $\mathcal{H}_g$, is defined as $\mathcal{H}_g = \{Z \in \mathbb{C}^{(g,g)} : Z^* = Z, Im(Z) > 0\}$. Now to define the action of the modular group $Sp_{2g}(\mathcal{O}_{k_p})$ on $\mathcal{H}_g^{\frac{p-1}{2}}$, elements of $\mathcal{H}_g^{\frac{p-1}{2}}$ are listed as $\tau = \left(\tau_1, \tau_2, \cdots, \tau_{\frac{p-1}{2}}\right)$, where $\tau_1, \tau_2, \cdots, \tau_{\frac{p-1}{2}} \in \mathcal{H}_g$. Then the action of $Sp_{2g}(\mathcal{O}_{k_p})$ on $\mathcal{H}_g^{\frac{p-1}{2}}$ is defined as

$$\begin{aligned}\begin{pmatrix} \alpha & \beta \\ \gamma & \delta \end{pmatrix}(\tau) &:= (\alpha\tau + \beta)(\gamma\tau + \delta)^{-1} \\ &:= \left(\left(\alpha^{(1)}\tau_1 + \beta^{(1)}\right)\left(\gamma^{(1)}\tau_1 + \delta^{(1)}\right)^{-1}, \right. \\ &\qquad \left. \cdots, \left(\alpha^{\left(\frac{p-1}{2}\right)}\tau_{\frac{p-1}{2}} + \beta^{\left(\frac{p-1}{2}\right)}\right)\left(\gamma^{\left(\frac{p-1}{2}\right)}\tau_{\frac{p-1}{2}} + \delta^{\left(\frac{p-1}{2}\right)}\right)^{-1}\right)\end{aligned}$$

for all $\begin{pmatrix} \alpha & \beta \\ \gamma & \delta \end{pmatrix} \in Sp_{2g}(\mathcal{O}_{k_p})$.

Now a Siegel modular form of genus g over k_p is defined as follows:

Definition 179. For $\ell \in \mathbb{Z}$, an analytic function $f : \mathcal{H}_g^{\frac{p-1}{2}} \to \mathbb{C}$ is said to be a Siegel modular form of genus g with weight ℓ over k_p if it satisfies

$$\prod_{j=1}^{\frac{p-1}{2}} \det\left(\gamma^{(j)}\tau_j + \delta^{(j)}\right)^{-\ell} f\left((\alpha\tau + \beta)(\gamma\tau + \delta)^{-1}\right) = f(\tau) \tag{8.10}$$

for all $\begin{pmatrix} \alpha & \beta \\ \gamma & \delta \end{pmatrix} \in Sp_{2g}(\mathcal{O}_{k_p})$ with $\alpha, \beta, \gamma, \delta \in \mathcal{O}_{k_p}^{(g,g)}$ and $f(\tau)$ has a Fourier series expansion of the form

$$f(\tau) = \sum_{\substack{N \geq 0 \\ N \in \Omega_p}} c(N) \prod_{s=1}^{\frac{p-1}{2}} e^{\frac{2\pi i}{\omega} tr(N^{(s)}\tau_s)} \tag{8.11}$$

with a suitable $w \in \mathbb{Z}$, where $N \geq 0$ means that N is a positive semidefinite matrix, each coefficient $c(N)$ is a constant and Ω_p is the set of all $g \times g$ symmetric matrices over $\delta_{k_p}^{-1}$. Here tr denotes the matrix trace function.

Remark 180.

(i) The set of all Siegel modular forms of genus g with weight ℓ over k_p is an algebra over $\mathbb{C}$, which is denoted by $\mathbb{S}_\ell^{(g)}(k_p)$.

(ii) Let $\mathfrak{P} = \{p \text{ odd prime} : \mathcal{O}_{k_p} \text{ is a Euclidean domain}\}$. Then by Proposition 2 of Rege [14], we know that $Sp_{2g}(\mathcal{O}_{k_p})$ is generated by the following matrices:

$$\begin{pmatrix} 0 & -I_g \\ I_g & 0 \end{pmatrix}, \quad \begin{pmatrix} I_g & T \\ 0 & I_g \end{pmatrix}, \quad T \in Sym(g; \mathcal{O}_{k_p}),$$

where $Sym(g;\mathcal{O}_{k_p}) = \{A \in \mathcal{O}_{k_p}^{(g,g)} : A = A^t = A^*\}$. (Here A^t denotes transpose of the matrix A.)

Note that the set $\mathfrak{P}$ is non-empty, as $3, 5, 7, 11, 13, 17, 19 \in \mathfrak{P}$ (see Lemmermeyer [12]).

From this point on, let us assume that $p \in \mathfrak{P}$.

8.5.1 *Determination of Siegel Modular Forms over k_p*

In this section, theta series of genus g is defined for the lattice $\Lambda_\mathcal{C}$ induced by a byte error-control code $\mathcal{C}$ over $\mathbb{F}_p$ and is further related with the byte weight enumerator in genus g of $\mathcal{C}$. Further, Siegel modular forms of genus g over k_p are obtained by substituting certain theta series into byte weight enumerators in genus g of self-dual codes over $\mathbb{F}_p$.

Definition 181. Let $\mathcal{C}$ be a byte error-control code of length bn and byte length b over $\mathbb{F}_p$, and let $\Lambda_\mathcal{C}$ be the lattice induced by $\mathcal{C}$. Then the theta series of genus g for the induced lattice $\Lambda_\mathcal{C}$ is a function $\Theta_\mathcal{C}^{(g)} : \mathcal{H}_g^{\frac{p-1}{2}} \to \mathbb{C}$, defined as

$$\Theta_\mathcal{C}^{(g)}(\tau) = \sum_{x \in \Lambda_\mathcal{C}^g} e^{\pi i Tr_{k_p/\mathbb{Q}}(tr((x^*x + \overline{x^*x})\tau))},$$

where

$$Tr_{k_p/\mathbb{Q}}\left(tr((x^*x + \overline{x^*x})\tau)\right) = \sum_{s=1}^{\frac{p-1}{2}} tr((x^*x + \overline{x^*x})^{(s)}\tau_s).$$

(Here each element of $\Lambda_\mathcal{C}^g$ is represented as an $bn \times g$ matrix whose column vectors belong to $\Lambda_\mathcal{C}$.)

In the following lemma, a transformation formula for the theta series $\Theta_{\mathcal{C}}^{(g)}$ is derived.

Lemma 182. *We have*

$$\Theta_{\mathcal{C}}^{(g)}(-\tau^{-1}) = \frac{1}{\sqrt{\Delta(\Lambda_{\mathcal{C}})^g}} \prod_{s=1}^{\frac{p-1}{2}} \det\left(\frac{\tau_s}{i}\right)^{bn} \sum_{x\in(\Lambda_{\mathcal{C}}^*)^g} e^{\pi i Tr_{k_p/\mathbb{Q}}(tr((x^*x+\overline{x^*x})\tau))}.$$

Proof. To prove this, let

$$Z = \begin{pmatrix} \widetilde{\tau_1} & & & \\ & \widetilde{\tau_2} & & \\ & & \ddots & \\ & & & \widetilde{\tau_{\frac{p-1}{2}}} \end{pmatrix},$$

where for any $e \times f$ matrix $A = (a_{ij})$, the matrix $\widetilde{A}$ is defined as

$$\widetilde{A} = \begin{pmatrix} a_{11} & 0 & a_{12} & 0 & \cdots & a_{1f} & 0 \\ 0 & a_{11} & 0 & a_{12} & \cdots & 0 & a_{1f} \\ \vdots & \vdots & \vdots & \vdots & \vdots & \vdots & \vdots \\ a_{e1} & 0 & a_{e2} & 0 & \cdots & a_{ef} & 0 \\ 0 & a_{e1} & 0 & a_{e2} & \cdots & 0 & a_{ef} \end{pmatrix}.$$

Now for a given $x_0 \in \mathbb{R}^{(bn,g(p-1))}$, let us consider $F_{x_0}(x) = e^{-\pi i tr((x+x_0)Z^{-1}(x+x_0)^t)}$ for every $x \in \mathbb{R}^{(bn,g(p-1))}$. Then for $v \in \mathbb{R}^{(bn,g(p-1))}$, let us define

$$\begin{aligned} \hat{F}_{x_0}(v) &= \int_{\mathbb{R}^{(bn,g(p-1))}} F_{x_0}(x) e^{-2\pi i tr(xv^t)} dx \\ &= e^{2\pi i tr(x_0 v^t)} \int_{\mathbb{R}^{(bn,g(p-1))}} e^{-\pi i tr(xZ^{-1}x^t)} e^{-2\pi i tr(xv^t)} dx. \end{aligned}$$

As $\tau_s = \tau_s^*$ for $1 \le s \le \frac{p-1}{2}$, by Lemma I 3.3 of Krieg [11], we see that $\tau_s = P_s^* D_s P_s$, where D_s is a $g \times g$ real diagonal matrix and P_s is a $g \times g$ matrix satisfying $P_s^* P_s = I_g$. It is easy to see that $\widetilde{\tau_s} = \widetilde{P_s^*}\widetilde{D_s}\widetilde{P_s}$ for each s. This gives $Z = \mathfrak{Y}^t \mathfrak{D} \mathfrak{Y}$ and $\mathfrak{Y}^t \mathfrak{Y} = I_{g(p-1)}$, where

$$\mathfrak{Y} = \begin{pmatrix} \widetilde{P_1} & & & \\ & \widetilde{P_2} & & \\ & & \ddots & \\ & & & \widetilde{P_{\frac{p-1}{2}}} \end{pmatrix} \quad \text{and} \quad \mathfrak{D} = \begin{pmatrix} \widetilde{D_1} & & & \\ & \widetilde{D_2} & & \\ & & \ddots & \\ & & & \widetilde{D_{\frac{p-1}{2}}} \end{pmatrix}.$$

This implies that

$$\int_{\mathbb{R}^{(bn,g(p-1))}} e^{-\pi i tr(xZ^{-1}x^t)} e^{-2\pi i tr(xv^t)} dx$$
$$= \int_{\mathbb{R}^{(bn,g(p-1))}} e^{-\pi i tr(x(\mathfrak{Y}^t\mathfrak{D}\mathfrak{Y})^{-1}x^t)} e^{-2\pi i tr(x\mathfrak{Y}^t\mathfrak{Y}v^t)} dx.$$

Now on taking $X = x\mathfrak{Y}^t$ and $V = v\mathfrak{Y}^t$, we obtain

$$\int_{\mathbb{R}^{(bn,g(p-1))}} e^{-\pi i tr(xZ^{-1}x^t)} e^{-2\pi i tr(xv^t)} dx$$
$$= \int_{\mathbb{R}^{(bn,g(p-1))}} e^{-\pi i tr(X\mathfrak{D}^{-1}X^t)} e^{-2\pi i tr(XV^t)} dX.$$

Further, let us write $X = (X_1, X_2, \cdots, X_{g(p-1)}), V = (V_1, V_2, \cdots, V_{g(p-1)})$ with each $X_h, V_h \in \mathbb{R}^{(bn,1)}$. On substituting this, we get

$$\int_{\mathbb{R}^{(bn,g(p-1))}} e^{-\pi i tr(X\mathfrak{D}^{-1}X^t)} e^{-2\pi i tr(XV^t)} dX$$
$$= \int_{\mathbb{R}^{(bn,g(p-1))}} e^{-\pi i\left[\frac{X_1\cdot X_1}{d_{11}^{(1)}} + \frac{X_2\cdot X_2}{d_{11}^{(1)}} + \cdots + \frac{X_{g(p-1)-1}\cdot X_{g(p-1)-1}}{d_{gg}^{((p-1)/2)}} + \frac{X_{g(p-1)}\cdot X_{g(p-1)}}{d_{gg}^{((p-1)/2)}}\right]}$$
$$\times e^{-2\pi i \sum_{j=1}^{g(p-1)} X_j\cdot V_j} dX_1 \cdots dX_{g(p-1)}$$
$$= \det\left(\frac{Z}{i}\right)^{bn/2}$$
$$\times e^{\pi i\left[(V_1\cdot V_1)d_{11}^{(1)} + (V_2\cdot V_2)d_{11}^{(1)} + \cdots + (V_{g(p-1)-1}\cdot V_{g(p-1)-1})d_{gg}^{(p-1)/2)} + (V_{g(p-1)}\cdot V_{g(p-1)})d_{gg}^{(p-1)/2)}\right]}$$
$$= \det\left(\frac{Z}{i}\right)^{bn/2} e^{\pi i tr(vZv^t)}.$$

Therefore we have

$$\hat{F}_{x_0}(v) = \det\left(\frac{Z}{i}\right)^{bn/2} e^{2\pi i tr(x_0v^t)} e^{\pi i tr(vZv^t)}. \tag{8.12}$$

Now by Lemma 3.3 of Choie and Jeong [4], we see that there exists an embedding $L : K_p \to \mathbb{R}^{(p-1)}$. We further extend this map to an embedding $L : K_p^{(bn,g)} \to \mathbb{R}^{(bn,g(p-1))}$, defined as

$$L(v) = \left(L_1(v), \cdots, L_{\frac{p-1}{2}}(v)\right),$$

where for $v = (v_{hj}) \in K_p^{(bn,g(p-1))}$, $L_s(v) = (L_s(v_{hj}))$ for each s. Using this, the theta series $\Theta_{\mathcal{C}}^{(g)}(\tau)$ can be written as

$$\Theta_{\mathcal{C}}^{(g)}(\tau) = \sum_{x\in\Lambda_{\mathcal{C}}^g} e^{\pi i Tr_{k_p/\mathbb{Q}}(tr((x^*x+\overline{x^*x})\tau))} = \sum_{V\in L(\Lambda_{\mathcal{C}}^g)} e^{\pi i tr\left(VZV^t\right)}.$$

On applying the Poisson summation formula, we get

$$\Theta_C^{(g)}(-\tau^{-1}) = \sum_{V\in L(\Lambda_C^g)} F_{\mathbf{0}}(V) = \frac{1}{\sqrt{\Delta(\Lambda_C^g)}} \sum_{V\in L(\Lambda_C^g)^*} \hat{F}_{\mathbf{0}}(V),$$

where $\mathbf{0}$ denotes the zero matrix. Using (8.12), we get

$$\begin{aligned}\Theta_C^{(g)}(-\tau^{-1}) &= \frac{1}{\sqrt{\Delta(\Lambda_C^g)}} \det\left(\frac{Z}{i}\right)^{bn/2} \sum_{V\in L(\Lambda_C^g)^*} e^{\pi i tr(VZV^t)} \\ &= \frac{1}{\sqrt{\Delta(\Lambda_C^g)}} \prod_{s=1}^{\frac{p-1}{2}} \det\left(\frac{\widetilde{\tau_s}}{i}\right)^{bn/2} \sum_{v\in(\Lambda_C^g)^*} e^{\pi i Tr_{k_p/\mathbb{Q}}(tr((x^*x+\overline{x^*x})\tau))} \\ &= \frac{1}{\sqrt{\Delta(\Lambda_C)^g}} \prod_{s=1}^{\frac{p-1}{2}} \det\left(\frac{\tau_s}{i}\right)^{bn} \sum_{v\in(\Lambda_C^*)^g} e^{\pi i Tr_{k_p/\mathbb{Q}}(tr((x^*x+\overline{x^*x})\tau))},\end{aligned}$$

as $\det(\widetilde{\tau_s}) = \det(\tau_s)^2$ for each s, $\Delta(\Lambda_C^g) = \Delta(\Lambda_C)^g$ and $(\Lambda_C^g)^* = (\Lambda_C^*)^g$.

This completes the proof of the lemma. □

In the following theorem, Siegel modular forms of genus g over k_p are obtained from theta series of genus g for the lattices induced by self-dual codes over $\mathbb{F}_p$.

Theorem 183. *Let C be a self-dual code of length bn and byte length b over $\mathbb{F}_p$. Then the theta series $\Theta_C^{(g)}(\tau)$ is a Siegel modular form of genus g with weight bn over k_p.*

Proof. As $p \in \mathfrak{P}$, by Remark 180(ii), it is sufficient to show that $\Theta_C^{(g)}(\tau)$ satisfies (8.10) for the matrix $\begin{pmatrix} 0 & -I_g \\ I_g & 0 \end{pmatrix}$ and matrices of the type $\begin{pmatrix} I_g & T \\ 0 & I_g \end{pmatrix}$ for each $T \in Sym(g; \mathcal{O}_{k_p})$.

Now for matrices of the type $\begin{pmatrix} I_g & T \\ 0 & I_g \end{pmatrix}$ with $T \in Sym(g; \mathcal{O}_{k_p})$, we have

$$\begin{aligned}\Theta_C^{(g)}(\tau + T) &= \sum_{x\in\Lambda_C^g} e^{\pi i Tr_{k_p/\mathbb{Q}}(tr((x^*x+\overline{x^*x})(\tau+T)))} \\ &= \sum_{x\in\Lambda_C^g} e^{\pi i Tr_{k_p/\mathbb{Q}}(tr((x^*x+\overline{x^*x})T))} e^{\pi i Tr_{k_p/\mathbb{Q}}(tr((x^*x+\overline{x^*x})\tau))},\end{aligned}$$

which equals $\Theta_{\mathcal{C}}^{(g)}(\tau)$, as $Tr_{k_p/\mathbb{Q}}(tr((x^*x + \overline{x^*x})T)) \in 2\mathbb{Z}$ for all $x \in \Lambda_{\mathcal{C}}^g$ (see Choie and Jeong [4, Lemma 4.1]). Next for the matrix $\begin{pmatrix} 0 & -I_g \\ I_g & 0 \end{pmatrix}$, by Lemma 182, we have

$$\Theta_{\mathcal{C}}^{(g)}(-\tau^{-1}) = \frac{1}{\sqrt{\Delta(\Lambda_{\mathcal{C}})^g}} \prod_{s=1}^{\frac{p-1}{2}} \det\left(\frac{\tau_s}{i}\right)^{bn} \sum_{x \in (\Lambda_{\mathcal{C}}^*)^g} e^{\pi i Tr_{k_p/\mathbb{Q}}(tr((x^*x+\overline{x^*x})\tau))}.$$

Since $\mathcal{C}$ is a self-dual code of length bn over $\mathbb{F}_p$, by Lemma 4.1 of Choie and Jeong [4], we have $bn(p-1) \equiv 0 \pmod 8$, $\Delta(\Lambda_{\mathcal{C}}) = 1$ and $\Lambda_{\mathcal{C}}^* = \Lambda_{\mathcal{C}}$. From this, we obtain $\Theta_{\mathcal{C}}^{(g)}(-\tau^{-1}) = \prod_{s=1}^{\frac{p-1}{2}} \det(\tau_s)^{bn}\, \Theta_{\mathcal{C}}^{(g)}(\tau)$. Further, by the definition of theta series of genus g for the induced lattice $\Lambda_{\mathcal{C}}$, we see that

$$\Theta_{\mathcal{C}}^{(g)}(\tau) = \sum_{x \in \Lambda_{\mathcal{C}}^g} \prod_{s=1}^{\frac{p-1}{2}} e^{\frac{2\pi i}{2} tr((x^*x+\overline{x^*x})^{(s)}\tau_s)}. \tag{8.13}$$

Now for each $x \in \Lambda_{\mathcal{C}}^g$, it is easy to see that $x^*x+\overline{x^*x} \in \Omega_p$ and $x^*x+\overline{x^*x} \geq 0$. Further, on comparing equation (8.13) with (8.11) and on taking $N = x^*x + \overline{x^*x}$ and $\omega = 2$, we see that $\Theta_{\mathcal{C}}^{(g)}(\tau)$ has a Fourier series expansion of the form (8.11), which completes the proof of the theorem. □

In the following theorem, theta series of genus g for the lattice induced by a byte error-control code $\mathcal{C}$ over $\mathbb{F}_p$ is related with the byte weight enumerator in genus g of the code $\mathcal{C}$.

Theorem 184. *Let $\mathcal{C}$ be a byte error-control code of length bn and byte length b over $\mathbb{F}_p$. Then we have*

$$\Theta_{\mathcal{C}}^{(g)}(\tau) = BW_{\mathcal{C},g}\left(\theta_a(\tau) : a \in \mathbb{F}_p^{(b,g)}\right),$$

where for each $a \in \mathbb{F}_p^{(b,g)}$,

$$\theta_a(\tau) = \sum_{v \in a+\beta_p^{(b,g)}} e^{\pi i Tr_{k_p/\mathbb{Q}}\left(tr\left(\frac{v^*v+\overline{v^*v}}{p}\tau\right)\right)} \tag{8.14}$$

with

$$Tr_{k_p/\mathbb{Q}}\left(tr\left(\frac{v^*v+\overline{v^*v}}{p}\tau\right)\right) = \sum_{s=1}^{\frac{p-1}{2}} tr\left(\frac{(v^*v+\overline{v^*v})^{(s)}}{p}\tau_s\right).$$

Proof. By definition, we have

$$\begin{aligned}
\Theta_C^{(g)}(\tau) &= \sum_{x\in\Lambda_C^g} e^{\pi i Tr_{k_p/\mathbb{Q}}(tr((x^*x+\overline{x^*x})\tau))} \\
&= \sum_{c\in C^g}\sum_{x\in\frac{1}{\sqrt{p}}\rho_p^{-1}(c)} e^{\pi i Tr_{k_p/\mathbb{Q}}(tr((x^*x+\overline{x^*x})\tau))} \\
&= \sum_{\substack{c\in C^g\\ c=(c_1,c_2,\cdots,c_n)^t}} \prod_{j=1}^{n} \sum_{x_j\in c_j+\beta_p^{(b,g)}} e^{\pi i Tr_{k_p/\mathbb{Q}}\left(tr\left(\frac{x_j^*x_j+\overline{x_j^*x_j}}{p}\tau\right)\right)} \\
&= \sum_{\substack{c\in C^g\\ c=(c_1,c_2,\cdots,c_n)^t}} \prod_{j=1}^{n} \theta_{c_j}(\tau) \\
&= BW_{C,g}\left(\theta_a(\tau) : a\in\mathbb{F}_p^{(b,g)}\right).
\end{aligned}$$

□

In the following theorem, Siegel modular forms of genus g over k_p are determined on substituting $X_a = \theta_a(\tau)$ for each $a \in \mathbb{F}_p^{(b,g)}$ in byte weight enumerators of self-dual codes over $\mathbb{F}_p$.

Theorem 185. *Let C be a self-dual code of length bn and byte length b over $\mathbb{F}_p$. Then $BW_{C,g}\left(\theta_a(\tau) : a\in\mathbb{F}_p^{(b,g)}\right)$ is a Siegel modular form of genus g with weight bn over k_p.*

Proof. By Theorems 183 and 184, the desired result follows immediately.

□

8.6 Invariant Space and Siegel Modular Forms of Genus g over k_p

In this section, a space of polynomials invariant under a certain matrix group and containing byte weight enumerators in genus g of self-dual codes of length bn and byte length b over $\mathbb{F}_p$ is determined. Further, an algebra homomorphism is defined from this invariant space to a space of Siegel modular forms of genus g over k_p. Theorem 185 follows as a corollary of this result. To do this, let us first define the following:

Definition 186. Let $H_{p,g}$ be the group generated by the matrices $\frac{1}{\sqrt{p^{bg}}}M_{p,g}$ (with $M_{p,g}$ as defined in Theorem 168) and $N_{p,g}(T)$ for $T \in$

$Sym(g; \mathcal{O}_{k_p})$, where the (a,c)th entry of $N_{p,g}(T)$ for $a, c \in \mathbb{F}_p^{(b,g)}$ is given by

$$(N_{p,g}(T))_{a,c} = \begin{cases} e^{2\pi i Tr_{k_p/\mathbb{Q}}\left(tr\left(\frac{a^t a}{p}T\right)\right)} & \text{if } a = c; \\ 0 & \text{otherwise.} \end{cases}$$

Then a homogeneous polynomial $f(X)$ with $X = (X_a : a \in \mathbb{F}_p^{(b,g)})$ is said to be invariant under the group $H_{p,g}$ if $f(L \cdot X) = f(LX^t) = f(X)$ for all $L \in H_{p,g}$, and the space of all such homogeneous polynomials over $\mathbb{C}$ is denoted by $\mathbb{C}[X]^{H_{p,g}}$.

The following theorem determines an invariant space of polynomials containing byte weight enumerators in genus g of self-dual byte error-control codes over $\mathbb{F}_p$.

Theorem 187. *If $\mathcal{C}$ is a self-dual code of length bn and byte length b over $\mathbb{F}_p$, then the byte weight enumerator $BW_{\mathcal{C},g}(X_a : a \in \mathbb{F}_p^{(b,g)})$ in genus g belongs to $\mathbb{C}[X]^{H_{p,g}}$.*

Proof. To prove this theorem, it is enough to prove that $BW_{\mathcal{C},g}(X_a : a \in \mathbb{F}_p^{(b,g)})$ is invariant under the matrices $\frac{1}{\sqrt{p^{bg}}} M_{p,g}$ and $N_{p,g}(T)$ for each $T \in Sym(g; \mathcal{O}_{k_p})$.

Towards this, we first observe, by Theorem 168, that $BW_{\mathcal{C},g}(X_a : a \in \mathbb{F}_p^{(b,g)})$ is invariant under the matrix $\frac{1}{\sqrt{p^{bg}}} M_{p,g}$. Next to prove the invariance of $BW_{\mathcal{C},g}(X_a : a \in \mathbb{F}_p^{(b,g)})$ under matrices of the type $N_{p,g}(T)$ with $T = (T_{qr}) \in Sym(g; \mathcal{O}_{k_p})$, let us consider

$$\begin{aligned} &BW_{\mathcal{C},g}\left(N_{p,g}(T) \cdot \left(X_a : a \in \mathbb{F}_p^{(b,g)}\right)\right) \\ &= \sum_{c_1, c_2, \cdots, c_g \in \mathcal{C}^g} \prod_{j=1}^{n} e^{2\pi i Tr_{k_p/\mathbb{Q}}\left(tr\left(\frac{a_j^t a_j}{p}T\right)\right)} X_{a_j}, \end{aligned}$$

where $a_j = \left(c_j^{(1)}; c_j^{(2)}; \cdots; c_j^{(g)}\right)$ for $1 \leq j \leq n$ if $c_m = \left(c_1^{(m)}, c_2^{(m)}, \cdots, c_n^{(m)}\right) \in \mathcal{C}$ with $c_j^{(m)} \in \mathbb{F}_p^b$ for $1 \leq m \leq g$. In view of this, it suffices to show that

$$\sum_{j=1}^{n} Tr_{k_p/\mathbb{Q}}\left(tr\left(\frac{a_j^t a_j}{p}T\right)\right) \in \mathbb{Z}.$$

For this, let us write $c_j^{(m)} = \left(c_{j1}^{(m)}, \cdots, c_{jb}^{(m)}\right)^t$ for each j and m, and let us consider

$$\sum_{j=1}^{n} tr\left(a_j^t a_j T\right) = \sum_{j=1}^{n}\left\{\sum_{\ell=1}^{g}\left\langle c_j^{(\ell)}, c_j^{(\ell)}\right\rangle T_{\ell\ell} + 2\sum_{1\leq q<r\leq g}\left\langle c_j^{(q)}, c_j^{(r)}\right\rangle T_{qr}\right\}$$

$$= \sum_{\ell=1}^{g}\langle c_\ell, c_\ell\rangle T_{\ell\ell} + 2\sum_{1\leq q<r\leq g}\langle c_q, c_r\rangle T_{qr}.$$

Since $\mathcal{C}$ is a self-dual code over $\mathbb{F}_p$, we have $\langle c_\ell, c_\ell\rangle = \langle c_q, c_r\rangle = 0$ in $\mathbb{F}_p$ for each ℓ, q, r. This, under the identification of elements of $\mathbb{F}_p$ with those of K_p, gives $\langle c_\ell, c_\ell\rangle = \langle c_q, c_r\rangle = 0$ in K_p for each ℓ, q, r. This implies that $\sum_{j=1}^{n} Tr_{k_p/\mathbb{Q}}\left(tr\left(\frac{a_j^t a_j}{p} T\right)\right) = 0$, which further implies that

$$BW_{\mathcal{C},g}\left(N_{p,g}(T)\cdot\left(X_a : a\in\mathbb{F}_p^{(b,g)}\right)\right) = BW_{\mathcal{C},g}\left(X_a : a\in\mathbb{F}_p^{(b,g)}\right).$$

This completes the proof of the theorem. □

The following theorem relates the invariant space containing byte weight enumerators in genus g of self-dual codes over $\mathbb{F}_p$ with a space of Siegel modular forms of genus g over k_p.

Theorem 188. *For an integer $n \geq 1$ satisfying $bn(p-1) \equiv 0 \pmod 8$, let $\mathbb{C}[X]_n^{H_{p,g}}$ be the space of all homogeneous polynomials of degree n that are invariant under the group $H_{p,g}$. Then the map*

$$\psi : \bigoplus_{n\geq 1}\mathbb{C}[X]_n^{H_{p,g}} \to \bigoplus_{n\geq 1}\mathbb{S}_{nb}^{(g)}(k_p),$$

defined as

$$\psi\left(f_n\left(X_a : a\in\mathbb{F}_p^{(b,g)}\right)\right) = f_n\left(\theta_a(\tau) : a\in\mathbb{F}_p^{(b,g)}\right),$$

is an algebra homomorphism.

To prove this theorem, we need to prove the following lemma:

Lemma 189. *For $a\in\mathbb{F}_p^{(b,g)}$, the following hold.*

(i) $\theta_a(\tau + T) = e^{2\pi i Tr_{k_p/\mathbb{Q}}\left(tr\left(\frac{a^t a}{p}T\right)\right)}\theta_a(\tau)$ *for each* $T\in Sym(g;\mathcal{O}_{k_p})$.

(ii) $\theta_a(\tau^{-1}) = \frac{1}{\sqrt{p^{bg}}}\prod_{s=1}^{\frac{p-1}{2}}\det\left(\frac{\tau_s}{i}\right)^b \sum_{\nu\in\mathbb{F}_p^{(b,g)}} e^{2\pi i Tr\left(\frac{a\nu^t}{p}\right)}\theta_\nu(\tau)$.

Proof. (i) By (8.5), part (i) follows immediately.
(ii) Following the same notations and working similarly as in Lemma 182, we see that the theta series θ_a of genus g can be rewritten as

$$\theta_a(\tau) = \sum_{v\in a+\beta_p^{(b,g)}} e^{\pi i Tr_{k_p/\mathbb{Q}}\left(tr\left(\frac{v^*v+\overline{v^*v}}{p}\tau\right)\right)} = \sum_{V\in L(a+\beta_p^{(b,g)})} e^{\pi i Tr\left(\frac{VZV^t}{p}\right)}$$
$$= \sum_{V\in L(\beta_p^{(b,g)})} e^{\pi i Tr\left(\frac{(V+L(a))Z(V+L(a))^t}{p}\right)}.$$

Now on applying the Poisson summation formula, we obtain

$$\theta_a(-\tau^{-1}) = \sum_{V\in L(\beta_p^{(b,g)})} F_{\frac{L(a)}{\sqrt{p}}}\left(\frac{V}{\sqrt{p}}\right)$$
$$= \frac{1}{\sqrt{\Delta(\beta_p^{(b,g)})}} \sum_{V\in L(\beta_p^{(b,g)})^*} \hat{F}_{\frac{L(a)}{\sqrt{p}}}\left(\frac{-V}{\sqrt{p}}\right)$$
$$= \frac{1}{\sqrt{\Delta(\beta_p^{(b,g)})}} \det\left(\frac{Z}{i}\right)^{b/2} \sum_{V\in L(\beta_p^{(b,g)})^*} e^{\pi i Tr(\frac{VZV^t}{p})} e^{-2\pi i Tr(\frac{L(a)V^t}{p})}$$
$$= \frac{1}{\sqrt{\Delta(\beta_p^{(b,g)})}} \prod_{s=1}^{\frac{p-1}{2}} \det\left(\frac{\widetilde{\tau_s}}{i}\right)^{b/2}$$
$$\times \sum_{v\in(\beta_p^{(b,g)})^*} e^{\pi i Tr_{k_p/\mathbb{Q}}\left(tr\left(\frac{v^*v+\overline{v^*v}}{p}\tau\right)\right)} e^{-2\pi i Tr_{k_p/\mathbb{Q}}\left(tr(\frac{av^*+va^t}{p})\right)}$$
$$= \frac{1}{\sqrt{p^{bg}}} \prod_{s=1}^{\frac{p-1}{2}} \det\left(\frac{\tau_s}{i}\right)^{b}$$
$$\times \sum_{\nu\in\mathbb{F}_p^{(b,g)}} \sum_{v\in\nu+\beta_p^{(b,g)}} e^{\pi i Tr_{k_p/\mathbb{Q}}\left(tr\left(\frac{v^*v+\overline{v^*v}}{p}\tau\right)\right)} e^{-2\pi i Tr_{k_p/\mathbb{Q}}\left(tr(\frac{av^*+va^t}{p})\right)},$$

as $\det(\widetilde{\tau_s}) = \det(\tau_s)^2$ for each s, $\Delta(\beta_p^{(b,g)}) = \Delta(\beta_p)^{bg} = p^{bg}$ by (8.5) and $(\beta_p^{(b,g)})^* = \mathcal{O}_{K_p}^{(b,g)} = \bigcup_{a\in\mathbb{F}_p^{(b,g)}} (a+\beta_p^{(b,g)})$ by (8.5) and (8.6).

Now let us consider $e^{-2\pi i Tr_{k_p/\mathbb{Q}}\left(tr(\frac{av^*+va^t}{p})\right)}$ for each $v \in \nu+\beta_p^{(b,g)}$. Let $v \in \nu+\beta_p^{(b,g)}$ be fixed, and let us write $v = \nu+x$ for some $x \in \beta_p^{(b,g)}$. Then we have

$$e^{-2\pi i Tr_{k_p/\mathbb{Q}}\left(tr(\frac{av^*+va^t}{p})\right)} = e^{-2\pi i Tr_{k_p/\mathbb{Q}}\left(tr(\frac{a\nu^t+\nu a^t+ax^*+xa^t}{p})\right)}$$
$$= e^{-2\pi i Tr_{k_p/\mathbb{Q}}\left(tr(\frac{a\nu^t+\nu a^t}{p})\right)} = e^{2\pi i Tr\left(\frac{a\nu^t}{p}\right)},$$

as $Tr_{k_p/\mathbb{Q}}\left(tr(\frac{ax^*+xa^t}{p})\right) \in \mathbb{Z}$ and $Tr_{k_p/\mathbb{Q}}(q) = q^{\frac{(p-1)}{2}}$ for each $q \in \mathbb{Q}$. This gives

$$\theta_a(-\tau^{-1}) = \frac{1}{\sqrt{p^{bg}}} \prod_{s=1}^{\frac{p-1}{2}} \det\left(\frac{\tau_s}{i}\right)^b \sum_{\nu \in \mathbb{F}_p^{(b,g)}} e^{2\pi i Tr\left(\frac{a\nu^t}{p}\right)} \theta_\nu(\tau).$$

This completes the proof of the lemma. □

Proof of Theorem 188. Working in a similar way as in Theorem 175 and by applying Lemma 189, the result follows. □

Remark 190. By applying Theorems 187 and 188, Theorem 185 follows immediately.

8.7 Partial Epstein Zeta Functions and their Functional Equation

In this section, partial Epstein zeta functions for each element in $\mathbb{F}_p^{(b,g)}$ are defined. A functional equation for these partial Epstein zeta functions is also derived using Mellin transforms of the theta series defined by (8.14).

Definition 191. Let e, h be $b\times g$ be real matrices, and let Y be a $b\times b$ matrix of a positive-definite quadratic form. Then for $z \in \mathbb{C}$ with $Re(z) > bg(p-1)$, the Epstein zeta function associated with (Y, e, h) is defined as

$$Z_b(Y,e,h,z) = \sum_{\substack{v\in\beta_p^{(b,g)} \\ v+e\neq 0}} \frac{e^{2\pi i\sigma(hv^*)}}{\left(\sum_{s=1}^{\frac{p-1}{2}} tr\left((e+v)^*Y(e+v)+\overline{(e+v)^*Y(e+v)}\right)^{(s)}\right)^{z/2}}.$$

Further, if I_b denotes the $b \times b$ identity matrix, then on taking $e = h = 0$ and $Y = I_b$, we get

$$Z_b(I_b, 0, 0, z) = \sum_{\substack{v\in\beta_p^{(b,g)} \\ v\neq 0}} \frac{1}{\left(\sum_{s=1}^{\frac{p-1}{2}} tr\left(v^*v + \overline{v^*v}\right)^{(s)}\right)^{z/2}}.$$

The partial Epstein zeta functions are defined as follows:

Definition 192. For each $a \in \mathbb{F}_p^{(b,g)}$ and $z \in \mathbb{C}$ with $Re(z) > bg(p-1)$, the partial Epstein zeta function $Z_{b,a}(z)$ is defined as

$$Z_{b,a}(z) = \sum_{\substack{v \in \beta_p^{(b,g)} \\ a+v \neq 0}} \frac{1}{\left(\sum_{s=1}^{\frac{p-1}{2}} tr\left((a+v)^*(a+v) + \overline{(a+v)^*(a+v)} \right)^{(s)} \right)^{z/2}}.$$

In the following theorem, a functional equation for partial Epstein zeta functions is derived.

Theorem 193. *For $a \in \mathbb{F}_p^{(b,g)}$, the partial Epstein zeta function $Z_{b,a}(z)$ with $Re(z) > bg(p-1)$ extends analytically to a meromorphic function on the whole complex plane having a simple pole at $z = bg(p-1)$ with residue $\frac{2}{\Gamma(bg(p-1)/2)} \left(\frac{\pi^{bg(p-1)/2}}{p^{bgp/2}} \right)$, where Γ is the Gamma function. Let*

$$\Lambda_{b,a}(z) = \left(\frac{\pi}{p} \right)^{-z/2} \Gamma\left(\frac{z}{2} \right) Z_{b,a}(z) \text{ for each } a \in \mathbb{F}_p^{(b,g)}.$$

Then for each $a \in \mathbb{F}_p^{(b,g)}$, $\Lambda_{b,a}(z)$ satisfies the following equation:

$$\Lambda_{b,a}(z) = \left(\frac{1}{p} \right)^{bg/2} \sum_{\ell \in \mathbb{F}_p^{(b,g)}} e^{\frac{2\pi i}{p} tr(a\ell^*)} \Lambda_{b,\ell}(bg(p-1) - z).$$

Alternatively, if $a_0 = 0, a_1, \cdots, a_{p^{bg}-1}$ are all the elements of $\mathbb{F}_p^{(b,g)}$, then the functions $Z_{b,a}(z)$ $(a \in \mathbb{F}_p^{(b,g)})$ satisfy the following functional equation:

$$\Gamma\left(\frac{z}{2}\right) \begin{pmatrix} Z_{b,a_0}(z) \\ Z_{b,a_1}(z) \\ \vdots \\ Z_{b,a_{p^{bg}-1}}(z) \end{pmatrix}$$

$$= \frac{\pi^{-\frac{(bg(p-1)-2z)}{2}}}{p^{-\frac{(bg(p-2)-2z)}{2}}} \Gamma\left(\frac{bg(p-1)-z}{2} \right) M_{p,g} \begin{pmatrix} Z_{b,a_0}(bg(p-1)-z) \\ Z_{b,a_1}(bg(p-1)-z) \\ \vdots \\ Z_{b,a_{p^{bg}-1}}(bg(p-1)-z) \end{pmatrix},$$

where $M_{p,g}$ is the matrix as defined in Theorem 168.

Proof. For each $a \in \mathbb{F}_p^{(b,g)}$, consider the theta series

$$\theta_a(\tau) = \sum_{v \in a+\beta_p^{(b,g)}} e^{\pi i Tr_{k_p/\mathbb{Q}}\left(tr\left(\frac{v^*v+\overline{v^*v}}{p}\tau\right)\right)}$$

$$= \sum_{v \in a+\beta_p^{(b,g)}} e^{\pi i \sum_{s=1}^{\frac{p-1}{2}} tr\left(\frac{(v^*v+\overline{v^*v})^{(s)}}{p}\tau_s\right)}$$

On substituting $\tau = itI^{(g,\frac{p-1}{2}g)}$ with $t > 0$, we obtain

$$\theta_a(itI^{(g,\frac{p-1}{2}g)}) = \sum_{v \in a+\beta_p^{(b,g)}} e^{\pi i \sum_{s=1}^{\frac{p-1}{2}} tr\left(\frac{(v^*v+\overline{v^*v})^{(s)}}{p} itI_g\right)}$$

$$= \sum_{v \in a+\beta_p^{(b,g)}} e^{-\pi t \sum_{s=1}^{\frac{p-1}{2}} tr\left(\frac{(v^*v+\overline{v^*v})^{(s)}}{p}\right)}$$

$$= \theta_a(t).$$

Similarly, on substituting $\tau = -\frac{1}{it}I^{(g,\frac{p-1}{2}g)}$, we get

$$\theta_a\left(-\frac{1}{it}I^{(g,\frac{p-1}{2}g)}\right) = \sum_{v \in a+\beta_p^{(b,g)}} e^{-\frac{\pi}{t} \sum_{s=1}^{\frac{p-1}{2}} tr\left(\frac{(v^*v+\overline{v^*v})^{(s)}}{p}\right)} = \theta_a(1/t).$$

As $\mathbb{F}_p^{(b,g)} = \{a_0, a_1, \cdots, a_{p^{bg}-1}\}$, by Lemma 189, we obtain

$$\begin{pmatrix} \theta_{a_0}(t) \\ \theta_{a_1}(t) \\ \vdots \\ \theta_{a_{p^{bg}-1}}(t) \end{pmatrix} = \left(\frac{1}{\sqrt{p}}\right)^{bg} \left(\frac{1}{t}\right)^{\frac{bg(p-1)}{2}} M_{p,g} \begin{pmatrix} \theta_{a_0}(1/t) \\ \theta_{a_1}(1/t) \\ \vdots \\ \theta_{a_{p^{bg}-1}}(1/t) \end{pmatrix}. \quad (8.15)$$

From this, it follows that if t is very large, then $\theta_{a_0}(t)$ is asymptotic to 1 and $\theta_{a_h}(t)$ is asymptotic to zero for each h, $1 \le h \le p^{bg}-1$. Furthermore, when t tends to 0, $\theta_{a_h}(t)$ is asymptotic to $\left(\frac{1}{\sqrt{p}}\right)^{bg} \left(\frac{1}{t}\right)^{\frac{bg(p-1)}{2}}$ for $0 \le h \le p^{bg}-1$.

Now for each a_h, we define

$$\phi_{b,a_0}(z) = \int_1^\infty t^{z/2-1}\left(\theta_{a_0}(t) - 1\right) dt$$

$$+ \int_0^1 t^{z/2-1}\left(\theta_{a_0}(t) - \left(\frac{1}{\sqrt{p}}\right)^{bg}\left(\frac{1}{t}\right)^{\frac{bg(p-1)}{2}}\right) dt, \quad (8.16)$$

$$\phi_{b,a_h}(z) = \int_1^\infty t^{z/2-1}\theta_{a_h}(t) dt$$

$$+ \int_0^1 t^{z/2-1}\left(\theta_{a_h}(t) - \left(\frac{1}{\sqrt{p}}\right)^{bg}\left(\frac{1}{t}\right)^{\frac{bg(p-1)}{2}}\right) dt \quad (8.17)$$

for $1 \leq h \leq p^{bg} - 1$. From this, for each z in the half-plane with $Re(z) > bg(p-1)$, we get

$$\phi_{b,a_0}(z) = \int_0^\infty t^{z/2-1} \sum_{\substack{v \in \beta_p^{(b,g)} \\ v \neq 0}} e^{-\pi t \sum_{s=1}^{\frac{p-1}{2}} tr\left(\frac{(v^*v+\overline{v^*v})^{(s)}}{p}\right)} dt + \frac{2}{z}$$

$$+ \left(\frac{1}{p}\right)^{bg/2} \frac{2}{bg(p-1) - z}.$$

Now on taking the Mellin transform of the function $e^{-\frac{\pi t}{p}\sum_{s=1}^{\frac{p-1}{2}} tr\left((v^*v+\overline{v^*v})^{(s)}\right)}$, we obtain

$$\phi_{b,a_0}(z) = \left(\frac{\pi}{p}\right)^{-z/2}\Gamma\left(\frac{z}{2}\right) Z_{b,a_0}(z) + \frac{2}{z} + \left(\frac{1}{p}\right)^{bg/2}\frac{2}{bg(p-1) - z}$$

$$= \Lambda_{b,a_0}(z) + \frac{2}{z} + \left(\frac{1}{p}\right)^{bg/2}\frac{2}{bg(p-1) - z}. \quad (8.18)$$

This yields

$$Z_{b,a_0}(z) = \frac{(\pi/p)^{z/2}}{\Gamma(z/2)}\left(\phi_{b,a_0}(z) - \frac{2}{z} - \left(\frac{1}{p}\right)^{bg/2}\frac{2}{bg(p-1) - z}\right).$$

Since the integrals in (8.16) converge for every value of z, $\phi_{b,a_0}(z)$ is an entire function of z. Furthermore, the functions $(\pi/p)^{z/2}$ and $\frac{1}{\Gamma(z/2)}$ are entire functions, and $z\Gamma(z/2) = 2\Gamma(z/2+1)$ is non-zero when z tends to zero. Hence the function $Z_{b,a_0}(z)$ has a simple pole at $z = bg(p-1)$ with residue $\frac{2}{\Gamma(bg(p-1)/2)}\left(\frac{\pi}{p}\right)^{bg(p-1)/2}\left(\frac{1}{p}\right)^{bg/2}$.

Working in a similar way for each h, $1 \leq h \leq p^{bg}-1$, we obtain

$$\begin{aligned}\phi_{b,a_h}(z) &= \left(\frac{\pi}{p}\right)^{-z/2}\Gamma\left(\frac{z}{2}\right)Z_{b,a_h}(z) + \left(\frac{1}{p}\right)^{bg/2}\frac{2}{bg(p-1)-z}\\ &= \Lambda_{b,a_h}(z) + \left(\frac{1}{p}\right)^{bg/2}\frac{2}{bg(p-1)-z}.\end{aligned}$$

This implies that

$$Z_{b,a_h}(z) = \frac{(\pi/p)^{z/2}}{\Gamma(z/2)}\left(\phi_{b,a_h}(z) - \left(\frac{1}{p}\right)^{bg/2}\frac{2}{bg(p-1)-z}\right)$$

has a simple pole at $z = bg(p-1)$ with residue $\frac{2}{\Gamma(bg(p-1)/2)} \times \left(\frac{\pi}{p}\right)^{bg(p-1)/2}\left(\frac{1}{p}\right)^{bg/2}$ for each h.

Next we proceed to derive a functional equation for $Z_{b,a_h}(z)$ for $0 \leq h \leq p^{bg}-1$. For this, we see, by (8.18), that

$$\begin{aligned}\Lambda_{b,a_0}(z) &= \phi_{b,a_0}(z) - \frac{2}{z} - \left(\frac{1}{p}\right)^{bg/2}\frac{2}{bg(p-1)-z}\\ &= \int_1^\infty t^{z/2-1}\left(\theta_{a_0}(t)-1\right)dt\\ &\quad + \int_0^1 t^{z/2-1}\left(\theta_{a_0}(t) - \left(\frac{1}{p}\right)^{\frac{bg}{2}}\left(\frac{1}{t}\right)^{\frac{bg(p-1)}{2}}\right)dt\\ &\quad - \frac{2}{z} - \left(\frac{1}{p}\right)^{bg/2}\frac{2}{bg(p-1)-z}.\end{aligned}$$

On taking $t = 1/u$ and by (8.15), we obtain

$$\Lambda_{b,a_0}(z) = \int_0^1 u^{-z/2-1}\left[\left(\frac{1}{p}\right)^{bg/2} u^{\frac{bg(p-1)}{2}} \sum_{\ell\in\mathbb{F}_p^{(b,g)}} e^{\frac{\pi}{p}tr(a_0\ell^*)}\theta_\ell(u) - 1\right] du$$

$$+ \int_1^\infty u^{-z/2-1}\left[\left(\frac{1}{p}\right)^{bg/2} u^{\frac{bg(p-1)}{2}} \sum_{\ell\in\mathbb{F}_p^{(b,g)}} e^{\frac{\pi}{p}tr(a_0\ell^*)}\theta_\ell(u)\right.$$

$$\left. - \left(\frac{1}{p}\right)^{bg/2} u^{\frac{bg(p-1)}{2}}\right] du - \frac{2}{z} - \left(\frac{1}{p}\right)^{bg/2}\frac{2}{bg(p-1)-z}$$

$$= \left(\frac{1}{p}\right)^{bg/2} e^{\frac{\pi i}{p}tr(a_0a_0^*)}\left\{\int_1^\infty u^{\frac{bg(p-1)-z}{2}}(\theta_{a_0}(u)-1)\frac{du}{u} - \frac{2}{bg(p-1)-z}\right.$$

$$\left. + \int_0^1 u^{\frac{bg(p-1)-z}{2}}\left(\theta_{a_0}(u) - \left(\frac{1}{p}\right)^{bg/2} u^{-\frac{bg(p-1)}{2}}\right)\frac{du}{u} - \left(\frac{1}{p}\right)^{bg/2}\frac{2}{z}\right\}$$

$$+ \left(\frac{1}{p}\right)^{bg/2} \sum_{\substack{\ell\in\mathbb{F}_p^{(b,g)}\\ \ell\neq 0}} e^{\frac{\pi i}{p}tr(a_0\ell^*)}\left\{\int_1^\infty u^{\frac{bg(p-1)-z}{2}}\theta_\ell(u)\frac{du}{u}\right.$$

$$\left. + \int_0^1 u^{\frac{bg(p-1)-z}{2}}\left(\theta_\ell(u) - \left(\frac{1}{p}\right)^{bg/2} u^{-\frac{bg(p-1)}{2}}\right)\frac{du}{u} - \left(\frac{1}{p}\right)^{bg/2}\frac{2}{z}\right\}$$

$$= \left(\frac{1}{p}\right)^{bg/2} \sum_{\ell\in\mathbb{F}_p^{(b,g)}} e^{\frac{\pi i}{p}tr(a_0\ell^*)}\Lambda_{b,\ell}(bg(p-1)-z),$$

as $\sum_{\ell\in\mathbb{F}_p^{(b,g)}} e^{\frac{\pi i}{p}tr(a_0a_0^*)} = p^{bg}$. Further, working in a similar way as above, we note that

$$\Lambda_{b,a_h}(z) = \left(\frac{1}{p}\right)^{bg/2} \sum_{\ell\in\mathbb{F}_p^{(b,g)}} e^{\frac{\pi i}{p}tr(a_h\ell^*)}\Lambda_{b,\ell}(bg(p-1)-z) \text{ for } 1 \le h \le p^{bg}-1,$$

which completes the proof of the theorem. □

Remark 194. The functional equation in the above theorem can also be rewritten in terms of Epstein zeta functions as follows:

$$\Gamma\left(\frac{z}{2}\right)\begin{pmatrix} Z_b(I_b, \frac{a_0}{p}, 0, z) \\ Z_b(I_b, \frac{a_1}{p}, 0, z) \\ \vdots \\ Z_b(I_b, \frac{a_{p^{bg}-1}}{p}, 0, z) \end{pmatrix}$$

$$= \frac{\pi^{-\frac{(bg(p-1)-2z)}{2}}}{p^{-\frac{bg(p-2)}{2}}}\Gamma\left(\frac{bg(p-1)-z}{2}\right)\begin{pmatrix} Z_b(I_b, 0, \frac{a_0}{p}, bg(p-1)-z) \\ Z_b(I_b, 0, \frac{a_1}{p}, bg(p-1)-z) \\ \vdots \\ Z_b(I_b, 0, \frac{a_{(p)^{bg}-1}}{p}, bg(p-1)-z) \end{pmatrix}.$$

References

[1] K. Betsumiya, Y. Choie, Codes over $\mathbb{F}_4$, Jacobi forms and Hilbert-Siegel modular forms over $\mathbb{Q}(\sqrt{5})$, European J. Combin. 26 (2005), no. 5, 629–650.

[2] K. Betsumiya, Y. Choie, Jacobi forms over totally real fields and type II codes over Galois rings $\mathbf{GR}(2^m, f)$, European J. Combin. 25 (2004), no. 4, 475–486.

[3] Y. Choie and S. T. Dougherty, Codes over Σ_{2m} and Jacobi forms over the quaternions, *Appl. Algebra Eng. Commun. Comp.* 15, pp. 129–147 (2004).

[4] Y. Choie and E. Jeong, Jacobi forms over totally real fields and codes over $\mathbb{F}_p$, *Illinois J. Math.* 46(2), pp. 627–643 (2002).

[5] Y. Choie and P. Solé, Ternary codes and Jacobi forms, *Discrete Math.* 282, pp. 81–87 (2004).

[6] G. Cooke, A weakening of the Euclidean property for integral domains and applications to number theory I, *J. Reine Angew. Math.* 282, pp. 133–156 (1976).

[7] G. Cooke, A weakening of the Euclidean property for integral domains and applications to number theory II, *J. Reine Angew. Math.* 283/284, pp. 71–85 (1976).

[8] S. T. Dougherty, M. Harada, M. Oura, Note on the g-fold joint weight enumerators of self-dual codes over $\mathbb{Z}_k$, *Appl. Algebra Eng. Commun. Comp.* 11(6), pp. 437–445 (2001).

[9] W. Ebeling, *Lattices and codes. A course partially based on lectures by F. Hirzebruch*, Adv. lect. Math., Vieweg, Braunschweig (1994).

[10] H. Kojima, On the Fourier coefficients of Jacobi forms of index N over totally real fields, *Tohuku Math. J.* 64, pp. 361–385 (2012).

[11] A. Krieg, *Modular forms on half-spaces of quaternions*, Lect. Notes Math. 1143, Berlin, Heidelberg, New York, Springer (1985).

[12] F. Lemmermeyer, The Euclidean algorithm in algebraic number fields, *Exposition. Math.* 13(5), pp. 385–416 (1995).

[13] B. Liehl, On the group SL_2 over orders of arithmetic type, *J. Reine Angew. Math.* 323, pp. 153–171 (1981).

[14] N. S. Rege, On certain classical groups over Hasse domains, *Math. Zeitschr.* 102, pp. 120–157 (1967).

[15] A. Sharma and A. K. Sharma, Byte weight enumerators and modular forms of genus r, *J. Algebra Appl.* 14(6), pp. 1–25 (2015).

[16] A. Sharma and A. K. Sharma, Byte weight enumerators of codes over $\mathbb{F}_p$ and modular forms over a totally real field, *International Journal of Information and Coding Theory* 4(4), pp. 237–257 (2017).

[17] K. Suzuki: Complete m-spotty weight enumerators of binary codes, Jacobi forms, and partial Epstein zeta functions, *Discrete Math.* 312, pp. 265–278 (2012).

[18] L. N. Vaserstein: On the group SL_2 over Dedekind domains of arithmetic type, *Mat. Sb. (N.S.)* 89 (131), pp. 313–322 (1972).

www.ingramcontent.com/pod-product-compliance
Lightning Source LLC
Chambersburg PA
CBHW050639190326
41458CB00008B/2339
* 9 7 8 9 8 1 1 2 1 2 9 1 8 *